Televisuality

Televisuality

Style, Crisis, and Authority
in American Television

John Thornton Caldwell

Rutgers University Press
New Brunswick, New Jersey

A version of chapter 5 was first published as "Televisuality as a Semiotic Machine" in *Cinema Journal* 32:4 (August 1993), and is reprinted here with permission of the University of Texas Press. Portions of chapter 4 were first published as "Salvador/ Noriega" in *Jump-Cut* 37 (1992), and are reprinted here by permission of the editors of *Jump-Cut*. The photographs grouped in the illustrations appearing on pages 17, 161, 162, 164, 194, 195, 197, and 226 are from the Terrence O'Flaherty Collection and are reprinted here by permission of Arts Library—Special Collections, UCLA. All other photographs are by J. Thornton Caldwell and are reprinted here by permission of the artist.

Library of Congress Cataloging-in-Publication Data

Caldwell, John Thornton, 1954–
    Televisuality: style, crisis, and authority in American television / John Thornton Caldwell.
       p.   cm. — (Communication, media, and culture)
    Includes bibliographical references and index.
    ISBN 0-8135-2163-7 (cloth). — ISBN 0-8135-2164-5 (pbk.)
    1. Television broadcasting—Social aspects—United States.
2. Television broadcasting—United States—History.   3. Television programs—United States—History and criticism.   4. Visual communication.   I. Title.   II. Series.
PN1992.6.C22   1994
302.23'45—dc20   94-39017
                                                       CIP

British Cataloging-in-Publication information available

Contents

Preface

Holling is tormented by Koyaanisqatsi dreams until he goes out and does the wild thing with a young stag. . . .
—Synopsis from production company "bible," *Northern Exposure*[1]

The collision of cineastic taste and streetwise sexuality, auteurism and rap—couched by primetime producers in this industrial document—was actually rather commonplace in television by the early 1990s. In fact, for at least a decade American television had exploited the programming potential of visual style in episodes like the one described above. Scripts, especially in prestige dramatic offerings like *Northern Exposure*, made a weekly practice of engineering their narratives around highly coded aesthetic and cultural fragments. This was no longer simply ensemble drama, a form with a long generic history in television. It was also a kind of ensemble iconography and a highly publicized ritual of aesthetic facility. This book aims to interrogate the nature of such performance, and to do so from three perspectives: as a historical phenomenon, as an aesthetic and industrial practice, and as a socially symbolic act.[2] In ways that I hope to make clearer later, the self-conscious performance of style is not adequately posed nor fully explained by reference to postmodernism. Furthermore, the industrial practice of televisuality challenges several other dominant variants of academic media theory and suggests that a reexamination of high theory is in order. A coalition of very strange intellectual bedfellows—from across a wide range of theoretical and institutional commitments—now posit the very status of the image as a problem. The inability of this intellectual coalition to fully engage contemporary television suggests that a desegregation of theory and production practice is very much in order.

Interrogating the status of the image in recent American television brings with it several inevitable tasks: coming to grips with fundamental changes in the ways that television is made and addressing the ways that production practice problematizes theory. The route that I took to writing this book meant navigating both issues. I recall, for example, while working in on-line video postproduction suites in the early 1980s, that practitioners in this environment utilized theoretical frameworks that were utterly alien to the accounts of television dominant in academic film and television theory. Of course, I never expected practitioners to use the same lingo as academics, but the fundamental paradigms upon which the digital video effects (DVE) artists worked, sound designers mixed, and the editors cut bore no resemblance to the explanations about television spun out by theorists. Of course these practitioners might be "dupes," as radical critical theory always seemed to imply, ideologically enmeshed as they were in the dominant media. Yet, they were certainly not stupid about what they were doing. Far from it. There was a semiotic and stylistic capability here that had to be taken seriously. Working with cinematographers underscored the same lesson. Even those doing video industrials could bring to productions an aesthetic facility with film historical and lighting styles, a consciousness that provided clients a de facto menu of styles-on-demand. What had happened to the video production industry since the early 1970s, when an orthodox stylistic canon ruled and the unions and guilds had little interest in aspirants with theoretical understandings and academic degrees?[3] Clearly, television had changed: practice itself was actively theorizing television. Critical theory surely needed to look at videographic evidence a bit more closely.

In addition, the rise of what academics would later label postmodernism had an immediate impact on the trajectory of the avant-garde. I had cut my educational teeth, so to speak, on the tenets of artistic modernism and independent media practice. By the mid-1970s, "guerrilla video" and "alternative television" provided "progressive" working models for production. *Radical Software* and Enzensberger provided instigating texts and calls-to-arms for progressive media theory.[4] Visionary video and video art seemed plausible institutional concepts, with both activist and aesthetic potential. Something happened, however, that began to problematize the notion of oppositionality. In the years around 1980, the dominant media began to announce ownership of the very tactics of radicality once thought to be defining properties of the avant-garde.[5] Commercial spots shot on Panavision took formalist compositional ideals to new heights. Computerized frame-accurate editing allowed broadcast television a dynamic and rapid editing style impossible to achieve with the low-band control-track editing equipment available in media access and production centers. Digital video graphics devices, and their high-resolution images, made the analog graphics of the Paik-Abe video synthesizer on which I had been trained seem like video burlap or electronic noise. For those who thought oppositional art could come from aggressive word-and-image appropriations and recombinations, CNN mastered the art of collaging endless visual images and verbal text. Their performance of recombinant reality-art aired around the clock and around the globe. At the same time, MTV marshaled many of the looks and tactics of the avant-garde—the appropriation of mass-culture images, intertextuality, fractured subjectivity, and yes, even irony.

Even the attitude of the avant-garde, then, could be mass marketed.[6] Was nothing sacred? Given the ostensible radicalization of the mainstream, what was left for independents, for film/video art, apart from catalog and museum apologetics? Ideologies of genius? Certainly the technological gulf began to widen between mainstream television and its opposition, a factor that raised other issues. In what ways, for example, were the much-heralded Infiniti automotive ads, with their zen-like cross-culture air and minimalist atmospheric style, different from the important and influential works of Bill Viola? Would Tom Wolfe soon strike against the theoretically subsidized institutions of the media arts world with a corollary version of *The Painted Word* diatribe? Surely not. The game of radicality had rapidly shifted to the commercial world. Even pop-culture tabloid programs like *Entertainment Tonight* showed that they could deconstruct both celebrities (with insider back-stories) and production simulations (through mini "making-ofs") in ways that would make Jacques Derrida, Paul de Man, and Jean Baudrillard proud.[7] Certainly every previous framework for the avant-garde—whether formalist, Brechtian, Godardian, or dadaist—had become highly visible in some form in the corporate world of the new television.

This was a key moment, then, in the history of both American broadcast television and in the history of independent film and video, a time of redefinition for both. Creativity for many independents increasingly meant trying to figure out how to cover the astronomical hourly rates charged for the ever newer state-of-the-art production technologies—apparati that provided the very looks that the new television demanded. In some ways these new looks, codified by music videos, became a visual burden that not only independents, but also commercials, primetime dramas, and even local news would eventually be measured against.[8] Yet, despite such pressures, despite the contradictions that came with assimilation, the mythologies of the "opposition" and the notion of the "independent" would live on, even if they did so out of habit. The political exigencies of the Reagan-Bush era, military intervention in Latin America, the spread of AIDS, and the highly vocal rise of the right wing all provided public conditions by which the avant-garde redefined the very terms of opposition. Within this climate of political crisis, and shadowed by a highly styled primetime, cooperative and oppositional groups like Paper-Tiger Television, Deep-Dish Satellite Network, and the grass-roots, camcorder-activist PBS series *The Nineties* showed a renewed kind of oppositional power—one that came when video activists elided the aura of the sensitive artist-producer entirely and recast the cutting-edge as low-tech, grainy, and socially topical. It is no irony, then, that even as mainstream television aestheticized itself, its opposition survived by renouncing aesthetic pretense.

During the period of this inversion, I began to critically analyze the operations of the new mainstream televisual forms, and to try to come to grips with the ideological implications of those forms. The book at hand is a result of that analysis. As I have revised and prepared this manuscript over the past few years, several trends have caused me to rethink the pessimism that originally drove me to try to understand mainstream television in the first place. The symbolic appropriation of the avant-garde that characterized the period following 1980—dramatized by its rapid assimilation into mass culture—now appears historically inaccurate.[9] Although the prestige genres of performance and video art may have been cloned by television and MTV, many vernacular forms were not. Community-access video, media cooperatives, educational and development video, and small-format camcorder activism have all survived and somehow prospered under the shadow of the new mass-market television.[10] The moment of broadcast television that I examine in this book—an institutional and presentational crisis fueled by the emergence of cable—then, also roughly fits within another historical trough, one formed by the mass-market cloning of the avant-garde in 1980 and the reemergence of an activist small-format camcorder revolution at the end of the Reagan-Bush-Gates era in the early 1990s. Although their ideological blueprints could not be more antithetical, the institutional trajectories of American broadcasting and the avant-garde are also in some ways very much interrelated.

For any critical theory of television, coming to grips with the complexity, scale, and operations of the televisual industry can be debilitating. If all of the aesthetic pretense that I describe above is now a property of national media corporations, critical accounts are prone to either celebrate its dominance or to withdraw from it in cynicism and defeat. Given the capital-intensive nature of the new televisuality, what options are left for independents, academics, and alternative producers? If the globally constructed image is a product of the multinational enemy, then the proprietary discourse of the image is also to be attacked and abhorred. This was precisely the tactic many academics used in response to Gulf War coverage. Yet this form of critical separatism was shortsighted. I plan to revisit a case very much related to Gulf War coverage, the L.A. rebellion, in the final chapters of this book, in a section that more completely examines political and cultural aspects of crisis televisuality. More than any other recent event, the rebellion in Los Angeles and its aftermath demonstrated three important lessons: (1) that televisual stylization continues to be a favored and operative mode in television, (2) that televisual abilities extend far away from the prestige producers of mass-market television, and (3) that outsiders can marshal televisual skills to counter program and resist the dominant media. The renewed optimism that I referred to earlier is based on the fact that youths in urban multiracial environments can and have used televisual tactics to give voice to the unheard and image to the invisible. Televisuality is not just manufactured by CNN, commercial production companies, and the Pentagon. It is, rather, a new and emerging mode of communication, well suited even for those written-off as marginal. As a means of expression with a bite and an attitude, televisuality can also enable "outsiders" to survive in contemporary American urban culture. Children of almost any class or culture seem to take to televisuality like fish to water or like virtual-combatants to Nintendo.

The chapters that follow combine three different levels of analysis, three perspectives not normally conjoined in a field that seems hopelessly split between those that reify context and those that celebrate the text. Three chapters—3, 5, and 10—look at the industrial basis for televisual style. Although television's mode of production seldom plays an important enough role in either critical theory or empirical research, its consequences are far-ranging. Three additional chapters are structured as broader historical surveys. Chapter 2 examines television's penchant for aesthetic posturing in the decades preceding 1980. It demonstrates how American television worked through the cultural and ideological conditions—the cultural appetite, if you will—that would eventually allow for the celebration of stylistic exhibitionism. Chapter 9 includes a survey of the historical development and televisual consequences of two alternative modes of liveness—the live remote and portable tape. The economic and programming bases for recent developments in televisuality are examined in chapter 10, while chapter 9 focuses on the televisual audience. The five remaining chapters—4 through 8—comprise the center of the book and elaborate recurrent formal modes and aesthetic guises that make up televisual style. Each of these chapters includes a contextualizing survey of stylistically related shows, along with an in-depth case-study and close-analysis of individual—but symptomatic—program texts: the epic miniseries, the auteurist signature show, trash television, and the televisual documentary.

Even a cursory look at one of these guises—the signature show—demonstrates that their appeals to the audience are indeed richly coded. When on-camera cineaste Orson Welles addressed primetime *Moonlighting* viewers in 1985, he was laying out the conditions of distinctive viewing even as ABC was teaching critics how to spin must-see status in the press: "Nothing is wrong with your set. Tonight's episode is an *experiment*. So gather the kids, the dog, the popcorn, and grandma . . . lock them in another room . . . and enjoy tonight's show."[11] Eight years later, the networks no longer apparently even needed such wraparounds as primetime aesthetic buffers. In one episode of *Northern Exposure* in 1993, for example, film director Peter Bogdanovich came to Cicely to help establish an international film festival with the works of Orson Welles at its center, while *Dances with Wolves* actor Graham Greene, playing Native American shaman Leonard, simultaneously sought to "locate the white collective unconscious."[12] Bogdanovich speculated around the table at Maurice's and afterward about film aesthetics and industry, even as Greene reminisced on Ciccly's backwoods streets about high-stakes public debates in film criticism: "What do [I] think about Pauline Kael? *We butt heads over Bertolucci."*

Blink, however, and you'd miss these fleeting, clockwork doses of auteurism. Bogdanovich's psychobiography of Welles, Greene's speculations on literature and film, and Ed's frame-by-frame visual deconstruction of Greg Toland's cinematography in *Citizen Kane* simply evaporated into the phosphorescence. When aired, such theorizations came across as little more than throw-away lines, lost within a formally complicated episode that combined critiques of anthropology, oral histories, the cultural clash between Caucasian and Native American cultures, the psychological function of myths and rituals, and the tension between self-centeredness and community. The hybridized aesthetic that fused Bogdanovich/Welles/Toland/Greene with producers Joshua Brand and John Falsey, however, also dissolved into a much broader institutional flow—into a channel-rich sweeps evening populated by countless exhibitions of sensationalism and stylistic knowingness. By 1993 the viewer was being flooded by cinematic ecstasies, electronic flourishes, and artspeak alike. In less than twenty years, the flow had been overhauled. How and why this happened is worth considering.

Acknowledgments

The collaborative nature of the editorial process combined the pleasures of intellectual interrogation with outright acts of generosity. During the time that this book was written, many individuals provided assistance and support, including the first-rate editorial board and staff of Rutgers University Press. Nick Browne offered invaluable comments, encouragement, and provocative suggestions for developing the original proposal; George Custen was a continual source of wisdom and intelligence throughout the long process of writing and revision; and Leslie Mitchner deserves a special note of gratitude for her advocacy of the project, for her understanding of the theoretical terrain, and for her patience and editorial insight as the book slowly took form. Laura Starrett and Marilyn Campbell were immensely helpful in preparing the manuscript for publication. I am especially grateful to colleagues Janet Bergstrom, Teshome Gabriel, Stephen Mamber, Chon Noriega, Vivian Sobchack, and Peter Wollen, for allowing me the year of archival research and graduate teaching during which time the final draft of this book was written. The professional staff of the UCLA Film and Television archives—Andrea Kalas, Luana Almares, Steve Ricci, and Robert Rosen—provided essential resources and information unavailable in any other television collection. Birgitta Kueppers, of Arts Library—Special Collections, made possible access to hundreds of photographic stills needed to complete this research.

The Department of Radio, Television, and Film at the California State University, Long Beach, a program with a tradition of integrating theory and practice, provided fertile terrain for the development of this project. Robert Finney's media expertise and advocacy as chair, and Karen Burman's logistical assistance, proved invaluable, as did the endorsements of Wade Hobgood, Thomas Ferreira, Dee Abrahamse, and Karl Anatol. Numerous students in the classroom provided a very different kind of proving ground, along with a wealth of perspectives that suggested the subject of television might be approached differently. For allowing me to test out some of the notions from this book in seminar settings, graduate and doctoral students elsewhere deserve a special note of thanks. Luisela Alvaray, Daniel Bernardi, Gilberto Blasini, Maren Chumley, Anna Everett, James Friedman, L. S. Kim, Lisa Kernan, Susan Knobloch, Mildred Lewis, Edward O'Neill, Potter Palmer, and Beretta Smith will undoubtedly recognize discussions within the text and references to their work at the end of the book.

Works like this one do not arise in a vacuum, and a number of scholars were instrumental in the formation of these ideas. Mimi White has freely offered insights from her work and has been a continual source of inspiration and guidance since this work first took form. Chuck Kleinhans has been a model of accessible and committed media scholarship and demonstrates the value of keeping the social and political implications of theory always in sight. Historians and theorists Jack Ellis and James Schwoch provided invaluable assistance and rich suggestions, perhaps in ways that they will probably never fully appreciate. I benefitted earlier in my career from a number of scholars, artists, and teachers. John Baldessari, Mike Kelly, Eric Martin, and Jim Shaw demonstrated the elegant logic that emerges when one views the aesthetic for what it is: a very practical matter and a matter of practice. Joel Sheesley, John Walford, Alva Steffler, and Patricia Ward all provided tangible and moral support, without which my career would surely have taken a different tack.

Although most of the material in *Televisuality* is printed here for the first time, I would like to acknowledge the editors of journals that have allowed me to reprint and expand on material for this book. A different version of chapter 3 appeared in *Cinema Journal* 37 (Summer 1993); a portion of chapter 4 appeared in *Jump-Cut* (Spring 1992); and chapter 11 grew out of a paper presentation at the "Console-ing Passions" conference on feminism and television in April 1992. One recurrent theme of this book, the denigration of the image in French media theory, was first prepared as a paper entitled: "Non-verbal Semiosis in Film and Video" for the International Summer Institute for Semiotic and Structural Studies (ISISSS) in 1986. I thank seminar director Thomas Sebeok of Indiana University for his insights and suggestions on that occasion. This theme, the conspiratorial suspicion of the image in high-theory later comprised part of a dissertation on the same subject.[1] Some of the other material in this book was also presented in provisional form at various media conferences. For those occasions, I must thank a number of panel chairs and organizers, in particular Jeanne Allen, Margaret Morse, Lynn Spigel, Sasha Torres, Linda Williams, Mark Williams, and Brian Winston. I hope they do not run for cover when they see their names mentioned here, for their questions and comments proved immensely helpful in orienting this study.

Because film and video production has been a professional influence in the formation of this book and an analytical focus in the pages that follow, I must thank a number of figures who have assisted me along the way. The late Malcolm Arth of the Margaret Mead Film Festival, Kathy Huffman formerly of the ICA in Boston and the Long Beach Museum of Art, Manfred Salzgeber of Berlin, Tom Weinberg and Jamie Cesar of WTTW-Chicago, Bob Harris of Anthology Film Archives New York, John Minkowsky and Lynn Corcoran of Media Study/Buffalo, the late Lynn Blumenthal of the School of the Art Institute of Chicago and the Video Data Bank, anthropologist William McKellin of the University of British Columbia, John Lalnunsang Pudaite of Global Films, and Joyce Bolinger formerly of the Center for New Television all provided invaluable support for my work over the years. For the opportunities to push image-making into problematic or provocative areas that these contacts allowed, I am more than grateful. Of course, not all of the formative influences behind this book were professional. A number of friends and family members—Paul, Ruth, Stephen, Otis, Julia, David, Beth, and Josephine—all provided healthy doses of moral support and therapeutic distraction during the years in which this book was written. To Thekla, I am indebted most of all—for her patience and for her good humor. By encouraging me early on to think about televisuality in relation to American cultural and social history, she helped me in ways that she will never know. To her this book is dedicated.

Part I

The Problem
of the Image

1 Excessive Style
The Crisis of Network Television

Television is to communication what the chainsaw is to logging.
—Director David Lynch[1]

There isn't much out there that looks real.
—Director/cameraman Ron Dexter[2]

Disruptive Practice

On the Friday September 8, 1989, edition of ABC's nightly news, erudite anchor Peter Jennings bemoaned the advent of what he termed "trash television." Prefacing his remarks by reference to a previous report by ABC on the subject, Jennings described the phenomenon with a forewarning. Norms of quality, restraint, and decorum notwithstanding, the new and ugly genre would in fact shortly premiere. Citing H. L. Mencken's adage about not overestimating the intelligence of the American people, Jennings signed off that evening with an obvious air of resignation. The class struggle, one sensed, might soon be lost.

Within two evenings, Jennings's warning was fulfilled. The highly evolved intertextuality that characterized television of the late 1980s was about to witness one of its most extreme manifestations to date. On Saturday, September 9, independent station KHJ-Channel 9 of Los Angeles uncorked the one-hour premiere of *American Gladiators* in syndication to stations throughout the country. Two nights later superstation KTLA-Channel 5 of Los Angeles aired its own nationwide trash spectacular—a two-hour premiere version of a show named *Rock-and-Rollergames*. Later that fall, pay-per-view television made available to cable viewers nationwide a special called *Thunder and Mud*. This latter trash hybrid featured various low-culture luminaries, and included Jessica Hahn of the recent PTL–Jim Bakker sex scandal, female mud wrestlers, wild-man comedian Sam Kinison, and the all-woman heavy-metal rock group She-Rok. The *Los Angeles Times* labeled the spectacle "Sex, Mud, and Rock-and-Roll." The producers, however, preferred the derivative punch of *Thud* to the official program title *Thunder and Mud*. To the show's makers, *Thud* was a "combination female mud-wrestling act–heavy-metal rock concert–game show with some comedy bits thrown in."[3] If any doubts remained about mass culture's reigning aesthetic in 1989, it was certainly clear that stylistic and generic restraint were not among its properties.

Rock-and-Rollergames was slated, interestingly, to air against the widely popular and front-running network sitcom *Roseanne*. Such competition was formidable given the vertically scheduled and heavily promoted sequence of shows that followed *Roseanne* Tuesday nights on ABC. Given trash television's excessive and low-culture pretense, such competition was significant, since *Roseanne* was being celebrated as television's premier "low-culture" hit; a status it had achieved with both viewers and tabloids during the previous season. By late fall, KHJ-TV had shifted *Rock-and-Rollergames* to Saturday mornings, and had renamed (and reduced) the spectacle to *Rollergames*, still in wide syndication in 1993. *American Gladiators* found itself shifted later in the year to the weekend schedule, and in subsequent years to late weekday afternoon strips. Together with its primetime airings, *Gladiators* found a lucrative niche by actively extending its competition out into the audience. Open trials were held throughout the country in highly publicized gladiator competitions at places like the Los Angeles Coliseum. Although trash television did not turn out to be an overwhelmingly dominant genre in primetime, *Gladiators* and other shows continued successfully in production with much success through the next four seasons. Musclebound, steroid-pumped women gladiators like Zap continued to grace the pages of *TV Guide* and the sets of celebrity talk shows through 1993.[4] A medieval variant of the trash spectacular, called *Knights and Warriors*, entered the trash programming fray in 1992–1993. Nickelodeon hyped and cablecast its hyperactive trash-gladiatorial clones, *Guts* and *Guts: All Stars*, for the younger set throughout the 1993–1994 season.

Although the genre was defined from the start by its distinctive no-holds barred look, trash spectaculars were also symptomatic of a broader and more persistent stylistic tendency in contemporary television—one that was not always castigated as trash nor limited to low-culture content. That is, trash-spectaculars can be seen as a stylistic bridge between lower trash shows—like professional wrestling or *The Morton Downey, Jr.* shock-talk show (series that exploited very low production values to blankly document hyperactive onstage performances for the fan situated squarely in the stands or on the sofa)—and higher televisual forms that more extensively *choreograph* visual design, movement, and editing *specifically for the camera*. Even mid-1980s shows with higher cultural pretension or prestige, like *Max Headroom* or *Moonlighting* on ABC or MTV's manic game show *Remote Control*, frequently stoked their presentational engines with excesses not unlike those that characterized trash spectaculars. Although broadcast manifestations of the televisual tendency took many shapes, stylistic excess has continued to rear its ostensibly ugly head—even in the ethically pure confines of Peter Jennings's network news division.

Bells and Whistles and Business as (Un)Usual

> We don't shy away from the aesthetic nature of the business. We have one foot on the edge, and we have to keep it there.
> —Local station executive, WSVN-TV, Miami

Starting in the 1980s, American mass-market television underwent an uneven shift in the conceptual and ideological paradigms that governed its look and presentational demeanor.[5] In several important programming and institutional areas, television moved from a framework that approached broadcasting primarily as a form of word-based rhetoric and transmission, with all the issues that such terms suggest, to a visually based mythology, framework, and aesthetic based on an extreme self-consciousness of style. This is not just to say that television simply became more visual, as if improved production values allowed for increasing formal sophistication. Such a view falls prey to the problematic notion that developments in technology cause formal changes and that image and sound sophistication are merely by-products of technical evolution. Rather, in many ways television by 1990 had retheorized its aesthetic and presentational task.[6] With increasing frequency, style itself became the subject, the signified, if you will, of television. In fact, this self-consciousness of style became so great that it can more accurately be described as an activity—as a performance of style—rather than as a particular look.[7] Television has come to flaunt and display style. Programs battle for identifiable style-markers and distinct looks in order to gain audience share within the competitive broadcast flow. Because of the sheer scope of the broadcast flow, however—a context that simultaneously works to make televised material anonymous—television tends to counteract the process of stylistic individuation.[8] In short, style, long seen as a mere signifier and vessel for content, issues, and ideas, has now itself become one of television's most privileged and showcased signifieds. Why television changed in this way is, of course, a broader and important question. Any credible answer to the question is only possible after systematically and patiently analyzing representative program texts. By closely examining style and ideology in a range of shows and series that celebrate the visual, the decorative, or the extravagant a more fundamental reconsideration of the status of the image in television becomes possible.

Televisuality was a historical phenomenon with clear ideological implications. It was not simply an isolated period of formalism or escapism in American television or a new golden age. Although quality was being consciously celebrated in the industry during this period, the celebration had as much to do with business conditions as it did with the presence of sensitive or serious television artists.[9] Nor was televisuality merely an end-product of postmodernism.[10] The growing value of excessive style on primetime network and cable television during the 1980s cannot simply be explained solely by reference to an aesthetic point of view. Rather, the stylistic emphasis that emerged during this period resulted from a number of interrelated tendencies and changes: in the industry's mode of production, in programming practice, in the audience and its expectations, and in an economic crisis in network television. This confluence of material practices and institutional pressures suggests that televisual style was the symptom of a much broader period of transition in the mass media and American culture. Yet historical changes are seldom total. Six principles—ranging from formal and generic concerns to economic and programming functions—further define and delimit the extent of televisuality. These qualifications will be more fully examined through close analysis in the chapters that follow.

1. Televisuality was a stylizing *performance*—an exhibitionism that utilized many different looks. The presentational manner of televisuality was not singularly tied to either low- or high-culture pretense. With many variant guises—from opulent cinematic spectacles to graphics-crunching workaday visual effects—televisuality cut across generic categories and affected some narrative forms more than others. For example, the miniseries proved to be a quintessential televisual form, while the video-originated sitcom—at least with a few notable exceptions—resisted radical stylistic change. Conceived of as a *presentational attitude*, a display of knowing *exhibitionism*, any one of many specific visual looks and stylizations could be marshaled for the spectacle. The process of stylization rather than style—an activity rather than a static look—was the factor that defined televisual exhibitionism.

Bells-and-whistles TV.
Nightly zooms through
crystalline and chrome
digital packaging on
Entertainment Tonight.
(Paramount)

Consider *Entertainment Tonight*, for example, a hallmark televisual show that influenced a spate of tabloid, reality, and magazine shows during the 1980s. *Variety* hailed *ET*, a forerunner of tabloid horses *A Current Affair* and *Inside Edition*, as "the granddaddy of all magazine strips" for its "brighter look and provocative stories."[11] Having survived over three thousand individual episodes and having prospered nationally in syndication for over a decade by 1993, the show's executive producer explained the show's secret to success: "We continued to update our graphics and other elements of production, the bells and whistles. If you look at our show, let's say once a month for the last seven years, the only constants are the title, the theme, and John and Mary hosting the weekday show. Everything else continues to change. So we go through a continual process of reinventing the wheel."[12] *ET*, then, airing five days a week, year-round, defines itself not by its magazine-style discourse or host-centered happy talk, but by the fact that the viewer can always expect the show's style—its visual and graphic "bells and whistles"—to change. Televisuality, then, is about constantly reinventing the stylistic wheel.

2. Televisuality represented a *structural inversion*. Televisual practice also challenged television's existing formal and presentational hierarchies. Many shows evidenced a structural inversion between narrative and discourse, form and content, subject and style. What had always been relegated to the background now frequently became the foreground. Stylistic flourishes had typically been contained through narrative motivation in classical Hollywood film and television. In many shows by the mid-1980s however, style was no longer a bracketed flourish, but was the text of the show. The *presentational status* of style changed—and it changed in markets and contexts far from the prestige programming produced by Hollywood's primetime producers.

Broadcasting magazine, for instance, described the dramatic financial reversal of the Fox television affiliate in Miami's highly competitive market. The ratings success of WSVN-TV was seen as a result of the station "pumping out" seven hours of news "that mirrors the music video in its unabashed appeal to younger viewers—flashy graphics, rapid-fire images, and an emphasis on style." While the trades saw the economic wisdom of stylistic overhauls like this one, television critics marveled at the station's able use of an aggressive, wall-to-wall vi-

Cutting-through-the-clutter. Videographic televisuality ruled "live" coverage of the 1992 Barcelona Olympics, and included high-resolution virtual reality displays of Mediterranean geography and architecture, and expressionistic nightly music videos "commissioned" by the network and its sponsors. (NBC)

sual style to revive a dead station.[13] Even the vice presidents at WSVN theorized the journalistic success of the station in artistic terms, as a precarious but necessary form of aesthetic risk taking.[14] By marketing cutting-edge news as constructivist plays of image and text, even affiliate station M.B.A-types now posed as the avant-garde.

3. Televisuality was an *industrial product.* Frequently ignored or underestimated by scholars, television's mode of production has had a dramatic impact on the presentational guises, the narrative forms, and the politics of mainstream television. More than just blank infrastructure, television's technological and production base is smart—it theorizes, orchestrates, and interprets televisual meanings—and is partisan. The mode of production is also anything but static; it changes. The production base, then, is both a product of shifting cultural and economic needs and a factor that affects how we receive and utilize television and video. In order to talk adequately about television style or narrative, one needs at least to recognize that *television is manufactured.*

Technology, geographical issues, and labor practices were all important components in the formation of a televisual mode of production. There was, for example, a direct relationship between certain production tools—the video-assist, the Rank-Cintel, digital video effects—and popular program styles in the 1980s. Yet these tools did not cause television's penchant for style. Rather, they helped comprise an array of conditions, and a context, that allowed for exhibitionism.

Bravura lighting design and "feminine" boutique sensitivity in quintessential televisual showcase, *China Beach.* (ABC)

Digital video technology, for instance, has had an enormous effect on the look of mass-market television, yet it has not infiltrated all of television's dominant genres. Primetime workhorse Universal Television, for example, justified digital technology only in terms of specific narrative needs: "We're getting more and more into digital. There's no standard formula that's applied to every single show, and every show doesn't get the same services in the same way from the same facility. With sitcoms, we're not into the digital domain yet—not to say that it won't be happening soon."[15] Digital imaging, then, once thought to be a futurist preoccupation now shadows even conservative television genres in Hollywood. Yet far from the sitcoms and genres of Studio City and Burbank, the same effects became defining factors and prized properties. At a very different level in broadcasting during this period even local stations began to manufacture alternative identities around a technologically driven aesthetic. In "personnel upgrades" regional broadcasters like WRAL-Channel 5 in Raleigh, North Carolina expected key production personnel to have visual arts degrees and electronic imaging technologies skills. News and promotional producers as well were now expected to "be able to write and produce promos that sell and touch emotion, (and) *must be proficient in all aspects of production with an eye for visuals that cut through the clutter.*"[16] Even outside of primetime's prestige television, then, market competition was defined as clutter; and striking visuals and high-tech graphics as obligatory corporate strategies. New televisual tools had arisen to meet the dense on-slaught of programming alternatives.

4. Televisuality was a *programming phenomenon*: Showcase television in it-self was nothing new, but the degree to which broadcasters showcased to counter-program was distinctive. Television history offers important and influential precedents for quality television: the Weaver years at NBC in the 1950s; the MTM/Tinker years at CBS in the 1970s; the Lear era at CBS in the 1970s. Program-ming designed around special-event status was also not entirely new, although the kind of prestige and programming spin that special events offered threatened to dominate television by the late 1980s. Everything on television now seems to be pitched at the viewer as a special event—from nondescript movies of the week to the live coverage of some local catastrophe on the eleven o'clock news—so much so in fact, that the term *special* is now almost meaningless. Showcase and event strategies that used to be limited to sweeps now pervade the entire year. No pro-gramming confesses to being commonplace.

While event-status television offered programmers one way to schedule nightly strips, "narrowcasting"—a result of demographic and ratings changes starting in the late 1970s—allowed for a different kind of aesthetic sensitivity in primetime programming. Broadcasters began to value smaller audiences if the income-earning potential and purchasing power of those audiences were high enough to offset their limited numbers. Narrowcast shows that averaged ratings and shares in the low- and mid-teens in the late 1980s—like ABC's *thirtysomething* and CBS's *Tour of Duty*—would never have survived a decade earlier, given the higher ratings expectations in broadcasting at that time.[17] The audience numbers needed for primetime success continued to fall in the 1980s. Although the Nielsens were slow to change from their ideal of an average viewing family, advertisers, cable, and direct broadcast satellite systems (DBS) executives were obsessed with clarify-ing ever narrower niches tied to economic, racial, age, and regional differences.[18] This industrial reconfiguration of the audience, in the name of cultural diversifi-cation, helped spawn the need for cultural- and ethnic-specific styles and looks. Fox, Black Entertainment Television, TNT, and Lifetime each developed distinc-tive and highly coded looks that reflected their narrowcast niches and network personalities. Gender- and ethnic-specific groups do not, apparently, coalesce around content-specific narration alone. Stylistic ghettos continue to be manu-factured by cable and broadcast networks according to maps of their supposed niche potential.

5. Televisuality was a *function of audience*: While the audience was being redefined and retheorized from the outside by broadcast and cable programmers, the cultural abilities of audiences had also apparently changed by the 1980s. While trash spectaculars betrayed new stylistic appetites in what have traditionally been deemed lower-taste cultures, the networks during this period learned to cash in on yuppie demographics as well. Many viewers expected and watched programs that made additional aesthetic and conceptual demands not evident in earlier pro-gramming. Even if such demands came in the form of irony or pastiche, shows like *Late Night with David Letterman* on NBC and *The Gary Shandling Show* on HBO presupposed a certain minimal level of educational, financial, and cultural capital. Such a background provided viewers both with an air of distinction—as

viewers in the know—and presupposed enough free time to actually watch late-night programming, terrain once written off as fringe.

The fact that television was no longer anonymous, also presupposed fundamental changes in the audience. While many British directors in Hollywood—Ridley Scott (*Thelma and Louise*), Tony Scott (*Top Gun*), and Alan Parker (*Mississippi Burning*)—had started as highly respected television commercial directors before breaking into features, American film directors before the 1980s had typically been segregated away from television agency work. There was, and to some degree still is, an ego problem and an institutional wall between the advertising and feature-film worlds. This segregation, however, began to change in the mid-1980s. Heavyweight film directors now were self-consciously hyped as producers of TV commercials.[19] David Lynch (*Blue Velvet*) designed pretentious primetime Obsession ads; Martin Scorsese (*Mean Streets*) produced an opulent Georgio Armani cologne spot for a mere $750,000; Rob Reiner (*A Few Good Men*) and Richard Donner (*Lethal Weapon*) both directed big-production-value Coca-Cola spots; Francis Ford Coppola (*Apocalypse Now*) directed a sensitive and familial thirty-second road-movie for GM that was never aired; Woody Allen (*Annie Hall*) spun out a commercial supermarket farce; Jean-Luc Godard (*Tout Va Bien*) did an avant-garde and overblown European Nike ad; and the venerable Michelangelo Antonioni (*Blow-up*)—the closest thing that serious Western cinema has to an aesthetic patriarch—choreographed a psychedelic spot for Renault.

The clients of these figures were inevitably impressed with their auteurist entourage, at least before the spots were produced and aired. Agency directors who had dominated the field outside of Hollywood, however, were less than happy with this opportunistic invasion of showcased aesthetes on television. Clio Award–winning director Joe Pytka grumbled that such auteurs underestimated the difficulty of the fifteen- and thirty-second short-forms.[20] Nevertheless, the auteur-importing mode—popularized by Ridley Scott, Spike Lee, and David Lynch in the mid-1980s—remained a viable fashion well into the 1990s with CAA's strong-armed invasion of the ad world from Hollywood in 1993.[21] The auteur-import business raises important questions about television practice: who is patronizing whom? CAA now acts more like televisual Medicis—with directors Donner and Reiner their kept artists—than traditional point-of-sale admen. Aesthetic promotion now flows both ways in commercial television—to the client and the patron—in a corporate ritual seldom kept from the audience. Audience consciousness and facility makes this commerce of authorial intent possible. Marketing aesthetic prestige presupposes and strokes audience distinction and self-consciousness.

6. Televisuality was a product of *economic crisis*. Televisuality cannot be theorized apart from the crisis that network television underwent after 1980. Stylistic excess can be seen as one way that mainstream television attempted to deal with the growing threat and eventual success of cable. Stylistic showcases, high-production value programming, and Hollywood stylishness can all be seen as tactics by which the networks and their primetime producers tried to protect market share in the face of an increasingly competitive national market. No longer could CBS, NBC, and ABC—protected by the government as near monopolies since the late 1940s and early 1950s—assume the level of cash flow that they had enjoyed up to the late 1970s. Although the networks faced the first cable players in

1980 and 1981 with a smug and self-confident public face, this facade began to crack as each year took its toll on corporate profits. CNN and MTV were merely the first in a line of very profitable challengers to sign on to cable for the long haul. The trades gave blow-by-blow accounts of the precipitous decline in network primetime viewing. The networks had enjoyed complete dominance—an incredible 90 share—during the 1979–1980 season, but saw this figure plummet to a mere 64/65 share by 1989–1990.[22]

Complicating things further still, the new fourth network, Fox, was profitable by 1989 and was eating directly into shares of the big three. Financial analysts reported that 40–45 percent of Fox's additional revenues in 1993 would come at the expense of ABC, directly reducing the network's yearly revenues by $50 million.[23] Fox's growing appetite was a network problem, not just because there were fewer pieces of market pie to share with Fox, but because of the demographic stratification within those pieces. Fox gained its market toehold and survived by specializing in the hip 18–34 demographic, but now was expanding into ABC's 18–49 year old range. Escalating and unrealistic production budgets were also part of the network diagnosis. The trades explained the collapse of quality network shows like *thirtysomething*, *I'll Fly Away*, and *The Wonder Years* as a result of inflated budgets. Even half-hour shows could now regularly cost $1.2 million per episode—an unheard-of level even for hour-long series two decades earlier.[24] By the late 1980s front-page stories in the national press were loudly trumpeting the demise of the networks, who were "under attack"—besieged by an array of new video delivery technologies.[25] By the early 1990s, the networks were publicly wringing their hands, as victims of cable, of unfair regulatory policies, and of syndication rules. Government regulators characterized the market as driven by "ferocious competition that was unimagined ten years ago."[26]

In one decade, network viewership declined by a corporation-wrecking 25 percent. This hemorrhaging of viewership may seem ironic, given the fact that high-style *televisuality also emerged during the very same years*. I will argue in the pages that follow, however, that televisuality addressed the very same economic problem that hostile takeovers would tackle in 1986 and 1987. Stylistic exhibitionism and downsizing were obviously very different organizational tactics. One came from programming and encouraged budget-busting expenditures of capital; the other came from corporate management and brought with it widespread layoffs and fiscal austerity. Yet both, paradoxically, attempted to solve the same corporate crisis: the declining market share of the networks. In some ways, this was payback time for network television. The market incursion of network broadcasters in the 1950s had itself created an economic crisis in Hollywood that sent the film studios scrambling for excessively styled forms: cinemascope, technicolor and 3-D. Depending on how one looked at it, televisuality in the 1980s was either a self-fulfilling deathwish by extravagant producers, or a calculated business tactic that increased market share.

The Terrain

Televisuality can be mapped out along several axes: formal, authorial, generic, and historical. From a formal perspective, televisual programs gained notoriety

Television's inherent limitations? Epic casts and geography, deep space, and widescreen iconography typify cinematic televisuality in George Lucas's *Young Indiana Jones* in 1992–1993. (ABC)

by exploiting one of two general, and production-based, stylistic worlds: the cinematic and the videographic. The cinematic refers, obviously, to a film look in television. Exhibitionist television in the 1980s meant more than shooting on film, however, since many nondescript shows have been shot on film since the early 1950s. Rather, cinematic values brought to television spectacle, high-production values, and feature-style cinematography. Series that utilized this mode typically promised broadcasters and audiences alike television's big picture. Situated at the top of the programming hierarchy, shows like *Moonlighting*, *Crime Stories*, *Wiseguy*, and *Beauty and the Beast* fit well with the financial expectations of network primetime programming. They also inevitably drew critical attention by their very programming presence and cinematic air of distinction. It was as if the televisual producers packaged labels with their cinematic shows that read: *"Panavision Shows That We Care."*

Televisual programs that exploited the videographic guise, on the other hand, were more pervasive and perhaps more anonymous than cinematic ones, but were certainly no less extravagant in terms of stylistic permutation. In fact, for technological reasons, videographic shows made available to producers, at any given time, more stylistic options. That is, such shows had more embellishment potential given their origins in electronic manipulation. Far different from the bland and neutral look that characterized video-origination studio productions in earlier decades, videographic televisuality since the 1980s has been marked by acute hyperactivity and an obsession with effects. If MTV helped encourage the stampede to film origination in primetime, then CNN demonstrated the pervasive possibilities of videographic presentation. Starting in 1980—and without any apparent or overt aesthetic agenda—CNN created and celebrated a consciousness of the televisual apparatus: an appreciation for multiple electronic feeds, image-text combinations, videographics, and studios with banks of monitors that evoked video installations. Ted Turner had coauthored the kind of cyberspace that video-freaks and visionaries had only fantasized about in the late 1960s.

The "give 'em hell" look of MTV, on the other hand, popularized a mode of production that changed the very way television was produced beginning in 1981. By shooting on various film formats and then posting electronically on tape, indie producers were no longer rigidly locked into a single production medium. Earlier telefilm producers, by contrast, were required to produce a single conformed negative for broadcast and a positive print made from source material that was uniform in format and stock. With the MTV prototype, however, it no longer mattered where the material came from (stock, live, graphic material), what format it was shot on (super-8mm, 16mm, 35mm), or whether it was black and white or color. Once transferred in post-production to electronically recorded tape, almost any element could be combined or composited, mixed and matched, in useable configurations. Not only were labor and business structured differently in the new videographic worlds—CNN was a nonunion shop and MTV utilized an eclectic mix of production personnel (from studio technicians to independent producers to animators to video artists)—but the very technologies that gave MTV national audiences allowed for and encouraged a different kind of look and stylistic expectation.

By the time of its network airing in 1986, then, *Max Headroom* was merely a self-conscious and premeditated reference and homage to an existing industry-proven, videographic prototype. Like two other pervasive forms, primetime commercial spots and music videos, *Max Headroom* operated between the two stylistic modes by mixing and matching elements from both cinema and digital imaging. Purer forms of videographic televisuality were actually more common in cost-effective and low-end programming, like *Entertainment Tonight*, *Hard Copy*, *America's Most Wanted*, and *Rescue 911*. Live coverage—especially of crises like the Gulf War, the L.A. rebellion, and the yearly spectacle of the Super Bowl—has had an especially ravenous appetite for videographic exhibitionism. Although videographic series typically evoke less critical attention than prestige cinematic programming, they frequently undergo more tortured attempts to crank-out aesthetic embellishment. A fetish for effects rules the videographic domain. This emerging formal axis between the cinematic and the videographic can also be sketched out in historical terms (Table 1.1).While *Moonlighting*, *Crime Stories*, *The Equalizer*, and *thirtysomething* continued to push film-based primetime into more excessive directions, *Max Headroom*, *Pee-Wee's Playhouse*, and *Remote Control synthesized* the electronic and videographic lessons of MTV and CNN for new national audiences.

A second way to understand televisual programming is to consider it along an axis formed by relative degrees of authorial intent and manufactured notoriety. Commercial spots were certainly not unique in flaunting directorial auras to the

Table 1.1   The Historical Field Televisual Events

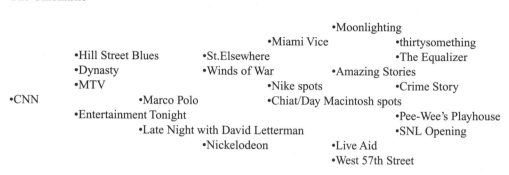

1980	1981	1982	1983	1984	1985	1986

The Cinematic

•Moonlighting

•Miami Vice   •thirtysomething

•Hill Street Blues   •St.Elsewhere   •The Equalizer

•Dynasty   •Winds of War   •Amazing Stories

•MTV   •Nike spots   •Crime Story

•CNN   •Marco Polo   •Chiat/Day Macintosh spots

•Entertainment Tonight   •Pee-Wee's Playhouse

•Late Night with David Letterman   •SNL Opening

•Nickelodeon   •Live Aid

•West 57th Street

The Videographic

NOTE: The *general* position of the titles laid out along the vertical axis suggests the *degree* to which the series utilized one of the two most privileged televisual modes plotted: the cinematic or the videographic mode. Although many of the positions are only approximations (since some shows might emphasize one trait in one episode and the other in another installment), some clear and direct lineages are evident: from CNN to *Live Aid* to the coverage of the L.A. rebellion and *Zoo TV* on the electronic, videographic axis; and from MTV to *Miami Vice* to *Quantum Leap* and *Cop Rick* on the cinematic axis. Note as well that

mass audience. Part of the emergence of the quality myth in 1980s television was that television was no longer simply anonymous as many theorists had suggested. Names of producers and directors assumed an ever more important role in popular discourses about television. While Aaron Spelling and Norman Lear were already household names, other producer-creators like Michael Mann and Stephen Bochco began to be discussed alongside their actors and series in popular magazines and newspapers. As with American film in the 1960s, authorial intent played an important role as an indicator and guarantor of aesthetic quality in primetime programming of the 1980s.

The centrality of this process can be demonstrated by considering televisual authorship within three organizing systems: marquee/signature television, mainstream conversions, and auteurist imports. Not only do the corporate and cooperative origins of television complicate singular attributions of authorship, but the various degrees of authorial deference within the industry show that televisual authorship is also part of calculated programming strategy. While many shows merit no authorial attribution by broadcasters, some demanding shows exploit it as a lifejacket, as a prop for survival, as a device used to weather programming clutter and to find loyal audiences. When *Brooklyn Bridge* suffered threats of net-

1987	1988	1989	1990	1991	1992	1993
						•Wild Palms
			•Cop Rock	•Northern Exposure		•seaQuest dsv
•Max Headroom		•Quantum Leap		•Homefront		•Tribeca
•Beauty and the Beast			•Twin Peaks	•I'll FlyAway	•Picket Fences	
	•China Beach			•Brooklyn Bridge		
		•Midnight Caller		•Seinfeld	•Young Indiana Jones	
	•War and Remembrance		•Camarena		•Hat Squad	•Homicide
	•Lonesome Dove					
•AT&T Spots		•Men	•Herman's Head		•Phillips' Clear Sound spots	
			•The Simpsons			•X-Files
	•Obsession spots		•In Living Color			•Animaniacs
•Tracey Ullman Show		•American Gladiators			•Mighty Morphin Power Rangers	
•Remote Control		•The Julie Brown Show			•P.D.I. Morphing spots	
•Max Headroom			•Hard Copy		•Roundhouse	
•A Current Affair		•Rock-and-Rollergames			•1992 Olympics	
	•Unsolved Mysteries		•America's Funniest Home Videos			
	•America's Most Wanted				•Honda VR spots	
				•I-WitnessVideo		
			•Gulf War		•Zoo TV/ Fox	
				•L.A. rebellion		•Bradymania

other shows, like *Max Headroom* in 1987, was influential on both ends of the stylistic spectrum: it flaunted the postmodern, dystopic cinematic looks of Ridley Scott's feature *Blade Runner,* but collaged them with endlessly dense vidographic and cybernetic configurations, surveillance, etc. Whenever possible, the series listed correspond to the years cited above them in the table, although space limitations mean that these are only approximations. Many of the shows and series laid out on this chart will be discussed in the pages and chapters that follow.

work cancelation in 1992 and 1993 due to poor ratings, the show's producer took stage before the press on a regular basis—proof positive that there was indeed a series of substance hidden under the very nose of the American people. Such showcase producers are manufactured by production companies and networks as banner-carriers, much in the way that Stephen Bochco and Michael Mann were showcased in 1981 and 1984. Other signature producer types, like Marshall Herskovitz, Diane English, David Letterman, and Joshua Brand, continue this showcasing tradition. As signature producers, each has functioned as a promotional marquee; a spotlit entré for programming seasons on their respective networks: ABC, CBS, NBC, and Fox (Table1.2).

While these marquee producer-creators are television insiders—honed, disciplined, and groomed through the ranks of the primetime industry—a second group of televisual authors come from the outside. Auteur imports from Hollywood and feature filmmaking typically elicit even more press than the Bochcos and Manns, but usually garner much poorer ratings. Among the great ironies of programming in the past decade has been, first, that television now attracts a wide range of film directors, and second, that most of these directors "fail" when brought on board to produce shows for networks. The list of auteur imports in primetime

Table 1.2   The Players   Primetime Televisuality

I. Showcase Producers Marquee Signatures Network Banner Carriers	II. Mainstream Conversions Acquired Mannerisms Embellished Genres	III. Auteur–Imports Cinematic Spectacle Visionary Emigres
Bochco, Steven Hill Street Blues, 1981–1987 L.A. Law, 1986–1993 Cop Rock, 1990–1991 NYPD Blue, 1993–	*Bellasario, Donald* Magnum P. I., 1980–1988 Tales of the Gold Monkey, 1982 Airwolf, 1984–1988 Quantum Leap, 1989–1993	*Spielberg, Steven* Amazing Stories, 1985–1987 seaQuest dsv, 1993–1994

Mann, Michael
Miami Vice, 1984–1989
Crime Story, 1986–1988
Drug Wars: Camarena Story,
   1990
	Cannell, Stephen J.
	The A-Team, 1983–1987
	Riptide, 1984–1986
	Stingray, 1986–1987
	Wiseguy, 1987–1990
	Silk Stalkings, 1991–
		Coppola, Francis
		The Outsiders, 1989–1990
		Faerie Tale Theater, 1986
		The Conversation, 1995–
		
		Lynch, David
		Obsession spots, 1990
		Twin Peaks, 1990–1991
		American Chronicles,
		   1990–1991
		Hotel Room, HBO, 1993

English, Diane Murphy Brown, 1988– Double Rush, 1995–	*Spelling, Aaron* Love Boat, 1977–1986 Fantasy Island, 1978–1984	

Letterman, David
Late Night With, NBC,
   1982–1993
David Letterman Show, 1993–
	Dynasty, 1981–1989
	Hotel, 1983–1988
	Twin Peaks, 1990–1991
	And the Band Played On, 1994
		Scott, Ridley
		Chiat/Day Mac spots, 1984
		
		Townsend, Robert
		Townsend Television,
		   1993–1994
		Parent' Hood, 1995–
		McDonalds Spots, 1995–

Kelly, David Picket Fences, 1992– Chicago Hope, 1994–	*Brooks, James L.* Mary Tyler Moore, 19780–1977 Tracey Ullman Show, 1987–1990	*Spheeris, Penelope* Thunder and Mud, 1989 Prison Stories HBO, 1991
Hewrskovitz and Zwick thirtysomething, 1987–1992 My So Called Life, 1994–	The Simpsons, 1990–	*McBride, Jim* The Wonder Years, 1990
Brand and Falsey St. Elsewhere, 1983–1984 Northern Exposure, 1990– I'll Fly Away, 1991–1994		*Lee, Spike* MTV Spots, 1986 Saturday Night Live, 1986 Air Jordan/Nike spots, 1990
		Lucas, George The Young Indiana Jones, 1992–1994, 1994–
		DeNiro, Robert Tribeca, 1993
		Zemeckis, Robert Tales from the Crypt, 1992
		Levinson, Barry Homicide, 1993–1995
		Stone, Oliver Wild Palms, 1993

Primetime's "auteur business" pitches the "signature" angle to the public in network public relations photos. Brandon Tartikoff sharing his vision with Eddie Murphy and company on the set of *Saturday Night Live* in 1983. NBC auteur Steven Spielberg imported for *Amazing Stories* in 1985 and for *seaQuest dsv* spectacular in 1993–1995. (Terrence O'Flaherty Collection, Arts Library—Special Collections, UCLA)

television since 1980 reads like a who's who list at Cannes: Spike Lee, Francis Ford Coppola, Steven Spielberg, George Lucas, David Lynch, Ridley Scott, Robert DeNiro, Barry Levinson, and Oliver Stone have all "done" TV. If the insider showcase producers are signature banner-carriers who give a network's seasonal offerings personality, then the auteur imports are aesthetic badges and trophies of distinction pure and simple. To the networks, any financial risk that comes with Lucas, Spielberg, and Stone is, apparently, worth it. What the networks get in return is a visionary aura of artistry and aesthetic challenge—an attitude they can toy with, at least until cancelation inevitably comes. Never mind the fact that much of the rest of programming is by comparison authorially mundane. Even if for a fleeting season, this imported class and visionary flash promises to work wonders for network programming—at least when aired and hyped in the right way; as when Levinson's *Homicide* was slated after the Super Bowl telecast or when Oliver Stone's *Wild Palms* barraged the viewer on multiple May sweeps nights in 1993.

So what, you say, Lynch, Levinson, and Stone were no more than bright but very curious flashes in the programming pan. This kind of import business in itself is no proof that televisuality dominated broadcasting during the 1980s. Yet, the auteurist import business is really just the tip of an iceberg. A third category of television producer-directors shows that authorship and excessive televisual style are linked in lower forms of television as well. The last fifteen years or so have also seen a marked stylistic change in successful but middle-of-the-road primetime producing figures. If stylistic exhibitionism was substantive and pervasive it should have affected these types of figures as well. Consider the marked stylistic differences between early and late works by Stephen J. Cannell, Donald Bellasario, Aaron Spelling, and James L. Brooks. In *Magnum P.I.*, *The A-Team*, and *Love Boat*, Bellasario, Cannell, and Spelling all succeeded by exploiting flesh and chrome rather than a stylized look. Such extra-cinematic objects were typically displayed in front of a neutral camera. The production apparatus was designed to show off and allow physical action and anatomy rather than pictorial or narrative embellishment. Later series by the same producers, however, show something very different. By the end of the decade Bellasario authored the quintessential televisual show *Quantum Leap*; Cannell choreographed *Sting-Ray*, a *Miami-Vice* clone; and Spelling banked on cinematic opulence in *Dynasty*. Other producers, like James L. Brooks—considered a quality producer by industry types and critics alike— actually gained their fame by making shows that were *visually uninteresting*. Brooks's early accomplishment in the rather bland-looking 1970s sitcom *Mary Tyler Moore* seems sedate by comparison to the presentational volatility and niche mentality of his *Tracey Ullman* show at Fox in the late 1980s. Even very competent, but middle-of-the-road producers, then, learned to value and exploit style for its own sake during this period. Who says television can't teach old dogs new tricks?

A third way of delineating the forms and functions of televisuality is to examine them within the framework of genre, since stylistic exhibitionism has not equally influenced all program formats. The extent of the trend can be better understood by comparing genres that favored televisual performance and those that did not. Television's bread-and-butter genres—where stylistic excess is an exception—include daytime talk shows, soap operas, video-origination sitcoms, nonprimetime public affairs shows, some public access cable shows, nonprofit public service announcement spots (PSAS), and late-night off-air test patterns. While PSAS and infomercials function as cheap but lucrative filler, the sitcom may have resisted televisuality for ideological reasons; that is, because of its inherently conservative cultural function.[27] In an ideological sense the sitcom, in almost every decade, always manages to reconfigure and update the nuclear family. In 1980s shows like *Full House* this meant awkwardly linking multiple parents of the same sex together as surrogate parental figures.[28] With the very myth, viability, and survival of the nuclear family as its chief creative task, primetime sitcoms had little need for the presentational possibilities—and the air of distinction—offered by stylistic exhibitionism. For a number of reasons, then, some genres simply do not care about style.

Many others, however, continue to share a marked penchant for stylistic exhibition. Although I have referred already to prestige film-origination genres—the miniseries, primetime soaps, and quality hour-long dramatic series—many other film-based shows market visual excess in broadcast and cable delivery systems: hyperactive children's television, archival syndicated programs, sitcoms and their parodies on Nickelodeon, feature film presentations on pay-cable channels, and a boundless number of commercial spots shot on film and aired across the channel spectrum. While many televisual program forms survived the industry's economic crash of 1989–1992, several flagship televisual genres from the 1980s—the miniseries, primetime soaps, and primetime dramas—were prematurely, and with some self-serving eulogies, declared dead by industry executives at the start of the 1990s. (The reasons for this eclipse and the ways that televisuality survived in hybrid forms are discussed in the final section of the book.)

Video-origination genres, by contrast, have continued to share a penchant for exhibitionism and include: network television sports shows, cable news, music television, magazine shows, most reality programming, home shopping networks, local commercials for cars and personal injury lawyers, and a veritable ton of interstitial and nonprogram material airing around the clock on almost all non-pay channels. Supermodel Cindy Crawford's videographic showcase, *House of Style* on MTV, provides but one explicit example of the exhibitionism that pervades television outside of primetime. Fashion- or anatomy-conscious thirteen-year-old white suburban girls or boys who watch her show are enticed by a type of performance that differs little from other televisual appeals made during off-primetime programming ghettos. Super-discount, high-volume auto-malls bankroll thirty- and sixty-second spot frenzies on the weekends; while ex-Fox CEO Barry Diller's QVC network teases home-shoppers with graphics-dense consumer bait. When downhome, regionally owned, right-to-work corporations like Wal-Mart produce national ads that look like a cinematographer's showreel, the implications are clear: mass retailing has made televisuality not just a passing production fashion, but a national consumer buying trend.

Modes and Guises

Popular TV critics have been accused by academics of wrongly isolating important and deserving shows out of the continual, redundant, and monotonous broadcast flow.[29] David Marc, for example, has argued against the critical and textual isolation of episodes, because: "The salient impact of television comes not from 'special events,' . . . but from day-to-day exposure. The power of television resides in its normalcy."[30] According to this perspective, individual episodes are rarely memorable, although series and their cosmologies are. Most academic theorists have followed this lead by attempting to elaborate fundamental structural and ideological conditions that comprise television's flow, the super-text, and the audience. I am arguing something very different here: that special television is a concern not just of critics, but of the industry and the audience alike. A great deal of television in the last fifteen years is significant precisely because it self-consciously rejects the monotonous implications of the flow and the conservatism of a slowly changing series cosmology. Whether or not televisual shows actually succeed in providing alternatives to this kind of stasis is not the issue. What is important is that they promote special status and pretend to both difference and change. Apart from a few important scholarly works, the very idea of special television has been undertheorized as an industry strategy and stylistic preoccupation.[31] Special television has historically played an important role in programming, and continues to do so today.

I have already indicated that many forms of televisuality have a difficult time raising their stylistic heads above the broadcast clutter, yet the obsession with distinction and with special status pervades both high and low forms of televisuality. *Whether deserving or not*, production technologies and writer-directors alike now continually angle for attributions of distinction. Many primetime televisual shows, for example, can be viewed as "loss leaders."[32] From this merchandising perspective, it does not totally matter if distinctive televisual shows—like *Homefront*, *Brooklyn Bridge*, and *Wild Palms*—score low ratings, return poor advertising revenues, and face cancelation. After all, most shows have low ratings and are canceled. The cancelation rate for new series is in fact overwhelmingly high and has been for some time. That is the very nature of television. This condition of turnover, an inherent part of program development, makes the critical emphasis on lauding select survivors a shortsighted fallacy. Since the 1970s, when shows like *All in the Family*, *Mary Tyler Moore*, and *M.A.S.H.* were designated and artificially isolated as distinctive, critics have ignored the vast majority of shows that come and go. Many shows that disappeared, ironically, made even more earnest formal and narrative claims to distinction than those critically privileged Emmy and ratings winners. Given the fact that most of the shows on television are ultimately ratings losers, which type of series should be deemed more symptomatic of a period, the few with high ratings and prestige, or the greater number with high prestige-claims but predictably low numbers? The ratings dominance in the mid-1980s of the conservatively styled *Cosby Show*, for example, does little to conceal the fact that almost everyone else up and down the ratings ladder was struggling to keep their signature looks above water. Distinction is an obligatory and pervasive programming tactic, not just a retrospective and limited critical attribution.

This kind of perspective—the cultural logic of distinction, of televisual loss leaders and special events—cannot be explained, however, without recognizing the fundamental role that style plays in facilitating distinction. More than just case-by-case formal taxonomies of televisual modes, then, the studies that follow aim to describe the favored guises of televisuality as part of a broader aesthetic economy. Coexistent with American mall culture, stylistic designations foreground television's obsession with merchandising and consumerism. The guises—boutique, loss leader, digital franchising, tabloid, trash, and ontological strip-mall—suggest both the programming logic of televisual forms and the types of presentational appeals and relationships televisual forms establish with viewers. Couching the performance of style in economic terms does more than just remove the discussion from the airless confines of formalism, it also demonstrates the fundamental industrial and cultural import of stylistic representations in television. Although my construal of an aesthetic economy may be open to criticism, the concept does not necessarily constrain or ignore the force of economic and political realities in the world at large. Television is part of the world at large and cannot be viewed apart from business conditions. This framework enables us to see televisuality as the industrial instrument and socially motivated ritual that it is. Unlike the fine arts, television aesthetics have never been locked into an intellectual netherworld of pure discourse.[33]

Those arrays of videographic signals and codes that are used pervasively in mass market television make up what might be called, in another context, its televisual language.[34] This language, furthermore, has emerged as part of a broader ideology of stylistic excess, one that pervades contemporary American television and mass culture alike.[35] Yet it is important to note from the start that even in the mass-produced industrial West, there is surely no singular ideology at work in mass culture.[36] For this reason, any paradigms that I refer to must be seen as part of a broader bundle of privileged views, some of which contradict each other. Paradigms can *compete, contradict,* and *coexist*.[37] Also, by "emergence," I hope to suggest that ideology, here taken to include even the way we think about art and imagery, involves an *uneven development*. Mythologizing takes place over time and so is inevitably partial or irregular in its presence.[38] A gloss of Thomas Kuhn's "paradigm shifts" in the history of science might suggest that revolutions in worldviews are drastic, comprehensive, and complete.[39] Last epoch's paradigms are, as it were, cleanly banished by new worldviews to the outdated ash heap of history. But this oversimplification of Kuhn in no way describes cultural change in the late twentieth century. Because there is no cultural pope to centrally organize and determine aesthetic culture, today's mass-media Copernicus must instead ply his or her paradigmatic wares on an open market—on a multinational electronic bazaar, only loosely regulated by the Federal Communications Commission (FCC).

For these reasons competing ideologies continue to coexist in broadcast and cable television. The visually and cinematically sophisticated *thirtysomething*, for example, was merchandized in book form even after its cancelation—a unique compendium of great and "sensitive" writing—as a way of "reliving moments with our favorite family."[40] Other shows, like Rush Limbaugh's syndicated right-wing shock-talk show, still invoke and use reductive studio production modes more typical of the 1970s than the 1980s. The question of which aesthetic paradigm governs television, in fact, depends as much on who you ask as on anything else. Robert S. Alley and Horace Newcomb, for example, claim that television is a "producer's medium"—but do so only after interviewing numerous television producers.[41] Jack Kuney argues that television is a director's medium—that the director "sets both the tone of the program and determines a show's impact on its audience"—after interviewing numerous television directors.[42] Whose medium do you think television would be if, instead, one interviewed editors, lighting designers, art directors or camera people; that is, any one of the hundreds of other people involved in primetime program production? The tension between aesthetic paradigms is not, then, limited to academic debates. Such conflict is an inevitable part of the television industry, as any one who has left a production "due to creative differences" can tell you. Given this context, then, the occasional presence of low-resolution or amorphous imagery within the present broadcast or cable spectrum does not disprove that a new aesthetic sensibility has emerged.[43] The trends and practices that I am theorizing are part of a *trajectory* of influences; notions that are bought and sold, aired and syndicated, cloned and spun-off. Even the ways that style is performed changes from season to season. Yet the widespread sensibility and urge to aestheticize and stylize suggests that televisuality is more than a passing fashion.

Mocking deconstruction of schlock footage on cable's *Mystery Science Theater 3000* in 1993, and on Dumont's *Window on the World* in 1949. Backstage reflexivity on *Father Knows Best* in 1955 and on *Seinfeld* in 1993. (Comedy Central, Dumont, NBC, NBC)

Nagging Theoretical Suspicions

Because so many recent trends in critical theory set themselves up in stark opposition to the aesthetic, the project outlined here may seem on shaky ground at best. John Fiske defines cultural studies, for instance, as a "political" framework in polar opposition to a study of culture's "aesthetic" products.[44] Why erase the aesthetic, in this way, as a theoretical and analytical category? What can be gained by this analytical retreat? Several important tactical assumptions, championed in high theory, work to hobble effective analyses of televisual style. Before examining, at the conclusion of this book, how a number of fundamental and strategic intellectual commitments have worked to conspiratorialize the image, it is important to consider how several more tactical schools in contemporary theory have impacted televisual analysis. Less antagonists of the image than heuristic complications, postmodernism, "deindustrialized" cultural studies, "glance theory," and the "ideology of liveness" myth all merit reexamination in light of television's penchant for exhibitionism.

Postmodernism. Given the traits of televisuality that I've already sketched out, one might ask, "Why not simply go to the postmodern theory as a basis for interpretation?" After all, the disembodied signifiers and textual extravagance that I describe here are central components in the postmodernist paradigm as well. This may be so. But stated simply, apart from textual description, postmodernism has little to offer broader explanations of American television. This is not because postmodern theory is wrong, only because the theory cannot be used easily to distinguish between what is postmodern and what is not postmodern in American television. Any systematic look at the history of television soon shows that all of those formal and narrative traits once thought to be unique and defining properties of postmodernism—intertextuality, pastiche, multiple and collaged presentational forms—have also been defining properties of television from its inception. Television history, unlike Hollywood film history, cannot be as neatly periodized into sequential stylistic categories: primitive, classical, baroque, modern, and postmodern. The hip gratifications that result from discovering loaded intertexts in *The Simpsons*, for example, are not necessarily unique to television in the 1990s, and tend to overshadow the fact that intertextuality was a central component in television from the start. Comedy-variety shows in the late 1940s and early 1950s—and not just Ernie Kovacs—repeatedly parodied and pastiched cultural conventions. Many other shows on both the local and national level combined intertextual fragments taken from various traditions—newsfilm, vaudeville, photography, radio comedy and drama—into single thirty- and sixty-minute program blocks. From a postmodernist point-of-view, 1940s and 1950s television had it all: self-reflexivity in *Burns and Allen*, intertextuality in *Texaco Star Theater*, direct address in *The Continental*, pastiche in *Your Show of Shows*; and social topicality—modernism's nemesis—in *I Love Lucy* and *The Loretta Young Show* (both made allusions to the Korean War for example).[45] Unlike classical Hollywood cinema, television had no centered gaze from the very start, and seldom had any seamless or overarching narrative. Multiple narrational modes issued from the same works, and audiences were constantly made aware of television's artifice and embellishment. In these ways, then, television has always been postmodern. Television has always been *textually messy*—that is, textural rather than transparent.

Consider, for example, two recent cable programs that gained critical notoriety in 1993: *Mystery Science Theater 3000* on Comedy Central and *Beavis and Butt-head* on MTV. Both bear all of the celebrated hallmarks of postmodernism and both utilize the same basic structuring motif: the series' "stars" sit and watch the same thing the audience watches, but make off-handed, on-camera comments that are either ironic, banal, hip, sexually loaded, or simply gross. Although *Beavis and Butt-head* is intended for a teen and preteen crowd that can appreciate a world numbed by too much glue-sniffing and *Mystery Science Theater 3000* is intended for jaded yuppies, both make an on-screen mockery of "found footage": *Mystery Science* deconstructs old low-budget trash and horror films; *Beavis and Butt-head* deconstructs heavy metal, music videos, and phantom video artifacts. Both are hip, ironic, and somewhat smart as well. Distinctively postmodern? Well, not quite. In the late 1940s Dumont Network's *Window on the World* used the same device. Dumont's host takes a breather from the show's on-stage action to mock and free-associate about bizarre turn-of-the-century archival footage unspooling in the studio's film chain. As beauty contestants in Atlantic City parade before the viewer, the announcer ironically mocks both their bizarre bodies and their ridiculous fashions. Like *Mystery* and *Beavis*, there is no laugh track. Like *Mystery* and *Beavis*, *Window* ironically deconstructs the newsreel for a knowing and hip *1949* audience. Like the audience for *The Simpsons* and for *Beavis and Butt-head*,

Dumont is playing its intertexts for those "in the know." Television has either always been postmodern, or its postmodern tactics are a part of a much different and less celebrated dynamic.

Apart from its descriptive capabilities however, postmodernism also speculates on the big picture behind such devices and attitudes. Fredric Jameson sees in such tactics late capitalism's logic and obsessive reenforcement of consumption. Jean Baudrillard is more fatalistic, and depoliticizes the universe even as he hallucinates about global spectacle. Jean-François Lyotard finds in postmodern practices evidence of the disappearance of distinctions between subject and object. But how do these explanations account for industrial and historical changes in American television? The descriptive tools of postmodern theory are powerful, but the theoretical grounding is frequently tautological. That is, once an account has committed itself to Jameson, Baudrillard, or Lyotard's logic, it tends to end up back at that logic even after exhaustive analysis. Postmodern theory can tell cultural analysts little more than the theory has already confessed to up front. Postmodern theory determines analysis, determines theory, determines analysis in an endless loop. I hope, in some small way not to prejudge this period in American television history by imposing tautological postmodern explanations. Such haste not only depreciates history, it also requires an analytical act of faith.

Deindustrialized cultural studies. Because of its underlying interest in popular culture and audience, this book shares in many of the tenets and objectives of cultural studies. Yet cultural studies, at least as it is sometimes marketed in academia and when it focuses on media, tends to gloss over one of the most important components of televisuality—the industry. This evasion is ironic given cultural studies current fascination with "technologies"—of gender, of cybernetics, of surveillance, of the body, of medical discourses. Even as television studies disappear into cultural studies, the field tends to ignore the extensive technological base of the subject itself: TV technology. It is perhaps easier to traverse multiple pop culture fields, than to account for workhorse technologies that comprise the dominant institution through which mainstream America consumes culture. The television industry may not be as flashy as VR and cyberpunk (tell that to Cindy Crawford, Billy Idol, and the folks over at *Liquid Television*), but it is, depending on one's perspective, surely no less problematic or ideologically complex.

As the academic turf called culture is taken on by humanities and arts colleges, media theorists now range freely and easily over the discursive and problematic turf once owned by sociologists, anthropologists, and political scientists. This recent and important intellectual overhaul, the leap to the culturally macroscopic as an antidote to disciplinary Balkanization, risks ignoring the need for more preliminary and extensive groundwork studies on questions of cultural and institutional *stylistics*. Even the development of a stylistic poetics of television is an important project.[46] It has become an academic fashion in recent years for intellectuals to go native, and so wed high critical theory with thoroughly vernacular forms from low culture. Recent anthologies suggest an institutional desire for validation at the hands of the everyday and the banal.[47] I am less interested in wielding the weapons of high theory to stake intellectual claim to the lowly than in explicating and questioning the ways that low culture itself performs and theorizes. All programming forms are complicated and mediated by style and technology even though "scientific" approaches, and many cultural studies approaches, tend to avoid or downplay this fact.[48] My call here, then, is not simply "back to the text," but back to the "televisual apparatus."[49] For after recognizing and accounting for the centrality and complexities of style, it is important to move beyond mere formal taxonomies. Describing how televisual technologies allow for, but also cut off and delimit, engagement with viewers, is surely an important concern for critical and cultural studies alike.

Glance theory: the myth of distraction. Glance theory, perhaps more than any other academic model, sidetracked television studies from a fuller understanding of the extreme stylization emergent in television in the 1980s. The myth's most cherished assumption? That television viewers are, by nature distracted and inattentive. Although its roots lie in the earlier work of Marshall McLuhan and Raymond Williams, John Ellis was the most forceful proponent of this definitive view. He argued that TV viewers not only lacked "intensity," but that they also gave up looking at all, by "delegating" their sight to the TV set.[50] This view—what I would call a "surrendered gaze theory"—while very influential, could not be a less accurate or useful description of emergent televisuality. Variations of the glance theory are however, commonplace: "We turn on the set casually; we rarely attend to it with full concentration. It is generally permissible to talk or to carry out other activities in its presence . . . (activities that) preclude absorption."[51] Ellis's position is an elaboration of Williams's earlier concept of the television "flow."[52] Within Williams's elaboration of the flow and McLuhan's rich speculations on media are keys to glance theory's flaws: *The mode of the TV image has nothing in common with film or photo.*"[53]

This extreme dualism between film and television—this mythology of "essential media differences" espoused by McLuhan—forms the categorical basis for many future speculations on television, including the glance theory. Once one assumes that there are innate experiential differences between the two media, then critical theorists are merely left to explain, post facto, the cultural and political reasons for those differences. Strangely enough however, Ellis and Williams deduced from this premise conclusions that were diametrically opposed to those of McLuhan. Whereas McLuhan argued that the low-resolution phenomenon fostered a highly active viewer response, one necessitated by the need to give conceptual closure to video's mosaiclike imagery, Ellis and Williams deduce just the opposite.[54] For them, the mosaiclike crudeness fosters inattentiveness and distraction, an inherent phenomenon that programmers try to overcome with the flow. Most contemporary critical works on television have followed the rationalizations of the later Williams-Ellis model of glance theory as an ideology, rather than the phenomenologically based and ostensibly naive futurism of McLuhan.[55] Ironically, the very same dualism—of essential media differences—gave one tradition an *active* viewer, and the other a viewer that *acquiesced.*

This distracted surrender gaze theory seems so far from an accurate portrayal of contemporary television consumption that one wonders whether glance theorists base their explanations of TV only on primitive shows produced in the early, formative years of the medium. When Ellis describes the "ignorance" and inability of TV viewers to know about the obscure and "inconsequential details" of television personalities, he seems to have mistaken TV viewers for what he describes as entranced and uniquely committed "cinephiles." Any cursory survey of the massive popular literature on television—including *Soap Opera Digest*, *TV Guide*, *People Magazine*, and others—will show a *extreme consciousness* by viewers of personality, marketing, and star promotion. Such literature also replicates on a mass scale a great amount of narrative detail in television, by summarizing a wide range of plot and character details in soaps and other genres.[56] The videophile—an impossibility according to Ellis—by the 1990s is actually a very informed and motivated viewer.[57] Contrary to glance theory, the committed TV viewer is overtly addressed and "asked to start watching" important televised events. The morasslike flow of television may be more difficult for the TV viewer to wade through than film, but television rewards discrimination, style consciousness, and viewer loyalty in ways that counteract the clutter. Whereas viewership for film is a one-shot experience that comes and goes, spectatorship in television can be quite intense and ingrained over time. Any definition of television based on the viewer's "fundamental inattentiveness" is shortsighted.[58]

The credence given glance theory in subsequent applications was due as much to television's inherent domestic context as to anything else. The notion of inattentiveness fit well the new emphasis on the home and on the "social use" and "object use" of the television set itself—an object that had to compete with other pieces of furniture "in a lighted room."[59] Again, the chief principles of glance theory are invoked: a distracted viewing context, a weak display, and a *very* unmotivated viewer.[60] Although some critics questioned the characterization of television's "regime of vision" as one where the viewer "lacks concentration," others extended the distraction model by shifting and overemphasizing the use of television sound rather than sight.[61] Even recent updates of the glance theory are based on the very problematic notion that television viewers are not actually *viewing* television but that the television is in the background while viewers are actually doing something else.[62] In an otherwise insightful analysis of video replay and video rental movies, one recent theorist attacks theory's aversion to low-culture viewing pleasures, while at the same time, continuing the orthodoxy that "there is a specific way" that television is watched. But if this particular TV viewer is *actually doing something else*, as the author says—while television is merely issuing-forth in the background—why extrapolate that this preoccupation with something else is symptomatic of the way television is *always* watched? Why not use an engaged and entranced viewer as the example upon which to build a theory of viewership?

Once the phenomenological basis for glance theory was academically sanctioned, however, more complicated ideological and psychoanalytic explanations were rushed to the fore. Since cinema spectatorship was commonly construed as psychologically regressive and hallucinatory—as a "totalizing, womblike, dream-state"—television, forever cinema's antithesis, was couched as just the opposite.[63] Following Freud, Jacques Lacan, and Christian Metz, film's viewing hallucination, like the dream, was described as an "artificially psycho(tic)" state that focused on the *pleasure of the image*. [64] Television's "more casual" forms of looking, by contrast, "substituted liveness and directness for the dream-state, immediacy and presentness for regression."[65] Because repeated television viewing presupposes *some kind* of pleasure however, psychoanalytic theorists sought the source of this pleasure and unity in places *other* than cinema's womblike state. From this point of view, television viewers, despite distractions and interruptions, achieve "an exhilarating sense" of power when they range across and "control" a wide variety of flow material. The ability to choose visually rather than to hallucinate visually is seen as a key to the pleasures of televisual distraction.

The fact that *some* TV viewers *are* deeply engaged in specific programs—and do find *pleasure* in entranced isolation while watching a show, star, or favorite performer—puts the validity of the psychoanalytic account into question. Since the conditions presupposed by the psychoanalytic glance theory are neither necessary or sufficient prerequisites for viewing, the applicability of the theory is severely compromised. Further problematizing psychologistic extensions of the glance are attempts to solidify the gaze versus glance, film versus video dichotomies into a model that explains male versus female gendering. It has become popular to see televisual distraction as a feminizing process and to extrapolate to polar conclusions that cast cinematic spectatorship as male and televisual spectatorship as female. The centrality of the housewife in the early decades of television does lend credence to this theory. There are, however, many pervasive and hyperactive forms of televisuality that can in no way be construed as feminine.[66] In fact, a number of hypermasculinist televisual tendencies have been an important part of television from the start.

Four correctives then, are in order: first, the viewer is not always, nor inherently, distracted. Second, if theorists would consider the similarities between television and film—rather than base universalizing assumptions on their "inevitable" differences—glance and surrender theories would fall from their privileged theoretical pedestals. Third, psychoanalytic and feminizing extensions of glance theory tend to put critical analysis into essential and rigidly gendered straitjackets. Fourth, and finally, even if viewers are inattentive, television works hard visually, not just through aural appeals, to attract the attention of the audience; after all, it is still very much in television's best narrative and economic interests to engage the viewer. Theorists should not jump to theoretical conclusions just because there is an ironing board in the room.

The ideology of liveness myth. Any effective analysis of televisual style must also shake itself of one other theoretical obsession: liveness. In recent American television, liveness is frequently packaged as an artifact. As often as not, pictorialism rather than realism, rules the context in which liveness is flaunted and seen. This practice of live embellishment flies in the face of some of the most cherished mythologies of television, ones that presuppose immediacy and nowness as the basis for television. Whereas glance theory focuses specifically on the nature of reception, the liveness mythology implicates production and cultural issues as well. The notion surely has its origins in the history of television production. In the pre-tape 1950s, television *was* live and broadcasters celebrated this distinctive fact. Yet, definitions of television and liveness emerged in phenomenological studies in the 1960s, in prescriptive aesthetics and manifestos in the 1970s, and in sophisticated poststructuralist analyses of the 1980s. Never mind that the technical medium itself has changed dramatically, and that it has done so several times. In high theory, the liveness paradigm simply will not die.

Sports iconography as stylistic templates for live embellishments of Gulf War conflict in 1991. Relationship of military to NFL's New York Giants spins one story. (ABC, CNN)

Although hardly ever cited by critical theorists, McLuhan laid the foundation for the academic myth of liveness, when he defined the medium as an "all-at-onceness" created by global television's erasure of time and space.[67] Other theorists expanded on the notion of television's innate liveness and nowness by examining the medium's broader appetite for currency and presentness even in non-live genres. Peter Wood argued from psychoanalysis that television's similarity to dreams caused it to evoke a "great wealth of familiar and often current material stored in the viewer's mind."[68] Horace Newcomb proposed a television aesthetic that was based in part on a commitment to the present, although he used the designation "history" to describe this form of currency. Newcomb alluded to the power of the present in television by arguing that "the television formula requires that we use our contemporary historical concerns as subject matter."[69] That is, sitcoms and historical programs alike elaborate contemporary cultural issues and current events. Within this tradition, then, everything in programming, *fiction as well as news and live coverage*, is a cultural and psychological operation defined by the present. Presentness and the past are inextricably related.

The myth of nowness also fed back into alternative production practice starting in the late 1960s. Video artists and electronic politicos alike embraced the liveness myth as a key to radical video production. Techno apologists argued that the instantaneous electronic medium altered "habitual ways of seeing" and transformed human experience. The technology-determined revolution was at hand.[70] Artists found in the "real-time" experience of live video, a fundamental force that could alter both personal experience and social practice.[71] Temporality was construed as magic; simultaneity as shamanism; and video art as altered consciousness. Video installations, environments, and small format tapes were designed and hyped around the concept.[72] Poet and modernist aesthetician David Antin, in a prescriptive treatise for media art, pursued what he termed the *"distinctive* features of the medium." Like Harold Rosenberg's advocacy of abstract expressionism before him (arguments that reduced painting to existential action), and Clement Greenberg's rationales for minimalism (ideas that reduced visual art to flatness and reflexivity), Antin attempted to reduce the phenomenon of television to its essence and its industrial obligation. Antin described video's fundamental and defining components as time and immediacy.[73] Postformalist critics and theorists like Rosalind Krauss later used simultaneity and liveness to demonstrate the centrality of psychosexual narcissism in the work of important video artists.[74] Works by Vito Acconci, Elizabeth Holt, and others exploited the medium's liveness through feedback loops and lengthy and indulgent performances, all as ways of exposing the human subject's "unchanging condition of perpetual frustration."[75] In an art world disciplined by rituals of specificity, video's immediacy myth was the perfect weapon for critical exclusion.

Even outside the limited institutions of video art, criticism, and theory, however, the liveness myth was snowballing. Basic production texts praised television's unique ontological burden for "realism and *authenticity.*"[76] Others contrasted the event-bound time of television to the innately sequential and objective time of film. Television's "most distinctive function [is] the live transmission of events. . . . The now of the television event is equal to the now of the actual event."[77] Mimicking the rhetoric of network television, then, production theorists argued that good television exploits the sense of nowness, since it is inextricably linked to an external event. Liveness, then, came to have completely different ideological effects, depending on the theorist who invoked it. For prescriptive modernist critics and video visionaries, on the one hand, liveness was a key to disruptive and *radical* artistic practice. Conventional production people like Alan Wurtzel and Herbert Zettl, on the other hand, argued that liveness was a fundamental quality of any good *mainstream* production work. The myth still held center stage, although the political implications of liveness clearly remained in the eye of the beholder.

During the 1980s television theorists who immigrated from film studies—a field that had spent nearly two decades deconstructing the ontology and ideology of realism—provided convincing explanations for the popularity of a related concept in television: the liveness ideology.[78] That is, liveness was seen to cover over the excessive heterogeneity and confusion of the broadcast flow by giving the medium a sense of abiding presentness.[79] Unlike many earlier critical theories, this view correctly noted that liveness in television is neither neutral, simple, nor unproblematic.[80] Other examinations also exposed liveness as a construction, but were even more explicit in *overstating the notion that liveness "pervades every moment of broadcast."*[81] As long as high theory continues to overestimate the centrality of liveness in television—even as it critiques liveness—it will also underestimate or ignore other modes of practice and production: the performance of the visual and stylistic exhibitionism.

More recent studies suggest that the industry has deontologized its own focus. Television now defines itself less by its inherent temporality and presentness than by pleasure, style, and commodity. Todd Gitlin's anthology of television criticism shows a renewed interest in "a common attention to the implications of form and style."[82] His own analysis is a highly visual account of the mood and tone of imagery and style in high-tech car commercials and the blank sleekness of *Miami Vice*.[83] David Marc takes as his analytical object a genre in television historically related to, and defined by, liveness. The comedy-variety show is linked to stand-up comedy and other overtly presentational forms of comedy that unfold in real-time. Presentational comedy, then, involves the traits one associates with liveness: improvisation, snafus, and spontaneity. Marc, however, describes the genre not around the notion of liveness, but as a spectacle of excess, a "framed" artform that "accepts the badge of artifice."[84] Margaret Morse's recent work analyses television within an American culture that invests heavily in rituals of distraction.[85] Television is linked to the ontology of the freeway, the shopping mall, and theme parks. These three recent approaches, Gitlin's *iconographic* view of television, Marc's explication of *artifice* in liveness, and Morse's *architectonic* analysis and critique, are all indicators of the importance *style and materiality*, *rather than temporality*, have come to play in contemporary television.

Yet, the ideology of liveness myth lives on, even if in modified form. A sophisticated analysis of catastrophe programming on television describes the ideology of time, and the sense of continuity that drives it on, as both a target and victim of broadcast catastrophes.[86] Liveness, at least when linked to death and disaster, is textually disruptive but ultimately pleasurable since its coverage works to assure domestic viewers that the catastrophe is not happening to them.[87] Television is again defined, even in this catastrophe theory, by its temporality and not by its image.[88] Yet, if catastrophic liveness *is* marginal and disruptive, then it is also an exception that proves the rule; it is an exception that indicates the dominance on a day-to-day basis of more conventional image and sound pleasures. If traumatic liveness induces extreme anxiety in the viewer, then hypostatized time and massive regularity comfort the viewer by providing a rich but contained televisual spectacle, an endless play of image and sound. The degree to which liveness and simultaneity still govern even recent theorizations is suggested by catastrophe theory's account of the new telecommunications technology: "The more rapid internationalization of television via the *immediacy* of satellites . . . replicates the *emphasis on transmission*.[89] Such an account suggests that liveness and immediacy will be even more important in global television than they are today.

This view ignores the fact that even satellite system broadcasts in Asia and Africa today seldom emphasize either immediacy or liveness. Star Network out of Hong Kong, for example, has become a quintessential packager of aged entertainment products—music videos, dramas, reruns—rather than a conduit for liveness, immediacy, or catastrophe. Very little, in fact, looks live or transmitted in international broadcasting. Even the domestic broadcasting of live and unscripted media events—like ABC's *Monday Night Football*, or major league baseball—are comprehensively planned, scripted, and rehearsed; are in fact highly regulated and rigidly controlled performances, fabricated to fit a restricted block of viewing time.[90] Now, as in McLuhan's 1960s, the resilience of the liveness myth still has as much to do with a vague notion (and hope?) of technological determinism as it does with anything else. As long as theorists look to the new technologies of television to prove the centrality of presence, simultaneity, nowness, or transmission, they perpetuate one of broadcasting's most self-serving and historical mythologies. Television has always boasted liveness as its claim to fame and mark of distinction, even though the programming that floods from its channels seldom supports this air of distinction and pretense of liveness.

In the spring of 1993, Mike Myers and Dana Carvey, stars of the recent hit film *Wayne's World*, hosted *Short Attention-Span Theater* on Comedy Central.[91] In the new world of cable, apparently, even *Saturday Night Live* alums could parody television's glance theory. Yet the trades discussed their appearance not as an indication of television's inherent distraction, but as just the opposite: the episode aimed to break the limitations of niche advertising by attracting a different audience. Niche economics on cable, after all, preclude the kind of inattentiveness that theorists celebrate as one of television's defining qualities.

Television and its performers have been no less conscious of the *stylistic* possibilities of liveness. David Koresh, founder of the Branch Davidian sect, proved that he understood the quintessential nature of televisual production when he forewarned: "The riots in Los Angeles would pale in comparison to what was going to happen in Waco, Texas."[92] Unlike his apocalyptic predecessor Jim Jones in Guyana—who suicidally fled to the afterlife rather than face NBC's approaching electronic news gathering (ENG) cameras—Koresh betrayed neither ontological subtlety nor televisual stage fright. The fires that raged when Los Angeles burned in 1992 provided not a sense of simultaneity or realism, but rather a powerful and codified template for stylized and horrific spectacle. The alienated televisuality of the L.A. rebellion could be appropriated and choreographed for the benefit of the mass audience, even by those in other places and with very different apocalyptic ends. Unfortunately for David Koresh and his followers, the ATF assault troops in Waco proved that the televisual spectacle, once unleashed, had an unforgiving mind of its own.

2 Unwanted Houseguests and Altered States
A Short History of Aesthetic Posturing

They're ripe for some avant-garde chatter . . .
—Dialogue from *Route 66*

On the premiere episode of *Beauty and the Beast* in 1987, viewers confronted an iconic map that defined the terms of the series: excessively visual sets, connotation-rich costuming, world-specific visual styles that demarcated the narrative realms above and below ground, and an array of culturally coded subterranean clutter. Taken together, this iconic configuration of neoclassical columns, paintings, sculptures, trophies, classic literary works and libraries of leather-bound books—all bathed in baroque lighting and all presented without verbal commentary or explanation—set out the terms of the narrative world and the conditions of viewing.[1] The overload and flurry of visual signs here raised important questions. To what degree were the aesthetic and high-cultural demands placed on viewers by the visual-narrative maps of *Beauty and the Beast* unique to this program or characteristic of late 1980s television in general? Was the preoccupation with this kind of visually defined alternative world a trait unique to primetime during this period, or was it an extension of some preexisting historical posture?

One way to understand the relative distinctiveness of televisuality is to study it within the context of the years that preceded the 1980s.[2] A close analysis of seventy archival television programs produced between 1948 and 1985—chosen because they all raised issues tied to the specter of art or its cultural trappings—illustrates that network television has always had an uneasy relationship with the aesthetic.[3] Two frameworks help clarify the aesthetic negotiations of such programs and show that televisuality was neither an abrupt invention nor one more example of primetime's bottom-line penchant for continuity and formula. The first perspective, an analysis of the explicit ways that programs framed themselves through those devices traditionally thought of as content—plot, dialogue, and characterization—shows how the very idea of the aesthetic was initially construed, subsequently contained, and eventually celebrated by American television. The second, a structural study, demonstrates how program form—the relationships between visual style, editing, and narrative—either effaced or foregrounded stylishness during the formative and classical eras of television. This particular approach looks at formal features from a range of television programs produced during four decades, with a special interest in symptomatic evidence of visual excess. The study of television's conceptual framing, on the other hand, considers not just the structural function of visuality in the formative years, but also the ways that programs themselves theorized art, the aesthetic, and the avant-garde

The icon-rich aesthetic netherworld in *Beauty and the Beast.* (CBS)

during that time. I say theorized, for it is important to establish, once again, that shows are not immanent art objects or empty forms of transmission that carry content. Programs throughout television history have also frequently provided viewing and interpretive guides—aesthetic self-theorizations, if you will—as part of their episodes, as part of the very structure and narrative of programs. Yes, Stephen Bochco, Pee-Wee Herman, and Max Headroom were not the inventors of television's aesthetic guise, but were rather functions of an industrial period that hyped and franchised that guise to a degree and for a purpose absent in the preceding years.

Rationalizing the Aesthetic
A Spatial Model for Culture, Class, Art

> I was in the fruit and produce business back in Philadelphia.
> —Contestant on *Ted Mack's Original Amateur Hour*

Having descended deep into the heart of the rural Ozarks at the start of one *Route 66* episode, Todd and Buzz put the Corvette in neutral, and muse on their near-term destiny. Should they opt for the simple life and go fishing, or pursue the "girls from the finest Eastern families, Boston, Philadelphia, and New York," who are holed up and suffocating in the fresh air at a nearby Dude Ranch? Buzz argues for the culture girls. "They're ripe for some avant-garde chatter over a dry martini—desperate for some contact with the outside world," he argues. "You and I—we're the outside world."[4] In one fell swoop, Todd has assigned the pair to the artistic vanguard, hyped their Eastern origins, and equated their avant-gardism with outsidedness and sexual potency. Like ripened fruit, these girls will fall from the tree in their male wake. Never mind the fact that the show's plot steers toward fishing instead of high culture and "white trash" girls rather than Eastern sophisticates. *Route 66* leads by playing its aesthetic card. Although many other formal aspects of the episode display a nondescript classical telefilm form, including the bland California lot on which it was shot, Buzz and Todd wear aesthetic distinction on their sleeves, even as they proceed to go slumming.

This conversational and geographical brokerage of art—with its awareness that American popular and artistic cultural was extremely segregated—pervaded even the earliest years of television. While television, replete with various rites of self-humility, did not always identify itself as art, it did constantly flag the presence of art in and around its periphery. There was, clearly, always an uneasy relationship to aesthetic quality in the early years of television. On *The Continental,* the male host was cast as a sophisticated, that is very European, womanizer who obsessed over romantic evenings with scores of beautiful women.[5] Like Todd and Buzz, almost two decades later, the spectacle of male sexuality here was conflated with the thick-accented Europeanness of the host. The show's sponsorship by a hosiery manufacturer suggests that art and alcohol, aestheticism and seduction, were being paraded for the pleasure of the mass-market housewife—not for the American literati. The *Ernie Kovacs Show* went one step further in equating high culture with acute and suspect forms of male sexuality. Kovacs's effeminate and lisping poet, Percy Dovetonsils, was one of the first television characters overtly to exploit gay stereotypes.[6] Described by later broadcasters as the "Alfred E. Newman of the gay set," astigmatic Percy sat in the studio's poet's corner and read precious but arty gems like "Ode to a Bookworm."[7] With genteel poise Percy sipped dry martinis even as he shared this intimate "literature" with the mass audience. In shows like these, TV teased the audience with glimpses of high-culture's aberrant dangers. Art was either sexually lecherous or aberrant, and it took place somewhere else, somewhere outside the safety or normalcy of home.

Other shows positioned the audience within an even more explicit relationship to high culture, art, and the east. *Ted Mack's Original Amateur Hour* made references to the working-class and low-culture origins of its contestants an obligatory ritual. In a typical episode, before and after each live performance, participants were made to describe their origins in a mode that was clearly confessional. "I am a metal lathe operator." "I am a forty-one-year-old machine tool inspector." "I was in the fruit and produce business back in Philadelphia."[8] The show repeatedly set up each contestant within a working-class milieu. Even child performers, groups of tumbling children, awkwardly express their class aspirations: "We wanna have a dancin' school some day." Having dutifully confessed, each performer was then allowed to act out his or her cultural aspirations. The lathe operator was exposed as a musician; the tumbling children were really dancers; and the produce worker, having packed his bags for New York (the closest thing that America had to Paris in 1949), actually belts out Italian opera. The amateur-variety genre clearly placed art and the East at the end of the American Dream. The awkwardness and flaws of both the confessors and their performances, however, make it painfully clear just how far these aspirants are from their cultural dreams. Pathos, a product of these recurrent and painfully flawed performances, makes it clear that high culture is actually always slightly out of reach.

If viewers in the late 1940s were somehow aware that they could not make it to high culture, television promised to bring high culture directly to them. Rather than make a spectacle out of the common man or woman, as when aesthetic culture cast its ominous shadow over contestants in the *Original Amateur Hour*, a Dumont Network series, *Photographic Horizons*, actually taught the viewer how to become an accomplished photographic artist at home.[9] Worried that you have the wrong zip code, live in the wrong neighborhood, cannot afford beautiful female photographic models? Don't worry, *Photographic Horizons* will provide beautiful models for you to "shoot off your television screen at home." Now *this* was aesthetic culture for the common man—for the homeowner, landlocked in the new 'publicly-sanctioned' suburbias.[10] No pressure to mix with bohemians and Madison Avenue–types here either. Simply mail your photos in, and Dumont will do the rest. Of course, shows like this had as much to do with the industrialization of snapshot photography and the rise of corporate giants like Eastman Kodak as with the importation of art. One need only recall Walter Benjamin's theory of this process—the democratization of photography as a "revolutionary" act—to sense the irony with which mainstream television was willing to meet the audience on its own cultural terms. For Dumont, "art in the age of mechanical reproduction" became a corporate sales tool.[11] The show's host actually came across more like a used-car salesman than Edward Steichen curating the "Family of Man" photographic exhibition at the Museum of Modern Art.[12]

As one of television's central industrial players in the late 1940s and early 1950s Dumont Network said it best on their series *Window on the World*: "Now we'd all *like to go to the opera—but it's a greater thrill for the opera to come to us*."[13] Now this was an aesthetic theory worthy of a corporation like Dumont. Dumont, after all, was diversified. It both produced programming and sold television sets on the mass market. We'll bring the opera to you. You'll see. It's artistically "better" that way. Furthermore, the TV-viewing audience depicted within the same show's ads, seemed clearly happy about the promise. Some musical variety shows were even more earnest about their educational function. Unlike the Dumont series, the *Voice of Firestone* did not just deliver high culture in reduced and user-friendly form to the homeowner, the show's direct address also made sure that the mass audience understood its position in the newly configured world of artistic culture. Each week the series stripped the alienating edge off of high-culture pretense, by turning on-camera classical performing artists into just plain folks.[14] Such attempts inevitably came across as heavy-handed and sometimes awkward exercises, but the need to buffer art for the masses was as pressing a task for Firestone as it was for Ted Mack and Dumont. Some, like *Ted Mack* and *Photographic Horizons*, pushed the populace up and toward culture; others, like Dumont and Firestone, awkwardly dragged culture on stage for the masses. The fact that the mass audience needed to be comforted during this period about this business of art, also meant that television inevitably imported only those aesthetic forms

Television's hand-holding operations. TV comes with prepackaged instructions about its use: Dumont program dramatizes positive affect on social life; *Honeymooners* teaches consumers about debt; *Letterman* crushes diegetic mic and sync sound with 10,000-pound Brechtian press. (Dumont, CBS, NBC)

that were safe, like opera and classical music. Modernism, and the many complications of its radical darkside, would simply have to wait on television's referential fringes for a more acceptable hour. Even after television moved to Los Angeles later in the 1950s and became a part of the Hollywood community, it tended to repeat the pattern established during this earlier period. Television deprecated itself—either unintentionally (like predecessors Dumont and Firestone) or by design (like *Ted Mack* and *Photographic Horizons*)—as a mere conduit to culture, not as culture itself.

Rigidifying artistic culture into one place (Europe, New York, Broadway) and the audience for television in another (the home) happened in other ways as well. An episode of *The Honeymooners*, "TV or Not TV," served as an explicit kind of viewer's guide to dealing with and accommodating the new television phenomenon.[15] Whereas Dumont's *Window* discursively taught the viewers in direct-address how TV programs and TV sets would make their lives better, *The Honeymooners* worked more obliquely. Enactment of the advent of television through comic narrative was another kind of hand-holding operation in prime-time. In this episode, Ralph and Alice argue at length about whether or not to get a television set "like everybody else." "Money is better than things," he says. She says, "I want a television set and I'm going to get a television set—I've lived in this place for fourteen years without a stick of furniture being changed!" More than just a marital spat, the argument that ensues sets out the possible terms for defining television within consumer culture. Although Ed Norton shows his male juvenile streak by needing a set in order to follow *"Captain Video,"* Alice positions television as a vehicle for more respectable pastimes: "I don't want to look at that sink, that stove, and these four walls—I want to look at *Liberace*!" Although Ralph seems content with the dingy working-class tenement that the couple have inhabited unchanged for more than a decade, Alice seems to know that culture exists "out there." This flattery and privileging of women as culturally aware is significant given the fact that advertisers at the time had promoted the idea that women were the chief purchasing agents in the American home. And while Liberace was not Horowitz, he was an acquired or learned taste, one that operated clearly above the lower-taste culture in which sociologist Herbert Gans would place the Kramdens.[16] Seen within the context of America's stratified economic classes, early primetime TV promised the middle-class a move up to opera and promised the working-class a move up to Liberace. Locked in their urban tenements the ticket to this cultural mobility came in the guise of furniture, and through the ritual of consumer debt.

Television was not framed paradigmatically in this show as mass communication, as dramatic narrative, as propaganda, or as a "window on the world." It was, rather, repeatedly theorized as a substitute and extension of the sink, the electric stove, and the water-softener—as a wood-cabineted "modern electrical appliance." One way to own culture was to own the appliance that transmitted it. In addition, these episodes also taught the viewer and the Kramdens how to manage consumer debt, the modern financial arrangement inextricably tied to the ownership of television and appliances. The show's discussion of buying on credit fits well the crucial postwar shift to mass consumer culture—a national agenda that promoted and legitimized consumer debt.[17] Learning how to borrow was not just Ralph and Ed's task, it was also the new task of mainstream Americans. Before being persuaded to purchase the set, Ralph castigates Ed: "Why don't you just go down and get it like you done everything else—on time!" "I can't," Ed responds, "they won't give me any more credit at the store. I got eighteen accounts already." By scheming, the pair finally acquire a set to share, and the show suggests that marital life will never be the same thereafter.

The *Burns and Allen Show* also celebrated and underscored the polar home-culture axis. The episode "Gracie Wins a TV Set" reduced television to the status of furniture even as it flaunted Burns's star status in television. After solving

Gracie's dilemma by buying her the set she thought she had won in a contest, Burns philosophizes in direct address to the viewer: "See what happens when you're married to a generous husband? You get a twenty-seven-inch television set. But why not, it's Gracie's happiness, it's a pleasure—and it's mostly her money."[18] To complicate Burns's act of generosity, the entire neighborhood follows his lead and goes out to get their own television sets. Television is pictured here as a consumer buying trend, with Burns as its instigator and apologist. Even at an early age, in shows like these, television saw the value in providing not only programming, but also explicit instructions about how to use that programming. Television graciously taught Americans how to use television.

Self-Flagellation: Art—The Unwanted House Guest

It must be nice to be an artist.
—Ben Casey, to patient in intensive care

Having established a spatial model for artistic culture, one that defined art as out there on the periphery, television proceeded to celebrate its low-culture origins and role through a systematic process of self-deprecation in the early 1950s. Both live television, especially in the form of the comedy-variety genre, and the sitcom mastered this form of artistic flagellation to great effect. Television in the 1950s was clearly not monolithic. Although many inside and outside of the industry associate the early 1950s with the "golden age," William Boddy has demonstrated how the live anthology drama was actually a function of a complex aesthetic alliance between network programmers and critics, a defense mechanism against Hollywood film producers, and a symbolic pawn in television's regulatory game with Washington.[19] While the aesthetic pretenses of the live-anthology drama have been dealt with elsewhere, I would like to show that a very different tactic was operating simultaneously in other genres.[20]

Introduced as "television's number-one star" in 1949, Milton Berle walked out in front of an audience on the *Texaco Star Theater* and repeated the announcer's statement that he "had just returned from California." Berle was setting himself up for a spate of jokes of his own, and television's, relationship not only to the more glamorous film industry, but also to the more respectable artistic worlds of Broadway and serious music. Hollywood was "God's country—Darryl Zanuck, cashier."[21] Broadway, on the other hand, was the home of great artists, one of which Berle introduced live from the studio audience, on network television. The aura of authentic artists pervaded the show, even if by tangential reference. Berle deprecated himself as "no Bob Hope," and guests, with comic interpretations, parodied real movie stars and performers. Meanwhile vaudeville acts, trampoline artists, and singers acted as awkward transitional devices. This show was hanging out at the seams—but America loved it. Why did Berle work so hard to shift cultural superiority and aesthetic legitimacy to a place outside the television studio, to deprecate the world of the comedy-variety show, and to place legitimate art in some other institution? Television was building audiences, and the construction of the performer as an agent with the point of view of the viewer would persist as a dominant form of audience address in the years to come. Denise Mann has shown how the comedy-variety show acted to mediate between the female fan at home and the stars that were imported to TV or that made guest appearances on television.[22] By using aging female stars as hosts on series like *The Martha Raye Show*, TV was using dual-purpose characters. Actors who still maintained some degree of status collaborated with the viewer by parodying or exploiting the presence of more serious and attractive Hollywood stars.[23] I am less interested here in looking at the way that television constructed and navigated a female audience than in seeing how this dynamic constructed an *aesthetic and moral mission* for television.

When Jack Benny helped produce a television show about himself on the *Jack Benny Show*, Rochester protested, "With all the celebrities in Hollywood—why'd they pick you?" Benny responded matter of factly, "I was the only one they wouldn't have to censor."[24] Several references like this one clearly define the film industry as licentious. Benny on the other hand—a *television celebrity*—is repeatedly characterized as impotent, uninteresting, and cheap. When seduced by a young blonde actress in a casting session, for example, Benny is physically unable to respond. His characterization as an uninteresting tightwad is significant as the show was clearly produced in Hollywood. A self-consciousness about this locale was communicated both by the show's narrative and style. Though Benny savors his simple origins as a boy from Waukegan, Illinois, his actions also simultaneously evoked the aura of Hollywood: the plot naturalized the ins-and-outs of everything from production casting calls to budget cutting to deal making. The show forms both a populist alliance with viewers who can empathize with uninteresting people, and a mythological showcase for the exclusive world available only in Hollywood.

Later in the 1950s, family sitcoms continued to develop and perpetuate this moral model of film and television art. In Waspish nondescript suburban towns like Hilldale and Mayfield, *The Donna Reed Show* and *Leave It to Beaver* kept the art-and-(im)morality specter of both high culture and Hollywood at arm's length. An episode entitled "Career Woman" makes art symptomatic of the very threat that suburban domesticity and the nuclear, gendered family faces in *Donna Reed*.[25] Donna's doppleganger, an old high-school girlfriend, is also a widely traveled and highly successful international high-fashion designer. This jet-setting style-setter is described as "irresistably stimulating," by Donna, but "a little bit . . . weird," by Donna's professional husband. The woman is, in short, everything that homemaker Donna had "wanted to be," but is not. The tension rises as Donna and her family undergo scrutiny from the eyes and in-home presence of this high-fashion mogul. "She's so chic," Donna anxiously remarks, "and all of her friends are so well groomed." Yet Donna eventually gives in neither to her own career aspirations nor to the urge to keep up with Paris and New York fashions. She eventually converts the designer and accepts her own lot by philosophizing: "Marriage means being a season behind in waistlines." Husband Alex softens the blow of this cultural resignation and acquiescence by noting that thirtysomething Donna still has what counts—the voluptuous body of a teenager. In another episode entitled "April Fool," the aesthetic-moral threat shifts from Paris and New York to the dangers of "show business." James Darren stars as a teen heart-throb rock-and-roll artist, who stays and wins the adolescent heart of Mary in the Reed home. When things get tense, son Jeff counsels Darren's character,

"They're not angry at you—they're angry at that sleazy 'show-business agent' of yours."[26] The episode really is a kind of cultural *sanitizing operation*: authentic artistry and the sexual restraint of the young male rock star is maintained, even as the manipulative aspects of show business are written off as crude and undesirable. As with the threat from high culture and fashion, the Stone family also finds a way of narratively and morally disciplining both show business and Hollywood.

Even *Leave It to Beaver* was no stranger to self-consciousness and aesthetic reflexivity. The Cleavers, like the Stones, also erected defenses against the loose morality of Hollywood from within the confines and comforts of their suburban home.[27] In an episode entitled "Beaver on TV," Beaver is selected to appear on a local television show called "Teen Forum." Faced with the possibility that Beaver might have been misled and taken off the show by producers, one of Ward's colleagues points to the "bright side" of his banishment: "Now suppose he'd been on the show, and came off great. The first thing you know, he's got show business in his blood. Then he's running off to Hollywood—and he wears the dark glasses, and runs around in one of those *foreign cars*."[28] Concerned brother Wally later repeats this linkage between art, lifestyle, and foreignness. "Then, he'll get a big head and go to Hollywood and drive one of those fancy cars like on *Route 66*." The clear and present danger in this narrative—in Beaver's appearance on a local television program—is that the impressionable child will be susceptible to "foreignness" and to the male ego-laden lifestyle excesses that accompany it in Hollywood. This suspicion of show biz is, of course, not unique to television, for the subject recurs in some feature films as well. Yet the threatening posture of the film industry as alien was more fundamental (and lucrative) for broadcasters, given television's greater economic dependence on the mythology of the home. The sitcom's moral defenses against art and culture in particular became, in practice, a dramatized ritual of primetime xenophobia.

But even as the primetime sitcom home was being fortified against outside cultural influences, other genres were emerging in the early 1960s that reflected a change in the perceived center of American culture. New series, like *The Defenders*, *Dr. Kildare*, and *Ben Casey*, reflected a shift in sensibility that historians have aligned with the altruistic social ideals and professionalism of the Kennedy era.[29] Other shows, like *The Fugitive*, *Peyton Place*, and *Run for Your Life* are characterized as "stories of substance . . . mature and sophisticated dramas that demonstrated a deeper concern over social issues and national purpose."[30] If the aging family sitcoms stretched the credibility of the dangers of aesthetic culture through their comic and moral preachments, then these new urban and professional shows promised a more intelligent and mature representation of those artistic forces. Yet the extremes to which such shows went to accommodate art were sometimes even more tortured than the suburban defenses erected in Hilldale and Mayfield.

One episode of *Ben Casey* entitled "A Certain Time, A Certain Darkness" begins with a violent automobile accident.[31] Having rear-ended a truck parked at a stoplight, a young woman's head is impaled by a two-inch thick steel pipe that crashes through the glass windshield in front of her. The reason? Her sin? The sequence leading up to the wreck showed her hallucinating in multiple-exposure montage—fixated on recurrent images of European modernist painter Paul Klee—

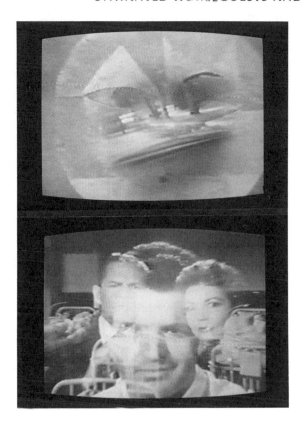

Dysfunctional woman hallucinates with images of modern art in *Ben Casey*. (ABC)

even as she dared a daytime drive through suburban Los Angeles. This dramatized opening is only the beginning event for what may be the most overdetermined female victim in 1960s television. Sensitive Casey slowly discovers that this woman is also an intelligent but multifaceted basket case: the alienated product of a dysfunctional but affluent family, a user and abuser of narcotics, an epileptic, a victim of psychotic hallucinations, a carrier of an unplanned and unannounced pregnancy, a victim of medical negligence, malpractice, and miscarriage, and—most damaging of all—an artist.

When stoic Casey teaches the patient to be honest with her family about her condition, she finally confesses to her young professional husband: "I'm an epileptic." In shock, and extreme close-up, the husband retorts "What!"—and quickly abandons her for the rest of her life. This is medicine at its best. In this process of primetime healing, the sensitive, Paris-trained, female artist-victim becomes sexually attracted to Casey's bedside manner. Rather than reciprocate romantically, Casey—obsessed as always by Hippocrates—shows her that life *can* be lived normally with epilepsy. So this is the lesson, the Kennedy-era appeal to relevance and social utility. The episode shows the woman as happy at the end of the episode. Yet the show also fails to tell her how to deal with her drug-addiction, her sexual desire, her artistry, her Eastern origins, or what it defines as her many other social dysfunctions. Casey's intelligence and sexual aura have, we are lead to believe, cured the pathological woman-artist of her expectations, but not of her dangerous artistic attributes: sensitivity, intuition, impulsiveness, expressivity. Multiply exposed hallucinations from Klee's avant-garde Der Blau Reiter movement could rear their ugly heads at any moment.[32] It is significant that this male, professional, Kennedy ethos on primetime conflates artistry with psychosis and femininity. After the show loads up this patient with all manner of sickness and aberrancy, Casey responds with resignation—and a straight face—"It must be great to be an artist."

An episode of *Run for Your Life* entitled "Hang Your Head Down and Cry" and starring Ben Gazzara runs on the same kind of conflation and model of aberrancy. This serious and adult series revolves around Gazzara's character Paul Bryan, a single corporate lawyer who discovers that he is afflicted with a terminal illness. As the episode begins, the character is still involved in his frantic cross-country attempt to compress "thirty years of living into one."[33] The mud-spattered bus that breaks down in the Arkansan heartland optimistically bears a lighted destination sign on the front that reads: "Los Angeles." Only a few years before *Easy Rider*'s eastward retreat from L.A. through America, *Run for Your Life* perpetuates the westward mythology of Los Angeles as a cultural Mecca where life's emotionally disfranchised migrate in order to "make up for lost time."[34] Into the male world of this series walks another young woman, also an artist and also a basket case. The story opens with a close-up shot of the artist-woman's stolen suede boots and an outburst of profanity, as she stumbles onto the bus. Postadolescent Kim Darby stars as a guitar-playing wandering artist and free-spirited drug-abuser. On first contact with the backwater riff-raff on this John Fordesque–Greyhound stagecoach, the Darby character celebrates her profanity and breeding as a legacy of her grandmother, who is now imprisoned at the Ohio State Penitentiary for concealing "raw opium in her corrective shoes."[35] Relegated to the back of the bus— an easily recognizable and still loaded social gesture in the mid-1960s—she queries Bryan: "Would you share a seat with a social outcast and a known dope addict?" Of course Bryan, an open-minded liberal like Casey, agrees. When gossipy passengers argue that she's "not normal" that "she's on the dope," the die is cast: artistry is construed alongside drug abuse and outsiderness. When the outcast woman addresses Bryan in French, and Bryan responds to her in French—a facility that betrays their downward mobility as a masquerade—it is clear that both have had socially privileged, internationally oriented educations. But having set up the threat in the episodic arc with this combination of cultural signs—art, drugs, Frenchness—the show delivers Darby's musical artistry not as that of the Yardbirds, The Velvet Underground, or some other psychedelic and underground group from the period. Rather, outcast artistry here is cast as the solitary heartfelt songs of a Joan Baez clone.

Run for Your Life was clearly not Antonioni's *Blow-up*, even though both attempt to negotiate similar terrain. Episodic television seems to blunt the edge of the aesthetic threat by defining it alongside dubious lifestyle choices, and by overhauling it with terms that better fit acceptable American mythologies. This woman-artist-victim, like Casey's, is oh so young, and oh so vulnerable, a pubescent and feminine Woody Guthrie. The location, the genre, and the music allow the problem to be seen within the context of the western, the road movie, and folk music rather than counterculture psychedelia. Bryan, the male-redeemer driven by destiny, heals the overdetermined sickness of the victim-artist, and so keeps the lid on the artist-as-aberrant–victim threat until next week.

By the 1970s the aesthetic began to emerge as part of a popular front against mainstream society and the status quo. Yet the terms under which the aesthetic was defined were still ambivalent. *All in the Family*, celebrated for its liberal engagement with political, social, and racial issues, also used art as a foil in its progressivist thematic soup. The controversial episode that depicted the threat of racial integration, entitled "Lionel Moves into the Neighborhood," begins when Gloria and Michael return from the city's art museum. They have just rented an abstract, stainless-steel sculpture for the Bunker's living room. As the horror of African-American homeownership begins to grow on Archie—represented as a fear of both declining property values and white flight—he turns and is shocked by the presence of modern art in *his* home. Faced with Archie's rage, Sally stumbles through a rendition of husband Mike's earlier philosophical justification for the work: "Why, the answers are all right here in the sculpture. This here represents the struggle of the races—man's inhumanity to man. Yet through it all, the shining hope of a new brotherhood." Unpersuaded by this critical maneuver, an enraged Archie then endures the Meathead's practical apologetic: "They've done sociological studies," when the first black families move into neighborhoods, "land values go up."[36] Only when art is theorized in terms of property values, can Archie get it. While this double-barreled affront—from race and art—is far too much for Archie to handle, the ploy does assign a socially redeeming significance to the modern aesthetic: like civil rights, art is apparently a component of liberalism and social progress. Unlike earlier liberal series, this show at least saw and described the aesthetic as a part of broader social and economic trends. Art was, Lear's world suggested, not just a social aberrancy or form of victimization.

Yet the show still managed to dance uneasily around the aesthetic. After all, the sculpture was only a model, not an original work; was being rented and not owned; and was being justified in a way that showed that its new caretakers did not *really* understand its significance. Mike and Gloria were also trying to convince themselves even as they theorized. The Bunker place, after all, was still a working-class home. While the surface of the plot explicitly justified and embraced the aesthetic, the characters' body language, social station, and ambivalence undercut it. These people knew they *should* understand art, even though they clearly might not. Art by now was, after all, a celebrated part of the popular front.

The positioning of art in terms of economic class continued in a number of other Lear sitcoms later in the decade. The most extreme case in this regard is probably *Mary Hartman, Mary Hartman*.[37] Seen at the time as the outgrowth of Lear's biting political satire and cynicism, the producer envisioned the show as one that would be both innovative and challenging; as a hybrid between the soap opera and the sitcom. Without a laugh track, the grim life of the housewife depicted here may have stylistically crossed the line with American viewers. In the first episodes, the main character's continuing concern with the "yellow waxy buildup" on her kitchen floor is dispassionately juxtaposed against the mass murder of an entire neighborhood family and "their animals." With the high-volume nightly production burden demanded by syndication and the resolute, jaded, and blank aura that dominated each episode, the show was not, finally, a ratings

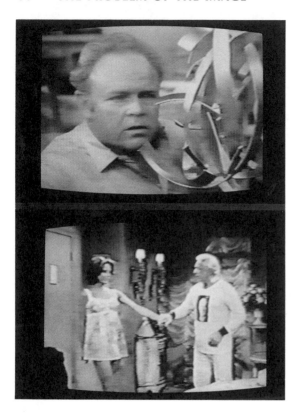

Archie, simultaneously dumbfounded by incursion of racial minorities *and* modern art in *All in the Family*. Mary on *The Mary Tyler Moore Show* tolerating the boys' fantasies about her in dream sequences whose presentational quality is that of cheap skits. (CBS, CBS)

success. Ironically, the series negotiated the aesthetic more by its glaring absence than by a celebrated presence. These characters were clearly lost in a cul-de-sac of low culture; a world of the assembly line, endless television advertising, and numbing alienation. The neighbor's wife is a Tammy Wynette–wannabe; Mary's husband is mired in sexual impotency and self-pity; and Mary herself is perpetually on the verge of a nervous breakdown. If pleasure came from this show, it came because the narrative regularly gave the viewer a place of cultural and aesthetic superiority. Comforted by Mary Hartman's oblivion and stroked for their intellectual openness and aesthetic awareness, Lear's viewers concede, "There but by the grace of God go I." In the Lear world, by this point in history, programming had begun regularly to place audiences in states of cultural superiority and distinction. Significantly, television did so without the buffering and hand-holding devices used in earlier television: the laugh track, the redefinition of threat in acceptable terms, and the use of classical narrative resolutions. *Mary Hartman, Mary Hartman* clearly shared a stylistic and conceptual sensibility with postmodernist shows that aired more than a decade later: *Roseanne, Married with Children*, and *The Simpsons*.[38] What Lear's shows did not have, however, was the blankness of intent that characterizes the later era. Lear's shows still had a political axe to grind, still believed in a recognizable package of progressive causes. Lear's liberal concerns and serious social agenda survived as a public response to the threat from the new right in the 1980s. From *All in the Family* to *Mary*

Hartman, Lear had created a superior position for his viewers, *by making art an absent force* in the lives of what could be seen as either working-class heros or mass-produced morons.

The position of *cultural and moral superiority* in Lear's world—an aura that stroked viewer intellect by contrasting it to the very uninformed and unsophisticated working-class dramatic worlds depicted on-screen—was clearly in place by the late 1970s. Yet all of the conditions for televisuality were not yet as fully configured. Two other factors—the primetime exploitation of stylistic *excess as a motivated program form* and the *technological competence* needed for televisuality—had not yet emerged alongside television's new cult of moral superiority. The Lear shows were still, after all, fairly monotonous exercises when it came to visual style.

The Structure of Televisual Form
Framing the Window (1946–1954)

> Which camera! . . . You're all moving around! Everybody's
> sooo . . . confused!
> —Jerry Lewis, screaming live to camera, 1951

Early television was frequently sloppy, a least from a classical perspective that values art for its unity and formal coherence. In many genres, like the comedy-variety show, television was formally excessive and heterogeneous. This was due in part to the fact that a programming battle was being waged at the time between several different framing paradigms and aesthetic traditions. While the struggle between the most viable of the artistic models—radio, vaudeville, film, journalism—has been documented elsewhere, the presentational status given excessive form through the early decades of television has received much less attention.[39] Having surveyed the *explicit rhetoric* by which the aesthetic was dramatized within television programs, it is also important to account for the changing *formal and textual economies* of style.[40] How, for example, was stylistic excess corralled, arbitrated, and eventually made into a dominant guise? How did television's programming modes change so that, by the 1980s, exhibitions of style were used as pervasive programming signatures in primetime television? Television did not just theorize the aesthetic, it reconfigured its place and value in both program narrative and programming flow.

Contemporary critical theory tends to prize and pursue ruptures and inconsistencies in programming texts. These breaks are said to reveal the "ideological contradictions" operative within texts and programs produced by the dominant culture.[41] An application of this framework to 1940s television shows, however, that ruptures and discontinuities are frequently the norm in many shows and genres rather than the exception. For example, *Window on the World* tried to control the way that television was viewed by providing both images and explicit directions about how to use those images.[42] The parade of performers and images that followed such instructions on the show, however, betrayed and undercut Dumont's fantasies of control over the viewer. Although a contented in-studio family watched submissively as newsreels brought the exotic world "home to you," the vast majority of nonfilmed material in the show carried with it a sense of textual and temporal *burden*. That is, live and lengthy filler surrounds what are essentially parenthetical and attenuated newsreel segments. This filler includes what seems like an endless succession of live second-rate acts: vaudeville and trampoline artists, and dancers who, like whirling dervishes, do number after number until they are sweating and breathless. In a ritual that predates power aerobics and Step Reebok by several decades, the static camera watches as the performers drive themselves toward exhaustion. It is significant in this hybrid show that television's imported modes—film and documentary—worked immediately and worked well. The program seemed more comfortable controlling and orchestrating filmed material rather than live performance material. Pulling off the show was, in a sense, a *textual struggle*. When, in the tradition of the circus, the performers raised their arms at the conclusion of a dance or trampoline act, the accomplishment celebrated was one more of physical survival than of refined entertainment or escape. Given the gap between what was promised and what was actually delivered inside the TV box—that is, physical struggle rather than effortless sophistication—the audience witnessed television grappling to control its own form. In this format, television was less a master of spectacle than a stand-up for second-rate showbiz. Vaudeville-based shows were not alone, however, in the struggle to come to grips with the new presentational burdens of television.

Imported from radio, *Help Thy Neighbor*, another 1940s program, also underestimated the visual-presentational burden of the televised image.[43] The host, a patronizing, self-serving local celebrity and philanthropist, drags out his ritual display of charity to excruciating ends. After speaking for and about himself, the host speaks at length for and about this episode's featured victim: a young, single, African-American teenager with an unplanned pregnancy. The show's visual image, however, forcefully betrays the intent of the humanitarian display, for it is clear that the pregnant teen is being force-fed scripted lines that she cannot adequately deliver. The cemented smile of the always-cheerful host is incapable of covering over the forced awkwardness of the scene. The visual image is simply too explicit. Emotional appeals like this one on *Help Thy Neighbor* may have worked on radio, given the perceptual openness of audio—a trait that encourages listener imagination and endless degrees of empathy. Here, however, the forced spectacle was formally all up front. The visual power of the image was like a loose cannon, an unruly force that disrupted the presentational goals of many early producers. The sum total of this show was artifice and awkwardness, surely self-defeating characteristics for television's new social missionaries and opportunists.

In addition to the sense that stations and producers were desperately struggling to find successful *performance formats* for the new visually defined medium, early television also showed off its *technical limitations*. In the *Original Amateur Hour* volatile performers constantly kept cameramen on guard, as they stretched the limited space and optical focus of the television stage At any point, a tumbling child threatened to roll off-stage left or into the lens of the retreating pedestal camera.[44] In *Texaco Star Theater*, Berle bantered and ad-libbed with the audience in direct address, and constantly played up technical mistakes and snafus. With Berle pointing out everything from lighting limitations to production problems, TV was cast not just as a moral and aesthetic second cousin, but as a severely limited tech-

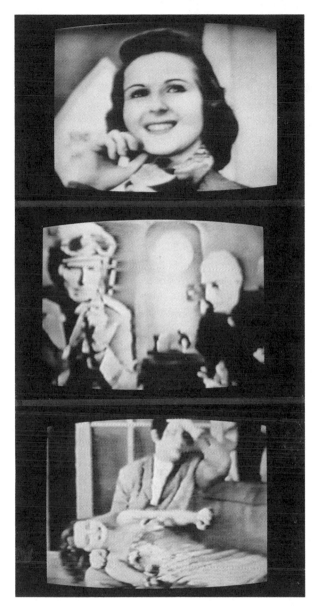

Photographic art for the homeowner (beautiful models included) follow Walter Benjamin's lead on Dumont's *Photographic Horizons.* Cheesy sets on *Captain Video* and controlled chaos oozing from Bakhtinian Berle. (Dumont, Dumont, NBC)

nical apparatus.[45] This celebration of technical limitations and performer volatility also showed up in *The Colgate Comedy Hour* starring Dean Martin and Jerry Lewis. Slamming a stage door during one of the live shows nearly brought the cheaply painted canvas flat to the floor. At another point, as one of the scripted gags breaks down, the ever-volatile and hyperactive Lewis has what appears to be a convulsive fit on stage. Running back and forth, and mugging for the camera, the manic Lewis outruns the entire technical staff, even as he squeals, "Which camera! . . . You're all moving around. Everybody's sooo . . . confused. You got no guts. You're running around. You're running. Which side do I go? Over here?"[46]

Staged and celebrated technical confusion is the operative concept here. Lewis constructed and exploited his uncontainable volatility by playing it against TV's technical inabilities. Marc comments that the "presentational forms" of comedy derived from vaudeville and burlesque inevitably involve excess.[47] Martin and Lewis clearly merit this attribution, but excess might better be described in these shows as chaos rather than anything else.

Other shows also acknowledged and celebrated the low-level artifice and limited technical prowess of television. Each episode of Dumont's *Captain Video* utilized cheesy sets, thrift-store costumes, football gear as space helmets, and perpetually static and monocular cameras.[48] One of the charms of the show was, in fact, its hopeless low-tech look, a quality that even Ed Norton could parody on *The Honeymooners*.[49] *The Continental*, for all its sexual and cultural appeals, was also a rather curious and awkward form of television production. With a few exceptions like the feature film *Lady in the Lake*, dramatic producers have rarely tried to use a first-person camera style to narrate an entire film or television program. The host of *The Continental*, however, looked continuously and directly into the camera-viewer's eyes—an awkward visual-narrative ploy that overtly positioned the audience as a potential date. In this guise, camera work was posed as a route to seduction. But this ploy in *The Continental* also drew attention to the severe limitations of the technical apparatus, and to the artificial nature of the interpersonal communication that was supposed to be taking place. An artificial production style and sloppy direction were not unusual in shows like *Beulah* as well.[50] *Beulah*'s one-liners, easy transitions, and comic asides in radio became, on television, awkward direct-address setups that were not smoothly integrated into the flow of the program.[51] Her lines were shot at different times, with clear mismatches in visual and aural continuity. The effect was a kind of textual segregation that mirrored the racial-occupational segregation in the show's narrative. The stylistic differences between *Beulah*'s shots on the one hand, and the more elaborately staged white family scenes into which they are cut on the other, created an acute sense of textual dis-integration.

Given television's own dramatic promises and narrative goals, one underlying effect of shows like these was that the television apparatus appeared technically limited and weak. Live performance, camerawork, and filmed source material constantly threatened to break out of their particular episodic constraints. In the face of this volatility and spilling over, television needed and presupposed decorum and restraint by its participants. A kind of unwritten contractual arrangement with guests was needed, an understanding that allowed performers to manage their volatility in order to work within TV's meager technical limits.[52] Even in serious, quality artistic offerings like *Marty*, technical limitations were not only evident, but they became a kind of badge of honor.[53] Poorly diffused pools of light and shadow abound in *Marty*.[54] Focus and depth of field were problems as actors moved in real-time; the sets were clearly inexpensive and minimal; and luminance levels changed noticeably as the actors walked from one side of the set to another. Even those generic forms from the period of "great" television art—like the live anthology drama from New York—flaunted explicit low-tech and antistyle airs as marks of distinction. There was indeed excess, in early television, but the trait did not involve pictorial refinement. Volatility, and the sense that television was coming apart at the seams, was instead an aura that reinforced television's obsession with liveness and immediacy. "The show that you are watching" could, as it were, fall apart at any moment. The technical apparatus was an essential part of the dramatic suspense. Only great and sensitive method actors could, apparently, navigate these perilous waters, or so anthology insiders and golden age historians would have us believe. Volatility and presentational ruptures were extensive and evident across a great deal of programming at this time. While local programs like *Help Thy Neighbor* unintentionally stumbled through attempts to control performer volatility and textual excess, prestige network dramas like *Marty*, on the *Goodyear TV Playhouse*, and *Requiem for a Heavyweight*, on *Playhouse 90*, played up the presentational threat intentionally.[55] The whole point, in fact, was to show how much one could do with very little. Live anthology drama wore the then-limited and meager televisual apparatus as a badge of dramatic honor and prestige.

Quality Control Telefilm: Late 1950s

Your mind is like a tape recorder—sometimes it plays back.
—Father Jim, after hallucination, on *Father Knows Best*

By the late 1950s, the pretense of television as a meager apparatus constantly on the verge of disintegration was more of a liability than a cause for celebration. Both live television from New York and telefilm production from Los Angeles were developed and mastered for corporate and economic reasons. NBC executive Pat Weaver had emphasized specials as an exclusive television form and as a signature aesthetic at NBC. Specials, after all, could only be produced through the centralized resources of a large national network.[56] Because of this budgetary appetite, specials became a weapon used to wrest control of programming away from single-advertiser sponsorship, an arrangement that had earlier given inordinate control of program content to sponsors. Having gone to bed with advertising in order to share the financial risks and development of early television, NBC now used special and showcase television to stab advertisers in the back. In addition to developing the modern programming theories of flow-through, hammocking, and counterprogramming, Weaver provided an early historical prototype for the way that a television of quality and distinction could be used in the 1980s.[57]

If Weaver saw the market benefits of showcasing and (over)loading television with expensive specials, Hollywood began to recognize the economic benefits of bringing to bear its own preexisting and underutilized stylistic resources: classical cinematic style and a stable of stars. While NBC practiced the financial risk of showcasing live specials in order to increase programming profits, Hollywood producers simply saw in television the opportunity to use the film studio's surplus production resources to make additional profit. MCA's Revue Pictures, for example, used the excess production capabilities of Universal Studios and its stable of stars to make nondescript film programs for series like *The General Electric Star Theater*.[58] An early episode entitled *Atomic Love* bears the Revue telefilm style: conservative single-camera direction, flat and high-key lighting, and the occasional cinematic montage to get from one plot point to the next.[59] As shorthand versions of feature B-films, these quickly made factory telefilms were really promos for second-rank film stars, and inexpensive cash cows for the studio. They fulfilled the same sort of economic function that a more modern genre, "making-ofs," provide for contemporary cable heavyweights HBO and TNT, and that shows like *Entertainment Tonight* provide for the industry in general.

If Revue was one of the first to see the value of promos *as* programs—almost thirty years before MTV was hailed by academics for the same tactic—then Disney Studios provided the prototype for the contemporary infomercial. Again, network and studio economics played a key role. Third ranked ABC saw in Hollywood a stylistic angle and production backlog that could give them a signature edge in the face of their network competitors at NBC and CBS.[60] Unlike Warner Bros., and other majors, Disney was more than willing to provide programming for Hollywood's archenemy television—on *these* self-serving terms: the *Wonderful World of Disney* was essentially a one-hour advertisement for both the studio's feature film arm and its newly established theme park.[61] Once the major film studios, like RKO, acquiesced to the inevitability of television, the large backlog of existing feature films flooded the networks.[62] Thirty-minute telefilms made for television were now joined by big-budget feature films aired nationally on the small screen. The stylistic impact of these changes cannot be overestimated. Cinema did not just import programs, it imported a way of seeing narrative and a distinctive way of constructing images. Although many telefilms were bland, expressive lighting and choreographed cinematography were not uncommon in late-1950s television. Along with the telefilms and features came practitioners and crafts people; 35mm Mitchell cameras and Mole Richardson lights; art direction and upright Moviolas; and the controlled stylistic world of the film-studio soundstage.

Even nondescript, generic family sitcoms showed the effects of this cinematic importation. A *Father Knows Best* episode entitled "Hero Father" transforms itself mid-episode into film noir.[63] When the father of one of Bud's friends leans on Jim for some underhanded and illegal influence with the zoning board, Jim's office literally turns dark and angular; the soundtrack turns dissonant. The home of the family sitcom may indeed be flat and high key, but the male world of work can be as dangerous and angst-ridden as any encounter with a hardboiled femme fatale. The series used expressionistic visual effects throughout the narrative of another episode entitled "Formula for Happiness."[64] Jim hallucinates, in this episode, about a recurrent nightmare, and phantoms appear and disappear in the glare of blinding lights. A television studio is displayed within the episode's narrative, as part of Jim's cathartic attempt to make his dreams rational. Sounding more like Timothy Leary than a suburban patriarch, Jim makes peace with his hallucinations: "The brain is like a tape-recorder—sometimes it plays back." In this particular episode, it plays back as an extended exercise in showcase cinematography.

77 Sunset Strip could easily marshal expressionistic and angular film noir lighting to trap its protagonist as well.[65] Critics immediately recognized the visual importance of the show. *TV Guide* praised the show's "excellent photography."[66] *Variety*—reflecting both film's arrogance toward the lesser medium of television and the trade's allegiance to the classical narrative model—qualified the praise, by pointing out that the cinematography was "surprisingly uneven, with the cam-

The studio big boys light late-1950s primetime with clean, hard compositions-in-depth and noir expressionism on *77 Sunset Strip.* (ABC)

era lingering on some shots and details that had absolutely no bearing on the storyline."[67] Nevertheless, the series for television represented a weekly exercise in proficient and capable feature film studio cinematography. If, as in *Father Knows Best*, a cookaloris was used to cast ominous venetian-blind shadows on the wall in *77 Sunset Strip*, it was done so as a special expressive statement.

Other telefilm genres, such as the half-hour westerns of the late 1950s, opened up the image in other ways. The westerns provided television with expansive landscapes, but these same images quickly became redundant as recognizable backlot settings through endless repetition. If, on shows like *Wanted: Dead or Alive*, Steve McQueen faced and fired at the viewer in stylized studio lighting, this dramatic motif was used merely as a punctuation point and framing device.[68] The stylistic agenda of the half-hour western that followed it could be described as mindless: a blank but efficient replication of the classical Hollywood style, complete with generic establishing shots, shot-reverse shots, reaction shots, and cutaways. The importation of cinematic style did have an effect on television style in the 1950s, but it was impact of a very muted and constrained kind.

Popular historians frequently point to the genius of Desilu productions in inventing and mastering the film-originated sitcom style for television in 1952. What such accounts seldom point out, however, is the extreme irony that Desilu rejected the distinctive cinematic and expressive edge that came with the very technology that it imported. Retrospective insider accounts of the studio and its stars, like the television special in 1992 on the "real" Lucy hosted by daughter Lucie Arnaz, celebrated the fact that German expatriate Karl Freund was hired to shoot the series and to develop a new four-camera telefilm style, but fail to point out that the proven artistic potential of Freund was stripped from the show.[69] Each episode of *I Love Lucy* during these years, in fact, shows a monotonous, flat, and generic lighting style.[70] Blocking and directing were nondescript vehicles for letting the comic actors show off. Was Freund here in name only? He was, after all, considered by many to be among the elite and most expressive cinematographers of all time: a 1920s product of UFA studios in Germany, a master of expressionist lighting, exposure, and the mobile camera, a co-conspirator with Fritz Lang, and one of the truly influential artistic forces behind the landmark film *Metropolis*.[71] Desilu liked Freund's fame and technical proficiency, but apparently was not interested in his distinctive and expressive stylistic abilities.

The early telefilm did indeed liberate film from the rote and disinterested recording task that television had first assigned the kinescope, but the expressive potential of film only slowly emerged in the years that followed its introduction. Television now used film, but confined it to presentational tasks of the most mundane sort.[72] Get the action on film; frame it wide; cut it to thirty or sixty minutes; and deliver it to the networks for broadcast. The 1950s telefilm either stripped its images of excessive style or, if flourishes were absolutely demanded by the narrative, corralled expressive images into special narrative boxes as "altered states."

Altered States: The 1960s

Intriguesville. . . .
—Cookie, *77 Sunset Strip*

I'm scared, Jesus.
—Sammy Davis, Jr., in a cinematic flourish, *The Mod Squad*

Since the 1930s in Hollywood, studios employed second-unit cinematographers and directors to produce montage sequences for classical Hollywood feature films. The sequences clearly had a secondary and supporting role within the overall form and narrative of the feature. In *Now, Voyager*, for example, optically layered montage sequences allowed the director to cover a great deal of temporal and geographic ground.[73] Thickened montage images of letters, faces, and locations allowed the main character to travel quickly between North and South America and to jump back and forth in historical time from different stages in the heroine's life. Multiply exposed images in Capra's *Meet John Doe*, on the other hand, are used to quickly establish historical time, the reactions of the press, and the ultimate despair and psychological trauma of Doe.[74] Difficult narratives like *Citizen Kane* used numerous montages to organize and structure complicated and out-of-sequence narrative scenes. In each case, the montage segments stood out stylistically from the narrative flow: as transitions, as parentheses, as comments, and as breathing spaces for the viewer. Television did not wait for the importation of telefilms from Hollywood to invoke the montage motif. Long before Ben Casey's patients hallucinated in multiple-exposure montage, live television in the early 1950s approximated the montage mode even if it did so in very approximate ways.

An episode of *The Goldbergs*, an early ethnic sitcom about an extended family in an urban tenement, is indicative of the way that live video achieved the status of the classical montage.[75] When Mrs. Goldberg is ostracized for not being able to dance, the director cuts to and holds on a multiple image closeup of moving feet on the dance floor. Since the multiple exposure of film using an optical printer was not even an option in live video, the director held the switcher midway on a fade between all three studio cameras, thereby simultaneously loading up the image with numerous pairs of dancing feet. The meaning of this display was obvious: time is passing, and Mrs. Goldberg is alienated from the other dancers who were all having fun. These visually thickened shots are unlike any of the other static and flat shots that dominate the program. They are also only a few seconds long, but quickly and efficiently give the audience a summation of Mrs. Goldberg's mental state and situation. Having established this state through visual shorthand, the narrative moves on to the next extended scene. These montage approximations are relatively excessive from a visual point of view, yet the show's narrative framework reduces them to the status of simple comments and asides. In this form of dramatic television, excessive signs are rendered and contained as passing asides.

By the time the films *2001* and *Easy Rider* had equated cinematic excess— optically printed special effects and rapidly cut sequences—with mind-altered states in the late 1960s, mainstream television had also recognized the presentational power of the extended and arty montage.[76] *Mod Squad* was quick to wear hipness and countercultural relevance on its sleeve. Each episode began with dynamic running shots, extreme camera angles, and whip-zooms in a rapidly cut opening. Heavy network promotions summed up this visual series signature as a progressive racial montage: "one black, one white, one blonde."[77] Although a static and predictable visual style dominated most of the bread-and-butter scenes in *Mod Squad*, special scenes became obligatory. In one episode entitled "Keep the Faith, Baby," a liberal-activist priest played by Sammy Davis, Jr., cries to God for help in prayer. Davis's Garden of Gethsemene crisis is motivation enough for a flourish of style: extreme angles, shaky handheld shots, and images composed with disruptive diagonals follow. Davis's character gets a second stylistic ecstasy when he has an abrupt flashback during the climax of the episode. This later hallucination erupted with an aggressive montage that would have made Eisenstein happy: crash cuts between excessive camera angles are embedded within flashbacks, flashforwards, and mind-altering backtracked music.[78] The Catholic priesthood apparently communicates to God with an ecstatic liturgy modeled after Stan Brakhage and Jonas Mekas.[79] It is important to note that both of the stylistic eruptions showcased in this episode are camera operator–based. That is, they result from simple *physical contortions* of the camera and abrupt editing collisions, rather than from the refined tonal control, lighting, and pictorial flourishes more pervasive in the 1980s.

The series *Julia*, another relevance show notable for its neutral lighting and directorial style, could also invoke the token experimental, film-school aesthetic when it needed to. One episode included a very filmic opening montage, cut aggressively to the beat of the theme music.[80] This rhythmic cutting and percussion worked to suggest both the show's ethnic overtones and to showcase dynamic camera work. The camera breaks a number of orthodox cutting and directing rules in this scene, as Julia's son, Corey, prepares orange juice. The sequence, for example, abruptly cuts in from a wide shot to a close-up without moving to another angle, thereby challenging Hollywood's artificially sacred 30-degree cutting rule.[81] When Julia returns home to find bloodstains from her now-absent son, the handheld camerawork is expressively isolated. It is important to note that the show typically showcased its female star, Diahann Carrol, with a rather conventional sitcom style, that is, with effaced direction and flattering high-key studio lighting that made her seem both very elegant and very white.[82] *Julia*, however, could also muster cinematic flourishes when called upon to do so, especially in the name of hipness and relevance or at moments of extreme anxiety.

Of all the self-consciously hip film-based shows of the late 1960s and early 1970s, *Kung Fu* most frequently promised to cross over into the realm of stylistic exhibitionism.[83] Formal excess was no longer relegated to the status of a few bracketed flourishes, but rather comprised one of the two dominant diegetic worlds woven together in the series. Kwai Chang Kane's contemporary nineteenth-century life as a wanderer in California was shot and directed in classical zero-degree telefilm style. The series included nondescript western locations, flat and frontal lighting, and traditional directorial blocking. Kane, however, was also half Chinese and was learned in the ancient ways of the Shaolin priesthood. Although typically meek and introspective, the warrior-philosopher-priest could also destroy men with his explosive martial arts skills, displays that were always depicted in drawn out slow-motion sequences. Apparently because of this cultural and psychological pedigree, Kane was prone to regular and frequent hallucinations and flashbacks to his monastic childhood in China. This retrospective narrative world, in contrast to the world of Kane's western present, was always heavily coded as an alternate and exotic reality. The show's cinematographers utilized every trick in their ditty bags to cook up these retrospective scenes as alien and Asian. Each shot in such sequences made heavy use of diffusion, color correction, and other optical filters. Such devices gave the imagery various levels of fogginess, off-Kelvin color (typically warm and golden), and dramatic-looking lighting effects. With *Kung Fu*, studio cinematographers in Hollywood were allowed to make stylistically excessive films for TV on a regular basis. But despite the fact that these exotic sequences comprised a greater percentage of the narrative than was typical on most primetime dramas, the sequences were still very much contained. They were motivated, however, not by formal brackets but by ideological attributions. That is, the excessive sequences remained televisual ghettos, as long as they were sanctioned and delimited by acute stereotypes of race, ethnicity, and Asian spirituality.

Shows like Rowan and Martin's *Laugh-in* promised to do for video-originated primetime shows what *The Mod Squad* and *Kung Fu* promised for shows shot on film.[84] Each series demonstrated the extent to which the performance of style could become a defining factor in the marketing of a program. Each threatened to break through the straitjacket of formulaic style inherited from the mass-produced conventions of three-camera shooting and the telefilm. *Laugh-in* was recognized immediately by critics as a revolution in television style. The frenetic and collage-style pace of the show produced "scripts that were measured by the pound," rather than by the page.[85] Each episode promised to pack as much television into one hour as was physically possible. Although the show was shot "live on tape," its rapid pace began to alter seriously the hierarchical relationship that existed between actors and presentational form; a relationship that had permeated live television since the early 1950s. The performers and stars simply took a back-stage to the textual flow of the program. With a density of jokes and visual gags this extreme, even bad jokes or poor performances no longer mattered. Richard Nixon's awkward visual and aural fragments were no better or worse than those of any other series star or guest star.[86] *Laugh-in* was one of the first series based entirely on rapid cutting, a preoccupation with image and sound bites, and the use of verbal graphics that were played against the grain of the image and flow.[87] On one national broadcast Governor George Wallace was asked to pick up his laundry in a scrolling textual graphic that continued, "Your sheets are ready." This personal communique had absolutely nothing to do with the nationally broadcast visual and dramatic image that it was keyed into.[88] This kind of televisual play provided a stylistic prototype for intentional text-image collisions—a deconstructive motif that would be celebrated in shows like *Saturday Night Live* in the 1970s and *David Letterman* in the 1980s.

Zero-Degree Television

No, no, no—I have limits. I will not get into that bed with you. I will only
go so far. Even if it is *your fantasy.*
—Mary, in *Mary Tyler Moore*

By the late 1960s and early 1970s the possibility of structurally inverting the dominant conventions of primetime had been suggested. While *Laugh-in* helped denarrativize television, it also reversed the previous dominance of performer over presentational text. In addition, the show made the videographic flow within programs a central component. *Kung Fu*, on the other hand, allowed visual style equal footing by structuring its plots around parallel narratives and historical worlds. The show provided a successful model for how programs could tie acute stylistic looks to alternative narrative worlds. The dominance of classical Hollywood style in the dramatic telefilm and naturalistic direction that historically dominated the studio show had been challenged. Yet, a revolution in style did not follow. The very *idea of hipness* that had spurred experiments with style during this period was transformed. Jethro's modish Hollywood masquerade on *The Beverly Hillbillies*, the countercultural restlessness and hippie-chick sensitivity on *Mod Squad*, and the psychedelic art direction and sock-it-to-me pastiche on *Laugh-in* all dated as quickly as any 1960s pop fashion.[89] Such things were, it could be argued, merely surface signs, naively imported to conventional generic forms as updates. As such, they would disappear rapidly when new and mindless fashions were available to replace them. The *structural* possibilities offered by late-1960s television, however, were covered over by a more fundamental change, one that relinked real-time drama and naturalistic narrative to the cutting edge.

Primetime redefined cultural relevance in the early 1970s as a product of social seriousness and intelligence, rather than lifestyle or fashion. The cutting edge was no longer a visual phenomenon. The much-acclaimed Norman Lear and

Tandem productions were instrumental in perpetuating this shift and in overhauling the very definition of quality television. Shows like *All in the Family, Maude*, and *The Jeffersons* assumed and reinforced the viewer's knowledge of social issues and intellectual problems.[90] It is not without consequence that these shows were also remarkably conservative in terms of style. Although Lear was celebrated for mastering the three-camera live shoot on videotape, critics ignored the fact that the lighting, camera work, and art direction were really throwbacks to a much more restrained period in Hollywood. Scenes were played wide, with a dominance of two- and three-shots that emphasize conversation. If close-ups were used, they were typically reaction shots, underlining a character's internal point of view. There were no flourishes, canted camera angles, videographic ecstasies, or even bracketed montages. The sets were just that: spaces where quality actors could perform live. Even in shows like *Good Times*, where race, ethnicity, and class might have suggested a different style-specific treatment, the three-camera Lear formula persisted.[91] African-Americanness was simply reduced to stereotypical surface trappings on the stock soundstage, while camera switching went off in rote fashion. The Lear shows typically won their Emmys for writing, acting, and direction, not for cinematography or lighting.[92] In Lear's hands quality television could also be identified by its antistyle.

In many ways this abhorrence of style was a throwback to the golden age of early live anthology drama, a connection that critics and producers were more than willing to tout.[93] The technical apparatus was in place only to allow the televised stage play to unfold. The fact that they were *just* stages added to the mystique of the quality drama on-stage. Unlike, for example, the flat cartoonish characters of Paul Henning's earlier rural comedies, *Beverly Hillbillies, Petticoat Junction, Green Acres*, Archie and Maude were clearly sophisticated and sensitive characters that demanded quality actors and solid writing.[94] The reemergence of serious drama and writing as center stage in television brought with it a renewed and dominant preoccupation with zero-degree studio style in television.[95] Television art, in effect, had been redefined as theater. Cinematic and videographic influences took a back seat as television entered an *ascetic* period during the first half of the 1970s.

Even shows that originated on film shared and perpetuated the zero-degree style. MTM, another quality production company, embraced a conservative live-on-film format for shows like *Mary Tyler Moore* and *Rhoda*.[96] Structurally, the aesthetic realm was offered to viewers as a literary experience. Although comic actor Moore ruled the stage, Grant Tinker's writer-producers James Brooks and Allen Burns came to be seen as the driving artistic forces. Like Lear's Tandem/TAT Productions, MTM developed its own signature comic-dramatic style, and, like Tandem, it won Emmys. As with Lear and Gary Marshall's enterprises, the MTM company style ruled its series. Even when *Mary Tyler Moore* did special episodes, the show tended to acknowledge its production style as a mere presentational game. In an episode entitled "Mary's Three Husbands," each of the three main male characters fantasize about what it would be like to be married to Mary. The three dream sequences that result use the same nondescript and flat studio lighting found in the nondream scenes, and each dream uses even more artificial and inexpensive sets and costuming than those that define the show on a weekly basis. Each of these meager fantasies comes across more like a sophomoric skit than a sexual fantasy. When Ted ("She's one hot mama") Baxter tries to "score" sexually with Mary, Mary draws the line, and steps out of character and protests: "No, no, no—I have limits. I will not get into that bed with you, even if it is your fantasy."[97] Series star Mary, in essence, takes control of her show, places the boys and their Vaseline-lensed fantasies in their place, and underscores the absolute flatness and artificiality of the dream scenes. Televisual style was just a half-hearted exercise in MTM shows like this one, and Mary's star persona was the beneficiary of this kind of stylistic effacement. Gary Marshall's productions *Happy Days, Mork and Mindy*, and *Laverne and Shirley* also perpetuated the zero-degree sitcom style on film, and each proved immensely popular.[98] If Mork got to use a cinematic device to clown around, it clearly functioned as a bracketed comment that only emphasized the set-bound, dialogue-driven world that dominated Marshall's sitcoms. Marshall's "family" of writers still ruled.[99]

It is clearly ironic that critics and historians have characterized this period in general as a second golden-age; an age when serious issues and quality writing and acting were again allowed on television.[100] Although the old aesthetic standby's—liveness, character acting, and sensitive writing—increased in programming value and stature during this period, many of television's stylistic capabilities were essentially ignored. Creators Tinker and Brooks, in fact, celebrate stylistic conservatism as MTM's company signature: simplicity, consistency, true characterization, compassion, and an approach that simply provided the steak rather than hyped the sizzle.[101] Horace Newcomb and Robert Alley have pointed out that Lear defined Tandem/TAT by its topicality, politics, and explicitness.[102] For both Tandem and MTM, then, company style was defined entirely as an issue of content, not form. This was clearly aesthetic television in only a very limited and attenuated sense. Like Herbert Gans's "higher taste-cultures," critics and learned viewers now could celebrate the birth of a TV art that was defined structurally by its sparseness and intellectual seriousness, not by messier formal excesses, kitsch, or camp. These liberal and ascetic tastes were reinforced by the wide popularity and financial success of the antistyle sitcoms. The zero-degree sitcom was a perfect vehicle for the mass-production needs of the primetime program producers and for the three networks that sought further dominance in the broadcast market. Mainstream production practice, especially in the 1970s sitcoms, tended to reject the medium's multiple and varied artistic precedents and to deny the televisual apparatus. In the process, the aesthetic was structurally reconfigured as a thinking and concerned place—available for escape or reflection.

Primetime dramas also settled into their own predictable zero-degree film-based style during the period. Episodic and dramatic-program producers, especially those associated with the Universal Studios lot, systematized film-based telefilm production with uniform settings, lighting, looks, and cutting in shows like *Columbo, Quincy, Delvecchio, The Incredible Hulk, The Six Million Dollar Man, The Bionic Woman*, and, later, *Knight Rider*.[103] These MCA-style shows were shot in single-camera 35mm feature-film style. Due however to television's inevitable pressures—the less than feature-scale budgets, rigid series scheduling deadlines, and union rules—such programs were notable for sharing and perpetuating a proficient, but very neutral, B-film style from the lot. The bread-and-butter production world of the studio and its personnel fit well the production needs of the networks, but had little to gain from drawing attention to the televisual apparatus or to televisual stylishness. The only thing that changed from week to week were stories, plots, and guest stars. Locations, on the other hand, were all recognizably southern Californian *and redundant.* MGM's zero-degree *CHiPs* lasted five years and produced 138 episodes using what appeared to be the same unfinished stretch of the 210 Foothill Freeway—a cheap and accessible stand-in for *all* of southern California.[104] The canned theme song, repeated throughout each episode, told the viewer that each chase was exciting, but the redundant chase imagery suggested deadening monotony; lighting was almost always flat and high key; the action seemed to take place between the flat and oppressive hours of ten A.M. and two P.M.—not exactly the cinematographers magic hour.[105] Cutting was formulaic.

Even movies of the week, originally envisioned as an alternative to importing feature films for broadcast on television, began to look exactly like episodic television. Both had roots, after all, in the same production mode. If visual flash *was* needed, these shows achieved it through props, action, and anatomy, rather than through tonal control, narrative manipulation, or visual stylization. Critics, for example, characterized the massively popular Aaron Spelling production *Charlie's Angels* as having "enormous visual appeal: three stunning women who often went braless."[106] Warner Bros.'s *Dukes of Hazard* was really not about the characters, but about the cousins' street-legal '69 Dodge Charger stock car. The show's obligatory and defining chase scenes went off in clockwork fashion.[107] In both its opening montage and dramatic scenes, *CHiPs* was also obsessed with the physical details of the male officers: chrome laden cycles, holstered weapons, metallic insignia, and the leather on Ponch and Jonathon's bodies. If freeway chases, fetishized tools, and weapons were not enough, wet T-shirt jiggle scenes with buxom starlets a la *Charlie's Angels*, could be pulled in front of the camera as filler, even in hour-long police dramas.[108]

In the homogeneous, studio-bound style of the 1970s dramatic telefilm, "stunning visuals" frequently meant placing loaded objects and libidinous bodies in front of the disinterested 35mm camera. Stylistic excess during this period had more to do with softcore fashion posing and automotive product photography than it did with painterly or expressionistic control over the image. The bread-and-butter Universal style of the 1970s, in fact, apes the classical Hollywood style of the 1930s and 1940s, but disregards even the modest and trended experimentation allowed in the earlier period. Without any strong independent competitors in broadcasting and with few alternative production looks available to viewers, this Los Angeles–based zero-degree studio telefilm style proved a cash cow for the monopolistic networks, the major studios like Universal-MCA, and a small group of major program producers like Lorimar. With factorylike efficiency, primetime programming assumed what André Bazin would call "the equilibrium profile of a river."[109] Mainstream television, centered in Hollywood, experienced little stylistic change in the mid-1970s. The zero-degree sitcom and telefilm styles worked and worked well, until, that is, an appetite for more distinctive programming and different geographical looks was developed outside of the lots at Universal, Warner, and CBS Television City in Los Angeles.

Monotonous landscapes, chrome, and flesh in front of disinterested camera of zero-degree, 1970s studio-style *CHiPs* (left, top and bottom). Exotic locations in *Roots* (top right), helped break geographic constraints of formulaic 1970s studio telefilm style, while acute designer contortions of picture plane in 1980s *Miami Vice* (bottom right) broke three-sided performance space and figure-ground segregation of earlier style. (NBC, NBC, ABC, NBC)

Unsettling the Studio Habit

Some critics complained because we showed a mountain peak in Henning, Tennessee, because that section of the country doesn't have mountains.
—*Roots* producer David Wolper

We've done something in making this that no one has ever done before. Let's show it in a way that no one has ever shown television before.
—ABC Network head Fred Silverman on *Roots*[110]

The emergence of a self-conscious aesthetic pose in the 1980s did not affect every genre. The antistyle tendencies of the Lear/Tandem sitcom mode, with its dependence on three-camera live taping or filming, continued through the 1980s despite some notable exceptions. Zero-degree sitcom style still forms the basis for contemporary shows like *Full House, Major Dad*, and *Coach*.[111] The massively popular *Roseanne*, for example, reflects the continuing power of the Lear-era antistyle. Its deferent presentational style works well to emphasize both the acting and the working-class topical focus of the series.[112] The effaced style is appropriate both for the anticultural pretense of the show and as a vehicle for its star, essentially a stand-up comedian/performer.[113] If *Roseanne* is excessive it is due to the star's unruly and disruptive presence on an otherwise highly controlled stage. Although stylistic flourishes occasionally occur in the conservative sitcom form—as when *Who's the Boss*, electronically keyed in multiple dream images during its Christmas episode in 1988—such episodes are exceptions.[114] Few sitcoms have worn artifice as a badge the way *Seinfeld* did by the 1992–1993 season.[115]

The homogeneous studio-bound look of film-based production, on the other hand, faced more immediate generic and geographic challenges. The rise of the miniseries and primetime soaps, in particular, undercut the conventional telefilm model by shifting onto the producers two new obligations: the temporal burden of a large cast and the textual burden of a large narrative. This shift in the scale of narrative and cast helped break open the confined and efficient texts mastered by studio major-telefilm producers in the 1970s. The ABC miniseries *Roots* demonstrated the power of nontraditional looks and non-Californian locations, filming in Africa, the rural South, and on the Eastern seaboard.[116] *Roots* was also massively and financially successful and demonstrated the profitable programming potential of expensive specials and nonepisodic programming like the miniseries.

The emergence of a new and more regional self-reflexive comedy genre was also instrumental in breaking the lock of the zero-degree styles: from England came *Monty Python's Flying Circus*, which was syndicated on PBS; from Toronto came the off-beat *Second City TV*; and from New York in 1975 came *Saturday Night Live*.[117] These shows gave new life to the now-eclipsed comedy-variety show—a genre that was still being done zero-degree style at the time by Carol Burnett in the very studios that had developed Lear's antistyle. Burnett's audience was made aware on a weekly basis that her show was coming from CBS's Television City in Los Angeles. *Python* and *SCTV*, by contrast, rejected studio-bound television style all together. Portions of *Python* were done underground, film style, on 16mm. *SCTV* productions from the period resist any attempts to describe the series within a single production category. Its episodes included some studio, some field production, and some film. With no live audience, and a disorienting collection of conventions and styles, *SCTV* also provided the viewer with little of the hand-holding available even on *Saturday Night Live*. If television had an avant-garde in the late 1970s, it came in the form of comedy produced outside of Los Angeles. The appearance and success of these new shows—the miniseries, the primetime soaps, and regional late-night comedy—had results. The new shows demonstrated the financial possibilities offered by a new set of primetime factors: exotic locations, sensitive regional worlds, self-reflexivity, and irony. The "outsiderness" that Buzz and Todd had only talked about in 1962 could be programmed with profit at the end of the 1970s.

House Styles and Emmy Bait

Let's be careful out there.
—*Hill Street Blues*

The value of zero-degree television depreciated in the face of several trends. First, even the majors increasingly began to place shows in regional locations. Unlike Los Angeles, places like Dallas (*Dallas*), San Francisco (*Hotel*), and Boston (*St. Elsewhere*) had not been stripped of their particular charms and cultural personalities through overuse and overexposure.[118] Geography, then, was one way that quality television opened up. Second, independent production companies like MTM now played a bigger role in engineering the look and aesthetic pretense of primetime shows. When MTM, Bochco, and Kozoll were "bought" and imported to NBC to produce *Hill Street Blues*, the network was consciously buying a look as well as an attitude. Third, the rise of serious, ensemble acting helped, ironically, to create the very conditions that encouraged stylistic exhibitionism. It is ironic because quality acting and writing emerged alongside Lear's antistyle, yet these serious relevance shows also created auteurs. Once producer-writer-director types like Norman Lear, James Brooks, Gary Marshall, Alan Alda, and others frequented the mass-marketed pages of *TV Guide*, the studio-bound zero-degree telefilm look was in trouble. That is, once the aura of artistry became a conscious part of industry hype, a *critical expectation for stylistic accomplishment* followed.

M.A.S.H. played an important role in bridging the gap between the homogeneous, nondescript studio styles of the 1970s and the styles of distinction that would follow.[119] By the middle years of the series Alan Alda had assumed control from Larry Gelbart as the show's creative force. Alda eventually became the only person ever to win Emmys for acting, writing, and directing. This publicized aura enabled the quality ensemble associated with the series to pursue tactics that became hallmarks of postmodernist television in the 1980s: including the masquerade of style, pastiche, and an extreme self-consciousness about the artifice of television. In the beginning, however, *M.A.S.H.* was simply a zero-degree sitcom. The show used a static one-camera shooting style, with master shots and close-ups but very few reverse-angle shots. Its simple history-specific sets were minimal enclosures that allowed actors to perform. The show was driven by the verbal wit and sequential rapid-fire ironies of the dialogue. There was little silence between or within speaking scenes and a very minimal use of background sound. The performance style was simply: *stand and deliver*. *M.A.S.H.* was then, in the early years, mostly a writer's vehicle.

Over its ten-year life, however, the series established precedents and terms that would be more extensively exploited in the excessive television of the 1980s. One notable episode was produced as a black-and-white documentary newsreel, with characters of the 4077th interviewed on camera by a reporter.[120] Another episode was directed first-person style, using the camera as the eyes of a wounded soldier.[121] Still another show was scripted and shot in real-time, à la Fred Zinnemann's *High Noon*.[122] The much-heralded final episode of the series, at the time the highest rated show ever, played as an extended psychotherapy session. The Alda-Hawkeye persona was set-up in the finale as the brooding and psychologically

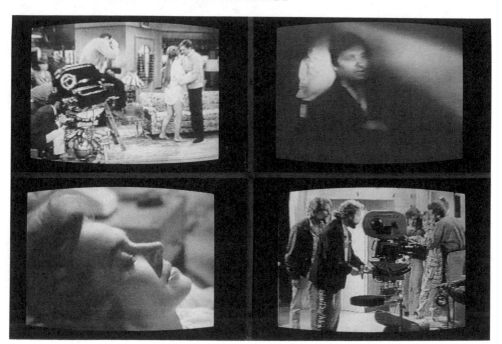

Dreaming about the evils and shallowness of Hollywood—from the *outside*—in the *Dick Van Dyke Show* (top left) and *The Donna Reed Show* (bottom left) in the early 1960s. Primetime personas as complex filmmaker-artists on the *inside* in series finales of *M.A.S.H.* in 1983 (top right) and *thirtysomething* in 1992. (CBS, ABC, CBS, ABC)

wounded creative force of the 4077th—a reflective subject that tried to make retrospective sense of the whole Korean experience.[123] Surely this was a fitting and author-izing role for one thought to be the creative center of the series. Each of these episodes was a critically acclaimed exhibition of sensitive storytelling, style, and expressivity. Yet, in light of the total 251 episodes of *M.A.S.H.* produced, these few showcases were clearly very limited exceptions and were not symptomatic of the show in general. They might better be viewed as Emmy bait—as special episodes targeted specifically at critics, the industry, and award juries of the Academy of Television Arts and Sciences. Promoting the occasional and exceptional showcase in this manner would develop into an art form in the decade that followed.

Other series helped sketch out the parameters of a new kind of stylistic playing field. In 1979, the massively popular *Dallas* began its long run on CBS. As a primetime soap set in Texas, it offered viewers a *regular* visual spectacle based in part on the lessons and precedents of the recent miniseries successes: the show had a large web of interrelated characters, regional locations and architecture, conspicuous lifestyles of consumption, and a *big narrative* that broke the discursive confines of a one-hour viewing slot.[124] Viewers were aware of the scale and spectacle of the show even if the show was rarely arty. *Dallas*, in fact, represented the zenith of the controlled, studio-bound, classical dramatic telefilm style, yet its expensive budgets regularly allowed for both a motivated kind of art direction and frequent forays into stylistically coded "altered states." The highly rated episode "Who Shot J.R.?," for example, costumed its characters and blocked its action in knowing and highly symbolic ways.[125] Sue Ellen, arrested for the murder, is garbed in designer Yin-Yang that divides her body into manichean geometric blocks of black and white, good and evil. Her jailhouse incarceration is composed-in-depth, with metal barred gates successively obstructing the viewer's vision. Spectacular, wide-angle shots of architecture regularly underscore the economic bed of the narrative. Several extended altered states are also included in the episode, as Sue Ellen's psychotherapist leads her back to the day in question through hypnosis. Not only was the hypnotic process expressionistically rendered, but the subject's drunken state, embedded within the flashback, was further stylized, with handheld camera and distorted audio. Again, the altered states appeared because they were motivated by the script, but the sequences were clearly more than just bracketed comments. Despite the high-key daytime look of the show, the series frequently indulged in such exercises, but did so only along the psychological faultlines of its "perfect" narrative and familial world. The producers' justification for Bobby's return to the series in the fall of 1986, for example, further underscored the artifice of the series and its world. By having Pamela, in the season premier, discover the long-absent Bobby showering at home, viewers were expected to believe that all of the previous season's episodes had been entirely dreamed. Any sense that Dallas was somehow about the concrete world of classical realism was undercut by this ploy, by this showcased deception. Even the prestige but zero-degree telefilm could, apparently, be discredited for gain by its producers and exposed as a play with artifice. This knowing celebration of the show's artifice, in fact, became a defining quality of the final episode of the series, in which J.R. sees visions and apparitions, and travels out-of-body to experience life in dimensions of the other characters.[126] The evolution of Dallas, then, demonstrates how even conservative, mainstream television genres could simultaneously perpetuate the conventional logic of flourishes, even as they knowingly and profitably exhibited the underlying artifice of dramatic realism.

If Dallas is notable for its gradual overhaul of inherited conventions, then Hill Street Blues stands as a self-made, and premeditated, stylistic breakthrough. The series proved influential both for writer-producers and for director-cinematographers. In the years since it aired, the show and its producer, Steven Bochco, have taken on an almost mythological stature, as a precedent for how great television art can happen again and prosper—that is, if the networks would simply practice patience. It has become an almost obligatory mid-season rite for critics and producers of low-rated (but quality) shows that face cancelation to whine that ABC, CBS, and/or NBC are no longer acting in good faith; that the challenging and serious show in question could find a long-term and valuable audience if the network number crunchers were not so obsessed with the impact of this week's bottom line. Bochco took artistic risks; enlightened president Fred Silverman trusted Bochco; and NBC weathered low ratings long enough for the show to develop a valuable demographics—or so the mythology goes. The result? Emmys and solid earnings for all. If only the current networks would embrace the same kind of forward-thinking vision. Like Marty and Chayevsky before them, Hill Street Blues

Table 2.1   Reconfiguring Quality TV   The changing relationship between defining series looks, showcased flourishes, and outright exhibitionism

	Emmy Bait 1973–1983 *M.A.S.H.*	**Faultlines** 1979–1991 *Dallas*	**House Look** 1981–1987 *Hill Street Blues*	**Designer Video** 1984–1991 *Miami Vice*
The Look	Zero-degree telefilm, but with special episodes that showcased style	Zero-degree hybrid telefilm, showcasing regional geography	Single, continuous, and distinctive series signature	Both a series look, and an obligatory weekly ritual of style
Showcase	TV as stage, stand-and-deliver acting, rapid wall-to-wall dialogue, and word-driven stories; flat lighting	Narrative complexity, with dramatic economy; exploits spatial and architectural showcases	Verité Panavision; volatile open space; chaotic visual image overwhelms dialogue; use of practicals for realism	Cinematic style; visual backgrounds inflect all dialogue; brooding men in stylized space; practicals as optical sp. fx.
The Logic	Unmotivated look with auteurist exceptions	Variant looks only if motivated by psychological fissures in characters	Motivated series look dominates episodes	Weekly performance of style and stylistic cycling
Flourishes	As episodic exceptions, and repertory specials	Only as special effects in script and altered states	As corporate, ensemble house-look	As episodic norm, and scenic obligation

and Steven Bochco could make news as prestige headliners. The commercial viability of this new consciousness and appreciation for aesthetic television was borne out by the long-term, unprecedented multiyear deal that auteur Bochco was given to develop future showcase projects.

While the backstory of the series influenced the way that the aesthetic was brokered in 1980s television, the presentational forms of the show were also highly acclaimed. From the start, Bochco wanted a gritty, messy visual look, so much so that the director of photography protested.[127] *Hill Street Blues* was modeled in part on Susan and Alan Raymond's 1976 low-budget independent video *Police Tapes*, a long-form low-resolution black-and-white verité look at police life on the streets.[128] The challenge for *Hill Street Blues* was to approximate this sort of gritty look using 35mm feature-quality camera technology. The now-famous roll-call sequence bears a look that recurred elsewhere in the series: handheld, jerky, verité Panavision; complete with minimally edited takes that include whip pans and overt follow-focus. The heavy use of practicals rather than just studio lights, faster film, and pushed processing all underscored the spontaneity and realism of this highly constructed image. Acting and directing styles added to the effect. The opening episode provided a directorial template for the episodes that followed.[129] Most of the episode's images were overloaded with actors, the soundtrack contained multiple and overlapping tracks, and the net effect was one of visual chaos. This compositional trait was scripted into the show, as violence constantly

Bellasario does *Altered States* and *JFK* on primetime in *Quantum Leap*; Oliver Stone does eighteenth-century Versailles, VR, and cyberpunk in postapocalyptic *Wild Palms*; and Lucas introduces *Young Indiana Jones* to his on-camera network mentors: Paul Gauguin, Henri Rousseau, Pablo Picasso, and Gertrude Stein. 1992–1993 was a good year for culture mavens. (NBC, ABC, ABC)

spilled over into the precinct offices. What the series consistently produced was a volatile open space that frequently overwhelmed the dialogue. Even in TV genres based on filmic realism, then, the status of the image takes on a new importance. In a structural and stylistic sense, *Hill Street Blues* was important for flaunting what might be called a house look. Earlier shows typically shared and shifted between inherited conventions and allowed for stylistic flourishes only as exceptions. *Hill Street Blues,* by contrast, created a wall-to-wall visual look that became a defining property both of each episode and of the ensemble of quality actors as well. This stylishness, furthermore, was not confined merely to the psychological faultlines as in *Dallas*, nor relegated to a parallel historical universe, as in

Kung Fu. The look and the attitude became a signature for the show and its producer.

The question of a stylistic signature in primetime increased in importance for series that followed *Hill Street Blues*, like *Miami Vice*. By 1984 everyone seemed to recognize *Vice* as a program distinguished by its obsession with high fashion and excessive stylishness. The producers not only instigated a basic house look coded to certain color schemes, costuming, and fashions, they also pushed beyond a defining series look and encouraged the very process of stylization as an almost autonomous ritual.[130] Starting with the premiere episode of the series, almost every aspect of the show—direction, cinematography, lighting, cutting, and sound design—was both expressionistic and, according to conventional production orthodoxy, excessive. Each production department regularly broke a cardinal rule of classical Hollywood or television. Rather than subordinating themselves to an overarching narrative, each craft drew attention to itself. From a traditional perspective, that is, the series was overphotographed, overcostumed, overmixed, and overcut. Unlike most programs in the history of television, it was not unusual for each narrative scene in an episode to have a different visual style. Tubbs's Brooklyn underworld was painted in browns and blacks, with smoke, neon, and backlit reflective light throughout. Crockett's marina digs were high key, pastels, and Armani. His ex-wife's place was cast in warm and diffused amber light. The architecture was embellished, reflective, Deco and postmodern.

If any recurrent scenes in the series were obligatory, they were the club scenes. These late-night cocaine-cavern sets allowed for the use of acutely spectacular lighting effects: strobes, neons, bar lights, smoke, motorized spots and floods. Such lighting sources could no longer be described as practicals in any traditional sense—they *were* the lighting design, the keys, and the performers. *Vice* was, in addition, a cinematographer's showcase and a veritable proving ground for production technologies and equipment companies. Any and all possible styles, after all, could be justified. In fact this rifling through visual looks became an obligatory burden, not just of each episode, but of each scene. In the end the perceptual universe and background were designed to overwhelm the brooding male characters in the foreground. Dialogue really did not even matter in many episodes. The influence of extra-Hollywood aesthetic tastes—MTV, Italian fashions, postmodern architecture, the club life and the consumer goods of yuppie culture—had entirely displaced the zero-degree lock that television had carried on its back for so many years. The fact that these artistic influences came from outside of Los Angeles is significant. The structure of television form had changed. Exhibitionism took its place alongside the earlier and more tentative orchestrations of form—the montage, the flourish, and the altered state.

Televisuality was not invented ex nihilo by a few forward-thinking creative players nor simply discovered by opportunistic programmers. A survey of historical programming shows that there are significant continuities as well as marked breaks between classical television form and the televisual exhibitionism of the 1980s and 1990s. The terms and conditions under which the aesthetic was eventually celebrated emerged as the result of on-going interactions between network programmers, creative personnel, and television's mode of production. In this chapter, I have tried to apply two frameworks for understanding the emergence of televisual style. The first focused on the ways that specific programs explicitly navigated the issue of art and rationalized the aesthetic. This historical negotiation by television of both the threat and the value of art played an important role in preparing the cultural ground and appetite for 1980s televisuality. The second framework examined the changing structure of television form. This perspective helps clarify the various ways that three of television's presentational components—the narrative, the dramatic, and the stylistic—alternately worked to support, to dominate, or to undermine one another.

If one looks at television's formal structure as a textual economy, a shift in cultural capital had clearly occurred by the early and mid-1980s, one that made stylistics a more valuable kind of programming currency. A structural inversion had taken place between the presentational functions of narrative and style. Excessive style was no longer confined to the margins in the form of bracketed montages, motivated flourishes, and script-justified altered states. Although zero-degree shows continued issuing from the production factories, many other shows no longer even made token attempts to integrate flourishes, music-video sequences, and graphics into their organizing narratives.[131] By the mid-1980s many shows were dominated by flourishes, and the look of the show frequently organized the narrative. Producers attempted programming toeholds in the competitive network markets by developing and hyping house looks and series signatures.

More importantly, the structure of television form had overhauled the very guise and meaning of quality artistic television. Artistry in the other golden ages—Chayevsky's 1950s and Lear's 1970s—had presupposed that television was a showcase for sensitive acting and a vehicle for writers. A reductive and visually monotonous antistyle was an inevitable part of this presentational hierarchy and focus. Although critics might be quick to put a Bochco or Mann into this aesthetic pantheon, they overlooked the problematic fact that many shows during the 1980s and 1990s had been reduced to high concept: *Miami Vice* as MTV cops, *Pee-Wee's Playhouse* as Mr. Rogers on acid; *Seinfeld* as a show about nothing. The 1980s were, after all, the decade that turned the two-minute pitch into an artform, a public offering, and a kind of performance art for M.B.A. types. It is no coincidence that the long-form television treatment-proposal fell into disfavor at the Writer's Guild during the period of the pitch. Along with the long-form treatment disappeared many of the ambiguities and sensitivities of serious and patient dramatic art. The complexities of method acting and existential dramaturgy simply don't fit the constraints of the pitch mentality, a concept-crunching method based on collaging eclectic generic motifs and stylistic fragments. There was still aesthetic television, but it looked nothing like Marty's or Archie's.

Packaging the Aesthetic Threat
Race and Sexuality as Cutting-Edge Commodities

Planting the seed deep . . .
—Beulah

Television's posturing and stylistic preoccupations were not simply apolitical formal practices. The changing ways that television rationalized the aesthetic and

its accompanying threats suggest that television has always had a serious investment in American racial and class politics. If we reconsider, first, the ways that
television defined the aesthetic, and second, the ways that it transformed the aesthetic into acceptable terms, the ideological implications of televisuality become
clearer. In the 1950s and 1960s, for example, art was frequently, though not always, defined as a threat, as an (im)moral force that came in many disguises. In
the early years of television, some genres worked hard, if not awkwardly, to
comfort the viewer by buffering high culture, classical music, and opera. Yet other
shows theatricalized art's many threatening edges. As this survey thus far has
shown, American television was more than willing to define the aesthetic as a
kind of dangerous sexuality, postured as either lecherous or gay; as an alien foreign influence; as an imposition of upper-class elite taste; as a pathology and a
mental disorder; as drug abuse; and as a racial threat. Before art was ever celebrated on television in the 1980s, then, it had long been ghettoized into what
mainstream America had considered marginal and aberrant social threats—as a
product of foreigners, of sexually active persons, of diseased patients, of drug abus-
ers, and of socially dysfunctional basket cases. This was not, clearly, a pretty picture, for xenophobia and paranoia appeared to rule the television homes, the
sitcoms, and the primetime dramas of middle America. It would be wrong, however, to misconstrue in this way television's aesthetic posturing as a mere reflection of existing cultural fears. Television was also active. It taught Americans about
these cutting-edge issues and aberrancies, in a way that betrayed the industry's
fascination and love-hate relationship with TV's unwanted houseguests: sexuality, race, and high culture.

The consistent historical conflation of art and race is perhaps the best example
of how television's posturing also functioned as a way to assimilate—rather than
to simply elide—its threats. Consider, for instance, how race and sexuality were
used to disintegrate and then overhaul programs. The series *Beulah* touted Academy Award–Winning black actress Hattie MacDaniel as its featured star.[132] The
character Beulah, however, proved to be more of a "backstage" token in a show
that placed more dramatic emphasis on the white family "frontstage." One episode showed the African American domestic help possessing innate forms of
rhythm and dance skills that were both hip and athletic. Out of pity, the black
help teach these exuberant arts to the awkward and alienated white son.[133] Now
in possession of this secret knowledge, the son then upsets the middle-class neighborhood with his new black bodily skills and excessive dancing. Despite the familial crisis that follows, the son is in the end proud and aware of his acquisition.
In the seclusion of a black-owned storefront, he then teaches the other white kids
the secret of black dance. In response, a bevy of young white girls line up to be
symbolically serviced by the now potent and Africanized white boy. Even in the
very earliest days of television, then, hipness and style were acknowledged, but
were placed somewhere out on the margins and conflated with dangerous aspects
of both race and sexuality. Television was simultaneously fascinated and repulsed
by black style and sexuality. In this particular episode, the character Beulah herself described the passing on of black style—in explicit sexual terms—as "planting the seed deep" in the white boy. The presentational forms of the show
then—segregated cinematic worlds: the black help in the back, and the white fam-

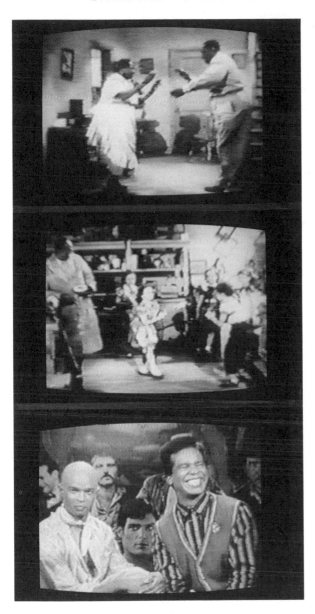

African-Americans as culture-givers and as mass culture's avant-garde. "Planting the seed deep" in *Beulah*, 1952, and *In Living Color*, 1993. (ABC, ABC, Fox)

ily up front—set up a racial and cultural distance that could only be crossed in authorized ways. In this case, only the white boy could import black hipness across the racial divide—since it had to be tamed and desexualized on white culture's terms.

MacDonald has shown that in later shows, like *Julia*, the African-American woman could only move freely in the white middle-class world once she was assimilated and packaged with a very white kind of sexuality and glamour.[134] "Try to be pretty," Julia's white boss gruffly demands. Julia's statuesque strides render white male character actors speechless from that episode on. This fear and accommodation of black sexuality persists and is aestheticized even in contemporary programming like *In Living Color*. Mildred Lewis has shown how the liveness on Fox's *Roc* is a direct result of the persistent conflation of live volatility, dangerous sexuality, and African-American masculinity.[135] L. S. Kim describes how television has demonstrated that fulfillment of the personal needs and sexual needs of the ethnic domestic *was possible*, but only in "legitimate" ways—typically by mothering and providing domestic consumer comforts to someone else's Caucasian children. Kane's Chinese spirituality and artistry in *Kung Fu*, on the other hand, were postured as hopelessly exotic and Asian. The *physical* skills of Kung Fu, however, were more easily imported—as weapons and sport—than the messy Asian cultural trappings that accompanied the warrior-priest. If style and artistry were not always *taken from* those racial sources, then white artists could be elevated by *displacing* the racial other back on those margins; that is, on the back of the presentational bus. Race-motivated substitutions like these suggest that people of color have always been allowed to operate as aesthetic repositories and consumer middlemen. When the ever out of fashion Jewish immigrant and mother seeks social acceptance from her sitcom peers on *The Goldbergs*, she goes, ironically, to the one source that can make her a hipper and more stylish American—the hot Latin dancers who teach her the "mamba, samba, and conga."[136] In shows like these, since the 1950s, the cutting edge of America was frequently defined by and around its racial margins. Television has historically idealized its racial characters as *culture givers*, and its Caucasian Americans as *culture takers*. The process of importing and transforming marginalized racial cultures continues to be a fundamental part of contemporary American culture. Outsider rap and hip-hop lived very short lives before being embraced and accommodated by white suburban mall culture. Race persists as television's avant-garde.

The fact that mainstream television caricatures dangerous forms of race and sexuality into adoptable aesthetic forms—jazz, rock-and-roll, hip-hop—is perhaps not surprising in itself. Television, after all, is an inherently conservative institution, and some African-American experts have argued that one underlying purpose of black music is "to mystify the white man and blow his mind."[137] What is remarkable, however, is the second half of television's persistent process of negotiation. Historically, television did not simply define and leave the aesthetic threat out there on the margins. It also actively worked to turn the aura into practical behaviors and commodities that "you too could own." In both ads and dramas, television consistently redefined the aesthetic as a *material product* related to family, home, and lifestyle. Dodge commercials in 1959, for example, went to great lengths to show how its new cars fulfilled a ten thousand–year design revolution that went back to the neolithic period. Loving close-ups of curvilinear chrome and Naugahyde upholstery underscored this latest cultural advancement.[138] With verbal instructions to the uninitiated, Todd and Buzz also postured contemporary industrial design as a hip awareness—a cultural sensitivity unavailable to inept white-trash gas station attendants who simply could not find the Corvette's gas tank.[139] Television dialogue like this fulfilled multiple functions: it was primetime dramatic fodder; it was mass-market advertising for Chevrolet; it was art appreciation. Both the Burns and the Kramden families were able to buy cul-

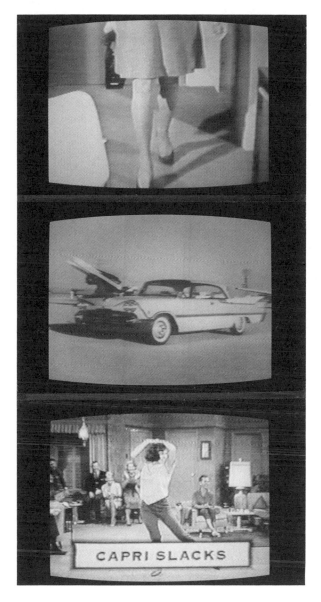

America's yearly commodity-substitutes for dangerous art; *Julia's* sexuality; Chrysler's cutting-edge automotive design; Laura Petrie's Capri slacks on *The Dick Van Dyke Show.* (NBC, Chrysler, CBS)

ture by buying a furniture of distinction: the television set. *Bachelor Father*, in a thinly veiled on-camera object lesson directed at delinquent youths, explicitly conflated lawyer Benchley's "hot digs" and fancy cars with upward cultural mobility.[140] The young and wayward men were no less impressed by the fact that Benchley's trappings and graduate education also brought him many articulate and sexy women. The Cleavers associated cultural dangers with films and hot sports cars.[141] Wally was clearly less ambivalent about such things than Ward. *77 Sunset Strip* somehow managed to put the hipster and beat countercultures— self-conscious bohemian undergrounds—up into L.A.'s brightly lit parking lots and into the flashy lifestyles of late Eisenhower era law and order proponents.[142]

Legitimate substitutes for art were not limited to sanitized forms of ethnicity and consumer products. The body could also be a cutting edge. Donna Stone and family automatically conflated aesthetic culture with dresses one could buy. When Alex Stone consoled his anxious wife Donna by arguing that her shapely adolescent body fulfilled a more important function than any fashion imported from Europe, the message was clear. The woman's body itself could become a symbol of aesthetic culture. Donna's body, after all, fit perfectly the postwar ideal of the feminine mystique. Her hourglass shape, thin waist and discretely draped torso, made her both highly sexual and maternal, a bearer of pleasure for all family members. Even this kind of female sexuality, however, was updated with changing fashions and social ideals. The *Dick Van Dyke Show*, for example, overhauled the 1950s moral skepticism of high fashion, by posing Laura Petrie as fashionable, not because of her vocation (she lacked one), but because of her hip and contemporary wardrobe.[143] Laura could somehow be pitched as the new woman of the 1960s: hip, domestic, *and* careerless at the same time. Through Donna and Laura's perfect bodies, the Reed and Petrie sitcom clans got to have their own substitutes for the avant-garde—their own acceptable and pleasurable bodily surrogates for a force still thought to be lurking out there on America's fringes.

In cars, dance, clothing, appliances, and the female body American television and Madison Avenue have long produced surrogates for the avant-garde. Commodities, both material and living, could become bearers of modern design and culturally distinctive collectibles. Better yet, these surrogate items, unlike art, could actually be used: around the home, on the golf course, in the subdivision. As long as art was seen as an aberrant threat, television worked to commodify it into acceptable forms. When art took television's center stage in the mid-1980s, however, commodification was no longer an ancillary process. That is, consumer culture was no longer something that television merely supported or pointed to outside of its aesthetic and dramatic confines. TV had also learned the value of repackaging itself as both *the* aesthetic experience and as a commodity ritual. The presentational guises under which this packaging took place will be more closely analyzed in the case studies that follow. Before doing so, however, it is important to move from a historical view of stylistics to an examination of the televisual mode of production. The performance of style was not just an invention of visionary producers and directors, it was also an aesthetic appetite tied to certain kinds of technology and production practice.

3 Modes of Production
The Televisual Apparatus

These radio news pictures projected from magic lantern slides onto the screens of the best picture theatres in the cities . . . [mean that] no newspaper can possibly put news events before the public as quickly as the theater can with radio news pictures.
—Technical proposal for theatrical television, *SMPE Journal*, May 1923[1]

Although some parts of the program technic may parallel the technics of the stage, motion pictures and sound broadcasting, it will be distinct from any of these. In effect, a new art form must be created.
—Modernist aesthetic espoused by RCA engineer R. R. Beal, 1937[2]

Television engineers have often acted as closet artists. From the very beginning, developers of production technology have seldom shied away from offering aesthetic theorizations about their new and constantly developing technologies. Even a cursory survey of the technical literature from the 1920s and 1930s shows that television might have ended up radically different from the form that we now know. Engineers hawked various visions of the artform: as a cinematic type of theatrical television, as a facsimile system, as radio photographs that produced paper prints, as a visual newswire, and as a video phone.[3] From the perspective of the 1990s, these early and alternative technical proposals, along with alternative economic proposals—like pay per view and a system of programming subsidized by TV set license fees—all seem incredibly forward-thinking. Each prototype, after all, now plays an important part in the contemporary multinational, multimedia environment. Yet, by 1950, each prototype had been written off as a failure.

This shepherding and attrition of technological and artistic prototypes suggests two things. First, media technologies are not easily dichotomized as either deterministic (forces that effect change) or symptomatic (phenomena that reflect cultural needs and ideologies), as Raymond Williams suggests.[4] As the above epigraphs indicate, prewar RCA and Society for Motion Picture Engineers (SMPE) aestheticians actively broached and bartered different aesthetic models in a give-and-take process of negotiation with stockholders, with government regulators, and with the supposed needs and tastes of the American people. William Boddy has shown incisively how government sanctioned monopolistic practice in the postwar era was a clear incursion into technological development—an exclusionary process of control that benefitted specific business interests.[5] To the deterministic-symptomatic model, then, one must also then add a third axis: the interventionist. History shows that mass-cultural processes are not always as subtle as some cultural studies suggest, nor are they always as ambiguous and contradictory as ideological criticism implies. Explicit aesthetic and theoretical discourses—as well as overt interventions of power—have always accompanied new media technologies and will continue to do so. Political and economic interests have never been queasy about publicly flexing corporate muscle to control paradigm shifts. These manifest tendencies in television's mode of production had an effect on televisual style as well.

Having surveyed the historical and ideological functions of aesthetic television, that is, the conditions and precedents for stylistic exhibitionism, I want to examine here some of the industrial conditions behind the emergence of televisuality. A consideration of the televisual mode of production—its production technology, methods, personnel, and organizational form—shows that the excessive looks of primetime television in the 1980s were not just illustrations of a stylistic or postmodernist sensibility, but were rather indications of substantive changes within the televisual industry and its production apparatus. To understand these changes, it is important to look at several key televisual technologies, their affect on production practice, and two major influences on televisual programming: the film-style look and the style-obsessed world of primetime commercial advertising. Chapter 9 will return to the issue of the televisual industrial apparatus with a more in-depth case study of two liveness modes: portable tape and the live remote.

Televisual Labor

Although students of cultural studies now flock to the audience and to the domestic living room in order to better explain television, few consider the practitioners or makers of what is transmitted over the TV a source for productive analysis. This academic oversight may be logical, given the problematic nature of much in the industry. The primetime production world, for example, is still in many ways an exclusionary old-boys network, fueled by patriarchy and by discourses that institutionalize anecdotes about "the ways things are done." There are those, for example, who celebrate the masculinist virtues of production technology. The latest full-page ads from Panavision in Hollywood show musclebound and loosely clothed male production studs erecting a tripod-mounted 35mm camera—all in a pictorial tableau clearly modeled on the bloody WWII Marine Corp flag-raising on Iwo Jima.[6] *Semper fi, Panavision.*

Film and television production accomplishment is, apparently, still very much fueled by testosterone. In recent and official how-to production publications by the Directors Guild, veterans teach television aspirants and newcomers the essential tricks of the directing trade: "Travel for [the director] becomes an adventure that no money can buy. Wherever he goes, he becomes involved with the natives at the location. . . . The relationship the director has with prominent people offers a different satisfaction than the one he has with the 'common' man. . . . I do not give brown envelopes with cash or cash in any form; I do not supply anyone with members of the opposite sex for any purpose."[7] In the old Hollywood, then, television commercial directing can be couched as a middle-aged forerunner of the "sex, drugs, and rock-and-roll" ethos. Straight-shooting, insider production books like this one cannot resist acknowledging the industry's important lifestyle functions and lures—travel, elitism, bribery, and illicit sex—as part of the director's professional package. This production patriarch has spoken, by substantiating each lifestyle concern with pregnant anecdotes from inside the production industry.

Compare this old-school Weltanschauung with an emblematic quote from a newer director, one more typical of the wave that entered television on terms very different from those laid down by the venerable Directors Guild of America: "I see my background in semiotic theory as the main reason why I've been able to cross over from such a radical avant-garde position to such a commercial medium. . . . Godard meets *Monterey Pop* is my ideal."[8] The references to semiotics and Godard are curious to say the least. These are not the references high theory expects to find in the *day-to-day workings* of the television industry. The remarks do show, however, that some parts of the industry are capable of viewing their work as a process of reflexive stylization and as a sign system. Semiotic and film historical consciousness counteract the way that the earlier industry overprivileged the maker's intentions as the key to a program's meaning. From a semiotic perspective, meanings are constructed by the program text through plays of image and sound, rather than predetermined by the program makers. The program, in short, is no longer construed as a neutral vehicle that transmits messages for its senders. Rather, the formal elements of the program themselves are the content. This particular homage to master-deconstructionist Godard, to the avant-garde, and to semiotic theory implies that some directors no longer dutifully accept the industry's received production wisdom and orthodox style. This degree of intellectual self-consciousness was not always a part of production.

Before the 1970s the production industry seemed to have little need for intellectual speculations on its form. After all, the industry's formal methods were self-evident, naturalized and codified through widespread use. Academic training, in fact, could be a definite mark against aspiring applicants to the primetime program production industry in Hollywood. The reason for this prejudice, of course, was that the only persons qualified to talk about or explain the industry were thought to be the practitioners and industry professionals themselves. The problem with this view, though, is the same one argued in philosophical aesthetics against the so-called intentionalist fallacy.[9] That is, critics of the fallacy argue that the meaning of an artistic work cannot logically be limited to the original *intent and expression* of the work's maker. If it is, the viewer is locked into a tautology, a closed loop. A work of art or film is not a work of art or film simply because their makers say so. Against the traditional view that privileged the originators of the artwork over its perception, Godardian consciousness presupposes the text as an (inter)active stylistic operation; that films and videos work as an array of signals and codes that are manipulated and orchestrated by producers and viewers. For a TV director to invoke Godard, and then to apply him to popular culture, is not just an exercise in the faddish, but empty, lingo of postmodernism. The behavior suggests instead a fairly widespread knowledge of film history, deconstruction, and stylistics.

I am suggesting that a new generation of production personnel in television is more stylistically *and* more theoretically inclined. Many production approaches in primetime are now premeditated rather than rote. One might counter this perception, however, by arguing that only those filmmakers tainted by the excesses of MTV or features actually evidence greater theoretical focus. After all, many in the industry tended to view with distaste the excessive style, hypervisuality, and self-consciousness of newcomer MTV. For example, *American Cinematographer* defined high-quality programs by setting them apart from MTV: "Pete Townshend's Deep End, *overcomes MTV special effects and overproduction and* exists on its own terms" (italics mine).[10] Notice here that excessive visuality—overproduction—seems to be a trait that producers should overcome. What the trade review argues for is stylish production, but only if it can be linked to a show's essential objectives, that is, to its own terms. The reference suggests that the industry is interested, not in excessive or empty style, but in stylistic originality and *motivated* style. Overproduction for *American Cinematographer*, then, might more accurately be described as special effects for special effects' sake, as canned looks that work apart from particular references or program objectives.

This example shows that even competing ideologies within the industry are reprivileging production style. On the one hand, the new generation of MTV-bred television and commercial directors play with the limits of style and radical theory. On the other hand, even the more conservative and older imagemakers associated with the feature film industry revalue style, but do so in a different way. *American Cinematographer* does not have an aversion to the stylish image—far from it. What it wants is *smart style,* rather than empty or unmotivated stylistic effect. Even this apparently reactionary critique against MTV stylishness is actually a call for stylistic heterogeneity. That is, it assumes that each work should be uniquely stylized according to its own objectives—neither wallpapered with special effects, nor forced to bear a standard and unmotivated industry look.

The industry's semiotic self-consciousness goes deeper than its various cutting edges and avant-gardes. Take for example the following trade account of production style in the popular primetime soap *Dallas*. Director of photography (DP) "Caramico sees his work as following the [series] look that's been established—of continuing the *tropes* laid down by previous cinematographers and by producer Leonard Katzman."[11] The reference to program tropes is not the kind of language one finds in earlier generations of books based on *prescriptive* rules and biblical production principles. Once one has viewed the look of a show through the lens of narrative theory—as figures of speech, as stylistic quotes, as smart references—form is no longer seen as a neutral vehicle. It is easy to foreground stylishness in unorthodox program genres. But here, even in the conservative and classical Hollywood style used in primetime dramas like *Dallas*, the look of the show presents itself as if derived from a well-stocked menu of possible stylistic figures.[12] Even if such programming does not invoke Metz's semiotic terminology, its utilization of tropes from a stylistic menu suggests that television is being theorized-in-practice as a visually coded formal system, as a semiotic smorgasbord. In this way, even conservative shows not originally pitched as stylistic can still be self-conscious about their visual poses.

The determining stylistic role that production personnel play is most evident at high-volume primetime production studios like Universal. With numerous episodic and serial shows simultaneously in production at any one time, visual style becomes both formulaic and very much tied to the tools assigned to each show. By 1993, Universal was still cutting conservative workhorse series like *Murder She Wrote* and *Columbo* on film. Flatbed film editing, after all, fits well the 1970s zero-degree telefilm style that is still used in both of these series. On more flashy and contemporary series though—*Law and Order, Crime and Punishment*, and the very hip and premeditated *Miami Vice*–like 1993 spin-off *South Beach*—the producers at Universal chose a much newer and flashier editing tool.[13] Universal

Studio executives praised the stylistic possibilities offered by the newer, nonlinear CMX-6000 editing system, but described the ultimate choice of editing technology as the series producer's call. Style, however, was not the only thing on the vice president's mind. Electronic editing also promised to be more cost effective—something the old-guard studios have always understood better than anybody else. Significantly, the new tools were tied directly to specific crafts people and to off-the-lot, third-party postproduction houses—all tied together in what the studio described as a new interpersonal relationship.[14] In Universal's eyes, then, televisual production technology cannot be distinguished from a new kind of producer and a new kind of extra-studio corporate relationship.

One of the most provocative illustrations of changes in televisual style comes from changes in the kind of background expected of directors. Television directing is no longer necessarily dominated by actor's directors. Veterans with decades of production experience, in fact, now argue that the best training for second-unit directing, a prerequisite for first-unit work, comes not from writing or producing, but from working as a cinematographer. "You have to be a director of photography before you can [direct] second unit, because you have to be able to shoot in any style. [As a director] you have to match what the main cinematographer does."[15] From the perspective of this director on the 1992 network series *Covington Cross*, then, one of the chief directorial tasks in primetime is to construct coherent stylistic worlds, on command, and from a wide variety of visual styles. Facility with rote, classical montage and blocking is no longer the issue. Requiring a cinematography pedigree for television directing, could not be more alien to the script- and acting-sensitivity celebrated by many earlier television directors. John Frankenheimer and Delbert Mann in the 1950s could not have cared less about the grade of fog filter used on the set or the characteristic curves of their primetime kinescope stocks.

Televisual Technologies

Simply looking at television's flow of ads, shows, and promos on any given night reveals the importance and consciousness of the televisual mode of production and its technologies. If many primetime shows now use all of their available "bells and whistles," then ads and music videos actually make production equipment a crucial part of their dramatic action as well. When director David Fincher, for example, used the new Raybeam, a lighting grid with thirty 1Ks on a recent Nike commercial, he liked the polished high-tech look so much that he included it as a prop in the background. From then on he had to re-rent the light as a prop to insure continuity among the other spots in the entire Nike television campaign.[16] The production tool became a fetishized toy in the hip urban world that Nike fantasized for its audience.

But new production tools not only influenced what was seen by viewers within television images, they also had a profound influence on how those images were constructed, altered, and displayed. It is important to see the emergence of stylistic exhibitionism in the 1980s alongside the growing popularity of six new technical devices in the televisual production world: the video-assist, motion control, electronic nonlinear editing, digital effects, T-grain film stocks, and the

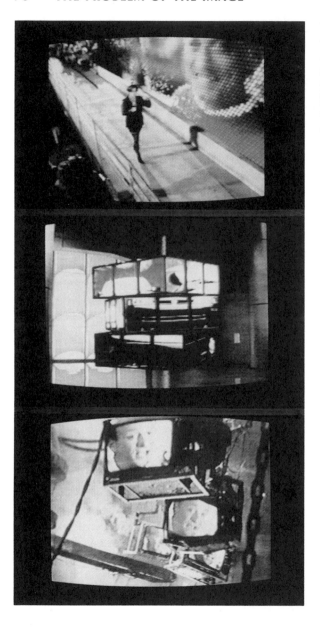

Television. Obsessed with electronic images and the technologies that create them, the televisual world is bathed in the light of the video image. (Bud Light, General Motors, ABC)

Rank-Cintel. The development and availability of digital video effects, for example has always promised (or threatened) to replace conventional production methods— a potential celebrated both by techno-futurists and production executives looking to save money in primetime by getting rid of real (and expensive) locations and sets.[17] Chapter 5 analyses this virtual world of excessive videographics in more detail. In addition, a case study at the conclusion of chapter 9 examines portable videotape, another influential industrial component in the rise of televisuality, and a technology directly implicated by new forms of electronic editing and digital videographics. Before considering the broader implications of these televisual technologies, however, a closer look at their importance in primetime production discourse is needed.

Video-assist. If one observes a film-based shoot for television today, one frequently confronts a production spectacle notably different from the way things used to be done. It is not uncommon today to see production personnel clustered around a glowing video monitor, entranced by the electronic image rather than by the actors or the action in front of the film camera. The video-assist makes possible this radical shift of the production group's gaze. In the old days, the nature of the image was really the business of only one or two people on a telefilm shoot: the director of photography and the camera operator. Although ultimately responsible for the show, the director was more concerned with acting and could only engage with the image as it was being shot, through the vague visual approximation offered by a neck-strung director's finder. Reinforcing the invisible nature of the television image during the shoot was the fact that the director and crew had to wait for projection of the film dailies—work-prints made from the camera original negative—which came back from the lab well after the day's shoot was over. As with feature filmmaking, sometimes the results were acceptable, sometimes they were not. For many years, television's 35mm Mitchell cameras did not even have reflex viewing. Camera people were forced to develop complicated "rack-over" systems to shift the camera away from the lens to enable the operator to frame a shot. This shift allowed the camera people to predict what the camera was seeing. Later, "reflex-viewing" on Panavision cameras allowed the operators to see the action exactly as it was being shot, but no one else at the time, including the director of photography, had any *certainty* about whether the shot worked, that is, whether it was exposed or framed correctly. The image was always in some ways a mystery, one that revealed its secrets only after a return journey from the lab's dark, chemical soup. This invisibility lent itself well to the mystique of the cinematographer's difficult craft and to the cult of professionalism.

With the introduction and availability of fiber optics, however, there was but a short and logical leap to the video-assist. Tiny fiber optics were tapped into the reflex viewfinder, fed out into a video pick-up device (a tiny video camera), and electronically fed to a monitor on the set. This device, then, allowed one and all on the set simultaneous and critical access to the once-mysterious camera image. What occasionally results from its use is a kind of team visualization, where every one of the key creative personnel has access to the composed image. This device, of course, saves money: a lot of takes never need to be printed, because flaws in acting or blocking become immediately apparent as the shot is being made or when videotape recorded from the video-assist is played back between takes. Yet the *displacement of the production gaze from the proprietary mysterium of the cinematographer*, to the public consumption of the entire crew makes everyone an expert on the image. The video-assist allows extreme precision during a setup, and saves money, but it can also be the bane of image makers when gaffers, actors, grips, and other "experts" offer suggestions about better ways to compose the image. Construction of the televised image before video-assist was based primarily on verbal commands between key personnel and mathematical calculations made by the cinematographer and his assistant(s). The video-assist, by contrast, allows everyone on the set to be highly conscious and concerned about visual quality. For better or worse, now everyone seems to be a master of the image.

Motion control. Another type of equipment that complements the video-assist, in both the *dispersal and intensification of the image*, is motion control. I include in this category not just the computer-controlled units that automatically program cameras to perform and reduplicate complicated camera moves, but also the Steadicam, Camrail, robotic-controlled studio cameras, and much less cybernetic devices like jib arms and motorized cranes. All of these devices are alike in one important way: they physically *take the camera away from the camera operator's eyes and move it through space in very fluid ways.* The resulting effect can be eerily nonhuman, as with the Steadicam—a body-mounted camera harness governed by a gyroscopic control that minimizes jerkiness and vibration. By taking out even the sensation of humans steps when the operator moves, the camera eye seems to float through space with a mind of its own. Popularized in excessively styled feature films by the Coen brothers (*Blood Simple, Raising Arizona*) and by Stanley Kubrick (*Full-Metal Jacket*), the Steadicam has been a workhorse in television commercial production, and an almost obligatory rental unit in music video shoots, for over ten years. Because of the Steadicam's extreme spatial fluidity in and around the body, the operator typically gauges his or her shot with the aid of the video-assist and a camera-mounted video monitor. Recognizing widespread demand, manufacturer Cinema Products has developed smaller versions of the Steadicam; lighter units specifically designed for television's video electronic field production (EFP) workhorse Betacam, and even the newer guerrilla format, high-8mm.

Everybody seems to want disembodied camera fluidity, not just feature filmmakers. The jib arm, a less sophisticated leverlike extension that mounts and pivots on the head of a tripod or dolly, also takes the camera eye far away from the operator's head. With video-assisted monitoring, television shots can now start far above a cameraperson's eye level and sweep laterally, vertically, or diagonally through a shot even as the camera rolls. Periscopic lenses on jib- or dolly-mounted cameras allow television cinematographers to shift from sweeping renditions of exterior action to snaking arterial moves through microscopic spaces as well. Programmed, computerized control of these moves allows directors to repeat identically the same complicated shots for one or one hundred–takes, all without the inevitable flaws and subtle differences that a human operator brings. Even outside of film-origination television, for example, at NBC network news and CNN, robotic control has had a profound effect on the way that live three-camera studio production is orchestrated. During John Williams's elegiac orchestral score at the start of NBC's nightly news, high-end robotic studio cameras glide subtly and smoothly around Tom Brokaw at center stage. One station technical director mused uneasily about his experience with the new automated studios: "Robocams—made 'em look so good, it's costing me my job."[18] As well as being stylish, then, television's robotic and autonomous eyes also have dire labor implications. Even the extensive and highly stratified camera crews in primetime production stand back and watch as a single composite operator (cameraman/DP/assistant cinematographer/grip) coaxes the Steadicam eye through its dramatic flight-like apparitions.

This family of motion-control devices all do one thing for the television image: they automate an inherently omniscient point of view and subjectivize it around a technological rather than human center. If anything reflects the onto-logical death of photographic realism in television, it is surely this gang of new and automated motion-control devices. The ideological effect of this basic televisual apparatus is one of airless and high-tech artifice. The televisual image no longer seems to be anchored by the comforting, human eye-level view of the pedestal-mounted camera, but floats like the eye of a cyborg.

Electronic and nonlinear editing: "Thirty-two levels of undo." If video-assist and motion control effected stylistic consciousness and disembodied fluidity *within the frame,* then a third group of technologies—electronic editing—helped shatter the *sequential and temporal* straitjacket necessitated by conventional forms of ed-iting. Electronic editing of videotape has been pervasive since the early 1970s, first as control-track editing, and then as frame-accurate SMPTE (Society for Mo-tion Picture and Television Engineers) "time-code" editing.[19] From the start, many telefilm producers despised these options. While video editing was acceptable for short network news stories, it proved impractical for longer forms, since video editing allowed no flexibility for change, modification, or revision once a suc-cession of shots was laid down. If twenty-five minutes of the final program had been edited and an early scene needed to be shortened or replaced, then the en-tire show back to that point had to be rebuilt and re-edited. This system of on-line editing meant that creative impulse had little place in the on-line suite. Directors and associate producers in charge of editing had to know the exact du-ration and sequence of each shot in the show. Mid-session changes of even a few shots could mean doubling the on-line editing costs. This was not, obviously, a creative or user-friendly system. Editing film, on the other hand, might be slower, but it allowed for numerous editorial reworkings. Telefilm editors could start cut-ting scenes in the middle of a program and work out from there with no negative effect. The total running time would simply expand as the editor added, deleted, or reversed scenes. The program's form grew and breathed with editorial changes. Telefilm editors were not locked into the rigid temporal sequences necessitated by videotape editing.

In the face of this uneven industrial reception by the primetime producers, major video equipment manufacturers—masters in the 1970s and 1980s of a high-tech industrial world obsessed with research and development—announced ever more highly sophisticated videotape editing systems. By 1993 even Sony's industrial and low-end broadcast editing systems had become proficient at loading up the television image with multiple simultaneous images and slow motion. Sony boasted that their "BVE-2000 editor connects to as many as 12 VTRs, control-ling up to 6 in any one edit."[20] This was the very kind of extreme visual facility that telefilm editors failed to find in the earlier variants of videotape editing. In some ways, the six-plus layered images simultaneously available even on this ba-sic industrial video system allowed for a denser and more complicated image than those of film. Telefilm producers, after all were limited to the two layers avail-able during any one edit (due to the A and B rolls in 16mm film negative; or the A roll and effects roll in 35mm). Videotape editing, then, has become hyperac-tive and visually dense with or without the endorsement of primetime telefilm producers. The production company for *Arsenio Hall* boasted that their hip and kinetic opening collaged visual fragments from a number of diverse sources: 35mm, 16mm, 8mm, black and white, and color. Anything and everything could be slammed together in the highly developed world of electronic videotape editing. Even before the wide acceptance of nonlinear cutting, then, videotape editing had provided practitioners with an apt tool for the new collage style favored in the 1980s.

The resistance to cutting on tape by the major telefilm producers began to loosen in the mid and late 1980s, with the development of newer random-access memory electronic editing systems, like Lucasfilm's Editdroid, the EMC2, the Montage Picture Processor, and the CMX-6000. All promised primetime program producers the ability to do film-style editing for television. The technological breakthrough that made this possible, was the increasing cost effectiveness and memory-storage power of newer recording media: electronic video discs and greatly expanded computer hard drives. As alternatives to videotape-based recording or editing, films or tapes were loaded into a computer's RAM, video discs or hard drives and any part of the original source material could be called up at any time within a microsecond. Shows cut on these systems could provide half-hour or hour-long screenings for producers even though no show actually or physically existed, either on tape or on work-print. The production footage was merely volatile, stored electronic information. The edited versions of shows were really software-driven computer files that pulled up scenes instantaneously and in a sequence on command. The ability to move scenes around endlessly was the promise of these devices.

At first, the claim to fame of this editing equipment was its ability to show your producer or client ten different completed versions of the same scene or program during the same screening session. The broader implications, however, also became clear: *nothing visual was set in stone.* Again, the majors at Universal and Warners were willing to consider the technology, not just because they wanted to ape the flashy style of MTV, but because it promised serious production economies. Grady Jones, vice president of postproduction at Walt Disney Television, rationalized about the nonlinear technology: "We're always trying to bring costs down. We don't have much padding anywhere. The only thing that we can do is try to control the length of time it takes to complete the editorial and sound effects cutting."[21] By 1993, one of the newest and least expensive nonlinear systems—the Mac-base Avid—received the kind of acclaim that indicated its new and extensive popularity. Avid was awarded an Emmy for technical accomplishment by the Academy of Television Arts and Sciences. Nonlinear Avids were, after all, being utilized everywhere: in New York–based commercial postproduction houses, in music videos, in primetime program production. The reason? The trades boasted about its ability to provide limitless reworkings. "The bottom line is that this system gives me enormous creative freedom. I can edit unlimited versions and save them all. The Avid has *thirty-two levels of undo* and that completely frees up the editor to *experiment.*"[22] Even mainstream television people in Los Angeles, then, saw and valued the dramatic experimental potential of the new systems. Forget orthodox editing wisdom, the whole point for editors now frequently is to demonstrate how far one can push the editing syntax on a project or scene, and how many stylistic variations one can showcase. After all, with nonlinear, there is no risk. Nothing is stylistically set in stone. Nonlinear encouraged, or fed, the televisual appetite for stylistic volatility and infinite formal permutations.

The two other crucial televisual technologies—the Rank-Cintel telecine and new high-speed film stocks—are best understood within the broader context of two emergent obsessions in 1980s television: the film-style look and primetime commercial advertising.

Playing with Limits
Self-Conscious Primetime Practice

Lighting for features, lighting for television, the light is identical.
—George Spiro Dibie, president, American Society of Lighting Directors[23]

On television, you can't be Vittorio Storaro. But what you can do is like music.
—Oliver Wood, director of photography, *Miami Vice*[24]

One of the central working concerns in television production in the 1980s concerned the formal potential of the television image, and especially the question of what can be done within the constraints and confines of the limited television frame. Consider the diametrically opposed views outlined above. Some DPs saw in primetime Bertolucciesque cinematic potential; others, melodic sensitivity. TV was inherently like film; TV and film were antithetical. Such contradictory answers—about what television can and cannot do, and what it can do best—abounded, but the question became more and more pervasive in the working and marketing discourse of the industry. Academic high theory, on the other hand, was working from two very different and problematic assumptions: first, that producers-practitioners could not be aware of the deep structure or ideological implications of their work, and second, that producers-practitioners used aesthetic criteria that were incomplete or naive. Evidence for this bias is found in the widespread penchant that high theory has for inventing its own frameworks, aesthetic categories, and critical terminology.

Even a limited examination of recent literature from the industry, however, shows that these assumptions and write-offs of the industry are misguided. Not only is television currently stylish, but it can be stylish in an extremely self-conscious and analytical way. While high theory was speculating on television as a distracting verbal-aural phenomenon, something very different was happening within the producing industry. There, in producer story sessions, in conversations between DPs and gaffers on sets, and among editors in postproduction suites, an awareness was growing of television as a style-driven phenomenon heavily dependent upon the visual.

Since the systematic approach to visual style is very much on the minds of some practitioners, and evident in the practice of many others, it is worth examining how media producers conceptualize this visuality.[25] Two areas of industrial debate—film-style programming and videographic programming—correspond roughly to a major generic programming division in television. That is, the division mirrors an institutional split between primetime dramatic and comedy series on the one hand—producer-dominant genres that use *film* pervasively—and the other extensive array of director-editor–dominant program forms and genres that are heavily dependent upon extensive *electronic* postproduction. Before turning to an indepth study of videographic embellishment in chapter 5—an electronic practice that dominates both off-primetime and interstitial material in primetime—a further clarification of cinematic televisuality is needed.

The Film Look

The issue of image superiority in the film versus video debate and questions about the merits of film-style production methods for television have received much attention in the 1980s and 1990s. Landmark work in this area was produced by Harry Matthias and Richard Patterson, and other works have followed.[26] I am not interested so much in the technical aspects of this debate as in the ways that practitioners interpret and explain the film look in television. To understand this kind of discourse, however, it is important to survey some of the technical issues that have become central for those television production people that make heavy use of film technology. A discussion of more fully electronic variants of televisuality will follow.

New film stocks. Predictions to the contrary, film origination in television has not been replaced by video imaging. Far from it, film origination has thrived and prospered in two major areas of television: (1) primetime programming (episodic shows, movies for television, and miniseries), and (2) commercial advertising for television. By the late 1980s, Eastman Kodak boasted about an increase, rather than a decrease, in film consumption in television. With "80 percent of prime evening time schedules for the three major networks made of programs originated on film . . . this [was] one of our best years ever in terms of original negative used for TV production."[27] Even as the aesthetic and formal possibilities offered by film increased, the popularity of shooting television on film stock also increased. Film-tape manufacturers argued that certain TV scripts in fact call for quality "production values that are more appropriate for film," especially any genre requiring "fantasy" rather than "immediacy."[28]

Not only did shows shot on film dominate television in the late 1980s, the quality of the film stocks allowed for a kind of visual sophistication impossible during the zero-degree telefilm years of the 1960s and 1970s.[29] In the early 1980s first Fuji, then Eastman, and then Agfa all introduced new lines of film negative stock with dramatic improvements in both sensitivity to light and graininess. Chemically engineered around new and less visible T-grain silver halide particles in the emulsion, the new stocks could be used in extremely low-light situations, could be easily "push processed" one or two stops, had more saturated color rendition, and provided a greater range of contrast and tonality within a single image than any of the earlier stocks. When primetime DPs boasted that "of course, not every television show is shot at ten footcandles," they were both showing off that they could shoot primetime at that unheard of level of darkness and also making stark contrasts to earlier, prestige production stocks.[30] Other primetime DPs tied the new speed and sensitivity to production mobility and freedom. "Technicolor was great but very slow—[it required] 900 footcandles at F/2 to get an exposure, and a camera and blimp so big and so heavy it was totally impractical to

The new film negatives. Rich tonalities, color saturation, and painterly light (in, top to bottom, *Young Indiana Jones, N.Y.P.D. Blue,* commercials) that would make Storaro envious. (ABC, ABC, Uncle Ben's)

use it in real locations. Nowadays we can turn up with a little camera and some film and shoot anywhere."[31] High-speed film stocks, then, also meant improved logistical mobility and more flexible production scheduling.

By 1990, Eastman provided a vast menu of professional negative stocks for telefilm producers, from grainless, color-saturated daylight stocks to nighttime stocks that could be pushed to 1000 ASA or higher. These were overwhelming options for television cinematographers trained in the 1960s and early 1970s, a period when one or two stocks were typically available and when high-speed negative was defined as a mere 64 ASA. Film stocks, therefore, had a direct impact on the ability of producers and cinematographers to marshal variant visual looks. The reasons were not just photochemical, however. With the increased use of computer technology—for design, engineering, and quality control of new emulsions during the 1980s and 1990s—film stock companies like Agfa, Eastman, and Fuji were now able to make potentially limitless numbers of stocks. Technical representatives from Rochester, New York, now continuously circulate among DPs and operators in their quest to formulate new stocks and discover new needs. If a stock with a particular look is not currently available for a commercial need, Eastman will now consider customizing one. This is not the development-lazy and cash-rich Fortune 500 company of the 1970s. Extreme international competition by Fuji and Agfa have changed the way that both business and engineering is done in the industry. But the facility with engineering endless photographic looks is not just an economic consequence, it is also an outgrowth of interaction between computerized emulsion engineering and the publicized stylistic needs of a new generation of primetime and feature cinematographers.

By 1991, filter manufacturer Tiffen had created a host of designer-color filters: "grape, chocolate and tropic blue—all available in three densities, and all available in half-color and graduated to half-clear." [32] Not only did the new film stocks have a direct impact on camera mobility, low-light sensitivity, and photographic tonality, they also allowed *a new level of visual detail in front of the camera.* Televisual sets and locations got visually denser and more complicated after Ridley Scott's *Blade Runner* and Chiat Day's Macintosh television spots in 1984, even as the ability to render such images by new film stocks—and higher resolution color television monitors at home—improved. In addition, the whole optical film industry was revolutionized by the improved resolution and chrominance abilities of television. Far from the days when primetime DPs were locked into the polar world of filters limited mostly to color and density corrections between tungsten light and daylight, the new filters and stocks were designed to render both wild variations and subtle nuances of color within a single image. When industrial players like Tiffen baptize their new "grape" glass as a "designer filter," the association is complete. The designer televisuality of Michael Mann's *Miami Vice* is matched by the designer optics of Tiffen's lense-mounted glass. When television cameramen now ask for more coral rather than simply more orange, the televisual revolution symbolically betrays its technical as well as producerly roots.

Rank-Cintel. The desire to infuse video with a visual style more typical of film was enabled by one technical development as much as by any other. In 1977, the first Rank-Cintel "flying spot scanner" was installed in the United States. Several generations of design improvements followed, and other companies marketed their own versions of the chip-based film-to-video transfer machines. Higher quality images could be rendered on videotape, since the Rank-Cintel was able to reproduce and take advantage of film's unique look and "incredible dynamic range." [33] Practitioners boasted that the Rank meant that "film provides an ability to record as much as a 400:1 range of brightness which is the difference between the brightest and darkest elements in any scene. This allows a talented cinematographer to use light to paint very subtle details which establish mood and setting." [34] Consider the not so subtle conceptual transformation that takes place here, from a purely technical description of contrast to an aesthetic theorization. Mere television transfer technique is redefined in terms of painting and cinematography. This simple verbal deduction undercuts the way that television image technology has traditionally been defined: as amorphous, low-resolution, flat, and crudely contrasty.

Television's marked shift toward using film negative was based in part on the promise that Rank-type transfer units could reduce electronic noise in the picture, while at the same time maintaining details in the darkest shadow areas of the television image. Rank-Cintel now offers producers a menu of various and distinctive looks depending on the type of film format being transferred: original camera negative, interpositive, dupe negative, or low-contrast projection prints all demand different setups and parameters on the Rank-Cintel. Each variety of stock affords the producer a different visual look. The producers of *Love Boat*, for instance, continued to transfer their shows from Eastman color *projection* print stock 5384, even though projection prints give much less subtlety and tonality than do transfers from negative.[35] For the producers of that show, the contrasty, saturated look of the print film gave the show a look that its fantasy needed and that its audience was thought to expect.

Transfer technology in the 1980s, then, did not just make the image *better* visually, it actually multiplied the various visual looks of television into discrete codes that could be tied to specific program ends. For *Love Boat*, projection-print Rank transfers *signified* fantasy. For other shows, transfers from negatives gave the subtlety associated with painterly chiaroscuro. Transfer technology, then, helped codify the look of television even before the artisans of postproduction were fully involved in a given program.

Once these two technical conditions existed in television (origination in film and transfer via flying spot scanner), pressures to change production aesthetics itself would intensify. Achieving the so-called film-look in shooting style became a production cliché in the 1980s. Originally, film-style video simply referred to the shift from three-camera live in-studio TV to remote single-camera shooting that became popular in the late 1970s. Electronic news gathering (ENG) changed to electronic field production (EFP) for more discriminating producer-practitioners. But this early shift in the way that video shoots in the field were organized—to ENG and EFP—was mostly logistical, since it had as much to do with camera and recorder placement as anything else. It was not until the 1980s that a more intensive change began to occur within the visual frame itself.

Improvements in lighting developed alongside EFP, and a subsequent shift to charged-couple devices (CCD) in cameras (rather than conventional tubes) afforded producers greater subtlety in visualizing their images.[36] It is, obviously, unlikely that technical changes alone intensified the image in television. Formal changes in genre and narrative greatly impacted the rise of televisuality. Industrywide though, there was an increased interest in transforming television images into complex, subtle, and malleable graphic fields. It is likely, given this shift, that formal and narrational changes provided the ideational resource required by an industrial transformation of this scale.[37] MTV and *Miami Vice* certainly were landmark programming developments that changed the way that television looked.[38] These changes have been discussed in detail elsewhere, so I will merely reiterate that distinctive programming forms and shows like these provided the conceptual framework—that is, the audience expectation and the cultural capital—needed to effect a shift in the televisual discourse. If film origination and Rank-Cintel transfers provided, in Brian Winston's terms, the "technical competence" needed for a change in the television industry, then the new highly stylized shows like *Miami Vice* and *Crime Stories* and MTV provided the ideational requirement for industry changes in the 1980s.[39] By the time feature-film director Barry Levinson was showcased on network television in 1993, the importance of televisual transfer technology was clear. Each episode of Levinson's *Homicide* opened with stitched-together footage shot on a primitive, spring-wound cast-off 16mm Bolex camera and feature-style images shot on 35mm. This mixing and matching of emulsion and format types, and the manipulation of colored filtered effects and black-and-white stock, were all possible because Rank-style transfers provided extreme options for stylistic control and reworking, even after the primetime footage was in the can.[40]

Program *Individuation*

While new stocks and transfer technologies reinforced and enabled one influential televisual ideal—film-style video—programming practice, acting, and promotional considerations encouraged a second industrial mythology: program individuation. ABC's *Moonlighting*, for example, frequently attempted dramatic scenes that pressed orthodox lighting and production methods to their limits. On one episode DP Gerry Finnerman attempted to shoot a complicated dialogue scene that involved many movements by the actors during a single take. To cover it without stopping the camera—with proper exposure, focus, and expressive lighting—meant staging the scene like a three-camera film show. Two people were assigned to the kind of dimmer control board that one associates with live theater and that allowed the crew to constantly move the key light and alter the background light even as the actors moved in front of the continuously rolling camera. In a blow-by-blow account, cinematographers marveled at the real-time skill and performance required by the crew to pull off these multiple moves, framings, lightings, and changes in composition: "Finnerman was using fill-light with Shepherd and cross-light with Willis. It wasn't easy, because Shepherd never looked in the same direction twice. The DP had special brackets on the camera that allowed him to use sliding diffusion. Sometimes he had two-shots with Shepherd covered by a Mitchell A filter and clear glass on Willis. Other times he used various different combinations."[41] This kind of primetime choreography, with filters, fresnels, and dollies flying, was more like ballet or performance art than single-camera film-style shooting.[42] It was, in a televisual guise reminiscent of the hometown NBA Lakers, "showtime" for the optical people. Like Indiana Jones, the not so implicit response in the competitive, production-oriented trades was: "These guys are good; very, very good." Bruce Willis, then, wasn't the only one hot-dogging on *Moonlighting*. The production enterprise itself seemed to embrace stylistic hot-dogging as its reason for being. *Style uber alles.*

It is commonly understood in industry parlance today that each show should have its own "look."[43] Even relatively traditional-looking and visually restrained shows like *Cagney and Lacey* postured an identifiable visual stance. The show's

DP explains how taking over the show also meant taking over the burden of its established look. "They wanted a more visual look, and to me that meant they wanted more contrast."[44] After choosing a film stock that gave them this contrast, the company also intentionally used long lenses in "over half of the shots" in the 1986–1987 season in order "to flatten space."[45] In addition to the practical effect of making downtown L.A. look like New York, both devices (long lenses and contrasty stock) helped to *stylistically individuate* the show. In competition with other primetime shows, this artificially constructed sense of place and geography were merely part of the overall effect that resulted from the show's distinctive stylization.

But shows like *Cagney and Lacey* only hint at the increased role that visual style played in other dramas. In the network dramatic series *Covington Cross*, the DP waxed eloquent about the historicity of their "flamboyant" signature style: "Smoke is the key element, because at that time, smoke was the source of energy for everything. At night, everything [in the show] looks as if it is lit by firelight, candlelight, or flambeaus. I'm using a quarter-fog filter on almost everything." He goes on to justify this excessive use of flammable and incendiary devices based on the assumption of an inherent physical need in television. In "TV where you have a smaller image, you need to go in stronger to create an atmosphere. I am also trying to light with a fair bit of contrast. People have always said you mustn't be too contrasty for TV."[46] In one fell swoop, then, this primetime cameraperson throws out the traditional view of the medium in its entirety. Precisely because the TV screen is smaller than that of film, *producers need stronger stylization*, not the weakened style that academic theorists have dichotomized.

Consider in addition the following breakthrough claim from the production of CBS's *Beauty and the Beast*: "We're very proud of *Beauty and the Beast* because the cinematography is really very important to it. The producers feel strongly enough about it to give the director of photography a credit at the beginning of the show rather than at the end—and *Beauty* is the only episodic TV show that does it that way."[47] Note that the programming breakthrough here is described as the process by which the shooter is given an unusual amount of creative power and a visible position in the credits along with other "above-the-line" personnel and talent. Not only was the show excessively stylish, but it was self-conscious about that trait. The show's unique style was centered around the heavy use of smoke, colored gels, Rembrandt lighting, fog- and halo-effect filters.[48] Writers for the show have discussed the unusually low ratio of total script pages to program running time.[49] With *Beauty and the Beast*, scripts frequently involved twenty to thirty pages of dialogue, rather than the forty to sixty pages typical of hour-long dramas. This sheer reduction in script verbiage challenges the most conventional wisdom about television style. When comparing their quality to highly visual film scripts, for instance, one frequently writes-off the quality of TV scripts as being too wordy, too explicit, or too expository. In worst case situations, television scripts can have dialogue that reads more like a director's cues, with overly explicit and redundant dialogue that repeats obvious visual information. In worst case scenarios, one actor may announce to another, "I think we should go to the door" rather than simply going to the door. In television, many producers have typically depended on the word to carry the story. Rationales for this verbal

Individuated looks, an obligation in the world of advertising. (Left: Miller Brewing Co., The Gap). Masquerade in primetime: *Casablanca* choreographs *Moonlighting*, *The Brady Bunch* possesses *Day by Day*. (Right: ABC, NBC).

privilege are frequently based on the argument that television's low resolution image is unable *by itself* to communicate essential narrative detail.

This orthodox wisdom about the centrality of verbiage and exposition in television scriptwriting now appears to be changing. Shows like *Beauty and the Beast* not only minimize talkiness, they also let an expressive visual style dominate the viewing experience. The fact that the producers of the show hired a poetry consultant as a member of the production staff suggests that even the nature of the script's written word has changed.[50] Script verbiage here is consciously addressed in a poetic and lyrical manner, rather than solely as an expository or action-oriented mode. This kind of shift suggests that even mainstream television can aim to gain viewership and win Emmys by foregrounding embellishment and expressive visuality. This stylization is not latent or subjugated by story either, as it would be in a classical context. Rather, the producers champion their visual style and iconographic accomplishment directly to viewers.

Masquerade

If some recent programs work by selectively intensifying their mise-en-scène around an identifiable look, others depend upon a third televisual mythology: a more eclectic and selective use of visual codes better termed "masquerade." That is, whereas *Beauty and the Beast* was known for specific photographic effects (saturated colors, directional lighting, smoke), other shows promote themselves by playing off or parodying cinematic styles. Film history itself becomes a playing field for many of today's television stylists. The award-winning and widely viewed series *Moonlighting* and *thirtysomething* on ABC were among the most visible of these network exercises in television and film history. Both shows toyed with numerous and eclectic visual styles in ironic and self-conscious ways. The knowing display of style became, for *Moonlighting*, an integral part of its performance. Viewers came to expect style references, and they got them in the various presentational guises that *Moonlighting* took on. Moonlighting did film noir; did MTV; did Orson Welles–Greg Toland deep focus; did Capra screwball comedy. In the later seasons of the series, the dramatic content of an individual episode was frequently tied to a specific visual style. What looked at first like aesthetic eclecticism though, became, through its presentational facility and range, a sign of connoisseurship. The boast is not just that such shows can do this or that visual style, but that they can cycle through a range of visual styles with virtuosity.

In 1989, the "Here's Looking at You Kid" episode of *Moonlighting* was nominated for "Outstanding Cinematographer in an Episodic Series" by the American Society of Cinematographers. The episode was designed around "shockingly exact replications of two classic films: *The Sheik* and *Casablanca*."[51] The crew was said to have achieved the black-and-white look of orthochromatic film stock throughout.[52] The production included a re-created airport, fog, black paint to approximate a wet landing strip, landing lights, and the heavy use of fog filters and nets. The set aimed to reduplicate Rick's Cafe in the original film "down to the palm fronds throwing shadows on the wall." While this penchant for retrostyling sets and locations was not totally alien to the high-production value system of classical Hollywood, the ways that the image is handled in shows like *Moonlighting* demonstrates an extreme awareness of the concept of visual simulation.[53] In this particular production, makeup was applied unevenly and heavily to artificially shift flesh tonalities in the same uneven way that orthochromatic stock did during the silent era. In addition, the film stock was push processed to achieve a contrasty, antique look. In the very ways that the texture and tonality of images was manipulated, shows like this one suggest that their audiences are neither aesthetically stupid nor solely interested in character or plot, as traditional dramatic theory assumes. Rather, regular investments and manipulations of style indicate that audiences must in some way find pleasure and engagement in a weekly aesthetic game that demands stylistic decipherment.

Moonlighting was not unique in this respect. In fact, many shows by the late 1980s had consciously utilized various methods of retrostyling and the explicit adoption and performance of visual style. By 1989, an episode of the sitcom *Day by Day* literally transformed itself into an episode of the 1960s *The Brady Bunch*. Five minutes into the contemporary program, a corresponding rupture in style occurred. Studio quality state-of-the-art sitcom video gave way to 16mm film origination, even flatter studio lighting, identifiable film grain, and poor camera registration. That same season, an episode of *thirtysomething* transformed itself into an episode of the *Dick Van Dyke Show*. *China Beach* went further in its stylistic transformation of one episode: it left the primetime dramatic style and tradition in order to mimic the interior, head-shot interview style of independent feature documentaries like *Seeing Red, The Good Fight*, and other oral-history films.[54] Stylistic references could be made, then, not just to the television and film history of mass culture, but also to style practices associated with higher or more marginal taste cultures, like independent film.[55]

These increasingly common practices suggest several things. First, primetime audiences by the late 1980s could apparently appreciate and *decode* self-conscious displays of cinematic and televisual form. Second, many shows now began to work not by simply making their mise-en-scène more excessive, but by making their *presentational demeanor* more excessive and sophisticated. In a sense, shows like those mentioned above positioned themselves as *impresarios* of style and aesthetic awareness. Maquerade shows revel in marshaling and displaying aesthetic systems, not just at making images more visual, which they also do. By doing this, by standing back from and acknowledging the form itself, the producers promote the television image as an image-commodity. If televisuality is about signs of excess, then its semiotic abundance comes not just from the frame that DPs and gaffers argue about, but also from the very broad cultural and pictorial traditions that practitioners can now bring to bear in producing shows.[56] For this reason, one cannot simply talk about televisuality's two-dimensional signs. One must shift from a compositional discourse to a pictorial and cultural one in order to understand televisuality's excesses. By manipulating pictorial sign systems, whether from film history or pop culture, television boasts to the viewer that it is a master performer of visuality, a master of stylistic masquerade.

In the past, television genres were defined by the fact that their narrative formulas were fundamentally static and repetitive, while only their situations changed from week to week.[57] Style was even more static than formula given the fact that style frequently came as part of the development package—it was dictated by the facilities and soundstages that housed the productions. This *static formula–dynamic situation* concept rang true of television in the 1960s and 1970s. Now, however, in many program and nonprogram forms, the stylistic and presentational aspects are the very elements that change on a weekly basis, while characterization becomes the medium's static and repetitive given from episode to episode. With *China Beach, thirtysomething, The Wonder Years, Quantum Leap, Northern Exposure*, and, yes, even less prestigious shows like *McGyver*, the viewer is now encouraged to speculate before each episode about what the program might *aesthetically transform itself into this week:* documentary, dreamstate, oral history, music video, homage to Hollywood, or expressionist fantasy.

Electronic Cinematography

All of the shows that I have described so far—*Love Boat, Cagney and Lacey, Beauty and the Beast, Moonlighting, Day by Day, thirtysomething, Covington Cross*, and *China Beach*—were relatively high-budget, primetime, and, in most cases, prestige shows. Each of the shows in this industry sample are primetime series that could afford to shoot with Panavision and 35mm film negative; a single-camera style that inevitably demonstrates that television's authority through cinematic distinction. Yet, the new mythology of film-style video, with its emphasis on visuality, has permeated non-primetime and video-origination programming as well. Even if producers do not have the resources to light and shoot television shows on film, many simply make their videos more cinematic and stylish by electronic means. The low-budgeted syndicated police show *The Street*—emblematic of the newly popularized genre, "reality programming"—was shot on video in urban locations and at night. At a frantic pace that enabled the crew to cover 125 script pages in five days, this show could in no way be described as prestige primetime telefilm material. Given this frenzied production schedule, the producers claimed, ironically, that they shot with a film aesthetic in order to achieve a "TV feel." Director of photography Rob Draper was hailed for treating "the camera as 'another film emulsion.' Ignoring the factory specified 125 ASA, Draper runs the Sony BVP-5 video camera at 800 ASA, and at +9dB. This results in a grainy feeling with electronic noise. Combining this technique with a spare-like lighting style gives the show its film look and documentary TV feel."[58] Although the producers argue that this gives them "a realistic look" for the show, what they actually get is an image far from illusionistic. In fact, by electronically boosting video gain (+9dB) to compensate for low-light actually only succeeds in filling the image with snowy electronic noise, or electronic grain.

The assumption here, then, is that viewers decipher noisy and low-resolution video images as both realistic and as somehow cinematic. Realism depends apparently on graphic opacity, rather than on representational illusionism. Far from clear or highly resolved, these images are forcibly videoized and degraded through the imposition of noise, but are somehow read as real nevertheless. So much for André Bazin's ontology of realism, a theory that constructed realism around a mode of visual and transparent replication, not around the graphic muddiness fabricated by this kind of television.[59] The frequent use of electronic degradation in "reality shows" does suggest that viewers can discriminate among the various presentational styles: from the film history masquerades choreographed in film-origination to the *ontological* obsessions of electronic origination. Both modes, the cinematic and videographic, are authorized by the narrative and generic assumptions of specific shows. In the case of *The Street*, all "film style" turns out to mean is minimal or nonexistent lighting, not elaborate motion-picture production value. Electronic noise, then, is considered as much a televisual code as are high-resolution transfers of richly toned film negative with their *absence* of grain. Each distinct look is tied to a specific referent, and shows are individuated by using either code. So powerful was the mythology of cinema's visual prowess, that even video noise could be conceptually retrofitted by television as a badge of stylistic distinction.

Commercial Advertising

Madison Avenue—defined and fueled by stylization—influenced the emergence of televisual exhibitionism as much as the family of cinematic mythologies: the film look, program individuation, and masquerade. If American television had an avant-garde in the 1980s it was surely primetime television commercial production. Commercial spots continue to be the most dynamic sites for visual experimentation on television. Packed into tiny temporal slugs of thirty and sixty seconds, advertising spots were probably the first type of programming to exploit the discursive and emotive power of hyperactive and excessive visual style.

Standard production wisdom says that a spot should focus on one major message in its short duration.[60] Given the limited potential for verbal discourse in short spots, then, nonverbal mechanisms are much more important in triggering the needed emotional appeals that drive home the spot's intended message to the viewer.

Well before MTV, in the late 1960s and 1970s, commercial spots learned the advantages of engaging viewers through the lower-sensory channels, through sight, sound, and tactility. Over several decades primetime advertising mastered a process in which the viewer is simultaneously flooded by a range of sensory signs. Visual style became visually excessive and temporally hyperactive on network television, one might argue, because ads must fight for the attention of distracted viewers during breaks from the program. Ad sequences with shot durations of one second or less now frequent both network and cable television. Ad cinematography, on the other hand, is frequently defined by its heavy use of designer filters (especially grads, diffusion, and colored effects). Primetime spots are, to use industry parlance, excessively "lensed." Cutting in contemporary commercial practice makes classic Soviet montage look lethargic. Since television really is about advertising, about selling viewers to advertisers, it is important to survey at least some of the favored televisual manifestations found in commercial spots during this period.

Digital Compositing

The clean European design and controlled studio product photography of an earlier period gave way in the early 1980s to ads that pushed television and its resolution to their limits. Producers sought to make video look like film, and talented newcomers left music video production for commercial advertising and program production. In the process, stylishness became a requisite for productions and products that sought memorability. The new commercial style infiltrated network programs as well. Title sequences in many programs began adapting the new frenetic visual style from spots. Even the segues to ads within programs became less overt as a result. A direct influence of MTV style showed in the *Saturday Night Live* title sequence of the 1986 season. Layer after layer of live-action imagery was artificially composited together. When broadcast, *SNL* cast members and digitized New York landmarks unscrolled past viewers on the screen. In this heightened performance of technical wizardry and hipness, no static frames were visible; and no shots existed in any traditional sense of the word. Instead, a realistically photographed, but graphically dense and continuous scroll unwound for viewers. The weeks that followed *SNL*'s use of the digital scroll saw widespread use of the mode by advertisers, who profited from the new and highly stylized visual effect.

Coca-Cola used the same effect to promote its newly announced product Cherry Coke. Significantly, they hired the Emmy-winning co-originator of the *Saturday Night Live* intro, John Kraus, to work on their piece as director of photography. While the final effect in both instances was dependent mostly upon the digital graphics capabilities of the new Charlex system and graphic Paintbox, the footage was all shot on high-end video. According to DP Kraus, the key to the success of the spot and intro was the use of a complicated film-style lighting scheme.[61] The lighting design aimed for directionality, 3-D modeling, and a com-

Commercials are to television what JPL is to NASA. Shaq stretches and tears the digital plane for Reebok. Time-traveling Paula Abdul composited with a young Gene Kelly for Coke. Breakthrough digitally composed scroll from *SNL* unrolls for Cherry Coke. (Reebok, Coke, Coke)

plex and shifting color scheme tied to specific objects or persons in the unfolding graphic scroll. All light was to be motivated in someway in order to fight the flatness and artificiality associated with most video effects. In addition, each of the multiple visual components and icons, filmed individually, was given a specific visual code or look. "We lit Marilyn [Monroe] with hard edge, 1950s light, and matched up two shots of her with and without her coat. When a werewolf, who pops out of a movie screen, turns into a handsome man after sipping Cherry Coke, and then turns up later on a rooftop with a lovely woman, we kept him in black and white to make that connection. We lit his date in a very rich light to convey glamour."[62] All of these "actions" actually happened only in the electronic ether. Cinematography here was essentially a process of *collecting* individual elements and fabricating virtual worlds through imagined light sources. Filmic and televisual composition, then, depends heavily upon electronic postproduction. Compositing demands and rewards directors who can skillfully choose from a wide range of specific visual lighting codes and styles. A style is no longer construed in a classical sense as a unifying formal element. Instead, in this type of compositing practice, styles are more like codified cards that are collected, layered, juxtaposed, and played with in a process of electronic postproduction. The frameless, digital environment that resulted in the *SNL* opening and Coke spots stands apart from the television image theorized in traditional media analysis. The penchant for visual density and the self-conscious orchestration of stylistic and lighting codes—evident in *SNL*/Coca-Cola's Emmy-winning compositing "breakthrough"—began to permeate the industry on other fronts as well.

The Anti-Ad

Whereas new videographic methods like digital compositing demanded of directors facility and skill in manipulating style codes—that is, the director needed to collect and combine of *lot* of imagery for a short amount of screen time—other commercial production practices foregrounded the issue of televisual style in a very different way. Visually aggressive "anti-ads" also became industry trendsetters in the 1980s. Consider the following industry explanation of the origins and aesthetic methods of the anti-ads: "Directors had been playing with Super-8/16mm black-and-white film for years, and these tools entered the commercial mainstream, as in Paula Grief and Peter Kagan's '*Revolution*' for Nike. Down and dirty film techniques made $200,000 spots look like home movies. Gritty was chic."[63] A self-conscious revolution had started, then, in and with the Nike ads. The "revolution" was not, however, based on high production values nor did it emphasize verbal messages. The visual stuff of the image itself—emulsion grain, flashframes, scratches, in short *the very elements that decades of production had sought to hide or disavow*—became in the emergence of televisuality part of the content itself. The down-and-dirty physical image that defined the Nike ad campaign was repeated numerous times with other corporate campaigns in the months that followed. What was it, one is lead to ask, about raw footage and acts of physical aggression against the image that inspired producers and attracted consumers? The great irony of this trend and the many anti-ad campaigns that followed, was economic. Commercial spot production budgets, which had only recently ascended to the quarter-million dollar range, now found themselves facing crude, and ostensibly inexpensive, forms of image degradation. Super-8 footage, amateur-looking but street smart, gave to the emerging televisual repertoire a new and influential code—a kind of "televisual povera."[64] What was being sold to American consumers of the 1980s was the street: the edgy urban environment, a raw and racially peopled existence that was as alien to Reagan's image of America as it was to the classical styles of earlier advertising. What was being sold, then, was an attitude, an ambience, and an image of America that was street smart, young, and raw.

An allied sensibility infused the Levi's 501 ad campaign produced by Foote, Cone, and Belding.[65] With the Levi's spots, the film stock and format were larger than Nike's super-8, and the distinctive look was colorized blue rather than black and white, but the visual codes were the same. Jerky handheld camera work, long lenses, and extremely abrupt cuts focused on hip Caucasian adolescents termed by the admen "urban cowboys." The message was really just a mood and a lifestyle. Levi's ideal buyer fantasized about hiply slumming it on the streets. No verbal discourse even survived in these spots. The "anti-ad" Levi's 501 campaign, like other high-end commercial productions during the period, was selling a specific lifestyle and attitude, not just a product.[66] The images and sounds were stylized, but self-consciously fleeting and ambient. In the evolution of advertising, the verbal strata—along with the physical product itself—was no longer even an requisite part of commercial spots.

The Documercial

Clio award–winning director Joe Pytka solidified spontaneity into a systematic visual code for use in what he coined "documercials"—a strategy that attempted to counter Madison Avenue's own flash with "authenticity." Visually confusing, aggressive, but thematically open ads by Pytka followed for the Wang business computer corporation. John Nathan invoked the same documercial codes for AT&T in its campaign. In Nathan's acclaimed "Washroom" spot for AT&T an executive panics when he discovers that the huge phone system he bought for his company is suddenly obsolete. "'You don't think they'll fire me?' he asks a colleague, who suppresses his gloat. The film is grainy, the lighting funereal. The camera whiz-pans between the two young men, desperately trying to record the conversation, a blur of paranoia, a career literally in the toilet."[67] Apparitions, paranoia, whip-pans to marked corporate men. These were advertising's hallmarked displays of authenticity and angst. Commercial director Nathan lauds his own innovations, "Now we're getting at what people are really saying and thinking." Thinking? Hyperactive camerawork, film grain, and visual and editorial desperation seem more related to fleeting sensation and apparition than they do to thought. Nevertheless, the director's hyped interpretion is important, for it shows that a systematic process was going on in the evolution of commercial production style to *find apt visual codes* for cultural preoccupations; to create and codify visual signifiers for the viewer—of thought, sensation, desperation.

In Pytka's commercial productions for Wang computers, the style is no less obscure and is intentionally disorienting. The viewer is never allowed, for instance, to see the whole picture or whole scene. In one spot, the camera is locked down on the back of a man's head. The concealed subject is talking to a vague, anthropomorphic shape that is pacing back and forth in the background. In another spot, two men listen to a speaker phone, but the viewer never sees their entire bodies, which are continuously and aggressively cropped. In a third spot, two shapes walk down a long hallway toward the camera and are entirely out of focus. These standing and shifting blurs of light and shadow finally dissolve to an all white logo, and the graphic message: "Call Wang."

What then, do these new and influential anti-approaches to television production style, found in both programs and nonprogram broadcast materials, have in common? Consider their formal operations. *The Street* added video and electronic noise to the image. Nike added extreme emulsion grain and contrast. Levi added jerky and disorienting camera work and unnatural colors. The producers of AT&T spots added washed-out funereal lighting and whip-pans. The *Washroom* spots for Wang used impossibly shallow depth of field and constantly obstructed the viewer's line of sight. These formal tactics have one general stylistic principle in common. They all take otherwise state-of-the-art imaging systems and *degenerate* them through technical and stylistic flaws. Levinson flawed his primetime show *Homicide* in the same way.

This kind of active and pervasive self-destruction and *flawing of the image* does not produce, in any conventional sense, a realism based on illusion. One might argue from this practice that the image-flawing trend belies my thesis that there has been an increased stylization in the television image. But this criticism only makes sense if visuality is defined by degrees of optical resolution. In fact, televisuality is not dependent upon higher and higher resolution. Instead, imagistic and stylistic violations continually draw attention to the television screen and to the *status of its image* as an image. Strategies of image annihilation are far removed from the goals of classical media image-making, precisely because they work to show-off such actions as stylistic marks and stylistic accomplishments. There really is no argument that these types of spots and codes were received as visual codes either. The industry press interpreted the trend immediately. One advertising producer, Stockler, explained the growing sense that such ads were overkill and trendy: "Handheld died when it began, critics said. Some work went beyond cinema verité, to video obscurité: Product was mystifyingly submerged."[68]

Stylistic fashions come and go, but it is worth considering the reasons that producers and agencies opted for self-annihilating visual tactics. Nathan explains that "the cliché-ridden vision of what goes on in the world—the domain of the TV commercial *until now*—has begun to pale and is perceived to be irrelevant to consumers."[69] Production people, then, counter the critique of the industry as clichéd by arguing that *their* cutting-edge methods and preoccupation with style are both relevant and interpretable to consumer-viewers. The image-destructive style perpetuated by anti-ads and documercials should really be seen, then, as a counter move, as a strategy to regain viewers. Production tendencies, even apparent antistylistic techniques, were hardened into marketable and reproducible displays that more accurately signified the thought and sensibility of America's changing consumers. Although the commercial advertising industry pretends to be a paragon of dynamic change and innovation, it also is a process that immediately hardens stylistic practice into an assembly-line succession of variant looks. Once made public, Madison Avenue's issuance of codes can then be taken over by other agencies, for different products and for different genres.

Smorgasbord

While jealous commercial directors—who made less than Joe Pytka's five-figure daily rate—wrote off the anti-ad and the documercial "shakeycam" as a passing fad, Madison Avenue practitioners were hoarding other looks in a veritable smorgasbord of acute aesthetic styles. Primetime spots in the month of December 1988 alone looked like a concentrated primer of art and film history. Mercedes cashed in on Greek classicism: although "form follows function" with all other car makers, Olympian plateaus presenced by Mercedes—a televisual world of white silk, hardened enamel, chrome, and stainless steel—was philosophized as a rare occasion; one "where form is free to pursue absolute beauty." Obsession made monochromatic Greek classicism its signature commercial style, with well-muscled, dispassionate, and minimally draped male and female bodies posed in pregnant tableaus. Shearson-Lehman's investment spots took the classical ideal of absolutism one step further by restraining form altogether. Its serious black-and-white spots posed heavy verbal adages over minimal textual epitaphs: "minds over money."

If elite automobiles and investment banking merited classical restraint, then breakfast cereals, cosmetics, and salad dressing used impressionism to tap the viewer's emotional surge. Kellogg's Muesilix was really an hallucinogen that packed the fleeting sensation of sun, wind, clouds, and magic-hour into its European "balance of grain nuts." Gloria Vanderbilt's perfume also swept the viewer up into fleeting slow motion and undulating optical reflection on the surface of water. Hidden Valley Ranch Dressing revived the landscape tradition of impressionism within each two-second shot in its painterly spots: each image was an overfiltered homage to the sensuous powers of soft, directional lighting. If domestic food products and feminine cosmetics exploited the easy sensory moments of impressionism, then products for "real men" demanded harder orchestrations of the avant-garde. Chevy's Camaro spots, for example, threw every tool of the cinematographer's trade at the audience in an expressionist eruption that would have made Matisse proud. Low-key and high-contrast photography collided with smoke, fog, and tire-squeeling urban streets. Chevy's impressionable twentysomething male learns that strutting women love men with hot rods, that is, as long as they cruise on overcomposed nocturnal streets bathed in neon reds and blues. For the male's morning-after, Gillette's Atra turned loose a combination of Russian constructivism and Robocop: overengineered high-tech edges and digital surfaces swarmed over facial stubble, thereby fulfilling everyman's need for a sensuous shave.

The dark-side of modernism, however, also had agency desciples. Following Lautremont, Breton, and Dali, Prudential Insurance found its holy grail in surrealism. The Prudential rock was really just a digital apparition that guided astral-traveling homeowners through comet-crashing glimpses of suburbia, family, and investment portfolios. Alfa-Romeo showed that surrealist metaphysician De Chirico and wide-eyed sentimentalist Spielberg informed every move in their promotional mantra for Milano: attitude, attitude, awe. J.C. Penney's cutting edge attitude was symbolized by minimalism and action painting. Would Jackson Pollack follow celebrities Cheryl Tiegs (Sears) and Jaclyn Smith (K-Mart) into this world of mass-marketed commodities? Spots for Verve showed that even the adolescent "femme-teen"—born long after pop and the summer of love—could appreciate the public possibilities of plastic, pastel, and Op art stenciled anatomical coverings. Swiss-made Swatch watches targeted the same adolescent crowd with hyperactive collages of clock-like human automatons jerking around like machine cogs. Was this postmodernism, futurist performance, or Marcel Duchamp's *Bride Stripped Bare by Her Bachelors*? While the teen crowd went for shallow and plastic, those who could afford Chanel went for the chic and self-satisfied world of erotic daydream. Fleeing her kept corporate world, and shadowed by phantom air-craft, Chanel's statuesque product model strode through wide-screen cinematic space in a compacted ritual of narrative ellipsis reminiscent of Bertolucci: "Come share the fantasy"—at least in its desert variation—was really an homage to what Maya Deren would have done if she had lived long enough to make a road movie.

This dizzying array of poses suggests that advertising brings to television not just a range of styles but an obsessive ritual and appetite for stylistic differentia-tion. *Advertising teaches television* in more ways than one: it is a hungry proving ground for new televisual production technologies; it is a leaky cache of creative personnel that denarrativizes television; it is an omnipresent aesthetic farm-system for primetime. Advertising's budgets, however, are far from minor league. In contrast to primetime's per-minute production costs, advertising budgets dwarf television's financial commitment. Advertising production, that is, gets more sty-listic "bang for the buck" than primetime or off-prime programming. For this rea-son, commercial spot production also underscores and reinforces one of the driving mythologies behind televisual exhibitionism: the idea that overproduction and sty-listic excess provide industrial leverage and corporate marks of distinction. *The televisual mode of production, then, is really an ad industry-proven mode of over-production.*

The Cult of Technical Superiority

Television's stylistic ability—including its penchant for overproduction—is directly related to a certain institutional privilege. During the licensing of any and all tele-vision stations, the FCC requires that potential licensees demonstrate to the gov-ernment that they are honest in character, financially stable, and providers of "diversity in the marketplace of ideas." But the FCC also demands that licensees be "efficient in operation . . . a technical factor involving the quality of the station's signal."[70] That is, stations must demonstrate that they are technically superior or as good as other potential licensees in the market. This burden of technical supe-riority has affected the mindset of broadcasters by perpetuating programming that exploits each technical and format breakthrough as a form of attention-grabbing authentication. The burden of technical superiority also tends to denigrate those civilians outside of broadcasting as amateurs. When television news covered Chicano hunger-strikers in Los Angeles in June 1993, for example, protesters and vocal critics alike were allowed hard-hitting on-camera political critiques. Yet the reporters—obviously oblivious to the political discourse they had just uncorked—regained control of the stories at the conclusion of each two-minute interchange. On-camera the reporters comforted viewers at home by paternalistically noting that there was nothing really political about the confrontations that had just been aired. These political discourses and confrontations were, after all, really just emo-tional and naive youthful outbursts.[71] "Back to you in the studio."

Affiliate stations also betray the cult of technical superiority when they send the televisual apparatus out on assignment to fulfill FCC-required local public service needs. One year after Los Angeles burned, major affiliate KCBS under-lined their Hispanic and Chicano social concerns by covering "Fiesta Broadway" in downtown L.A.[72] What was supposed to be a celebration of indigenous Mexican-American and Latin-American culture, however, was actually more like a low-budget Rose Bowl commentary that mixed fragments from numerous but picturesque non–English speaking cultures into one big, happy south-of-the-border family. Bored commentators faked enthusiasm from atop towering platforms. Anglo corporate representatives from Disney remarked at length about how much Hispanic people "like" Mickey Mouse. The station tokenized this particular weekend programming ghetto further by loading it up with as many Hispanic KCBS employees as it could lay its hands on. Cameras cut from one mindless overcomposed Betacam image to another, all centered around a billboard-like stage that proclaimed—not Los Angeles'—but *"KCBS's* Fiesta Broadway." Despite the fact that every videographic bell-and-whistle was being used, KCBS's production team seemed to be covering the event in its sleep. Even the awkward on-air banter indicated that this token crew of highly trained professionals could not wait to finish in order to go out and have a few cold ones. One sensed in this high-tech but sloppy production that the margaritas had been flowing for some time. An air of moral and technical superiority oozed out of the apparatus and enraged some viewers, even as KCBS's televisual "shock-troops" deftly fulfilled their FCC-mandated public service mission.

The networks are no less prone to wield their impressive televisual production apparatus as a threat to any and all would-be challengers. CBS's *48 Hours* did an hour-long critique of sensationalist talk-show programming and hopelessly wacko public access producers on an episode entitled "Talk, Talk, Talk."[73] Dan Rather's cynical ironies did little to conceal the fact that CBS's own magazine show differed little in verbal content from the talk shows and video populists that it demeaned. The *48 Hours* broadside, however, delivered from the technical safety of CBS's impressive production studio, was a televisual apparatus that provided its ambivalent broadcasters with continuous proof of their technical superiority. In the televisual apparatus, high-end production values not only separate the men from the boys—they distinguish the networks from lowly syndicators, affiliates, and cable activists. Big-time production values also help prove the moral and intellectual superiority of the FCC's chosen ones.

The construal of overproduction and technical superiority as indicators of cultural and moral authority does not, however, provide a comple picture of televisual exhibitionism. Styles, even in those cultured and aesthetic forms that pretend to autonomy, cannot be separated from the economic logic that sanctions them in the first place. The aesthetic has rarely, except in intellectual circles, been a realm of pure form or intellect. Television programming, in particular, fabricates the aesthetic as part of its industrial habit. The next five chapters examine what might be termed the aesthetic economy of televisuality, and elaborate in more patient detail the stylistic and presentational modes that typify televisual exhibitionism. More than offering exposition on the formal grammar or language of television in a Metzian sense, however, each chapter will first preface and situate a close presentational analysis—the case study of a symptomatic show, series, or genre—within a broader cultural context that motivates and favors exhibitionism. These close textual studies—of the epic miniseries, the broadcast hybridization of feature films, hyperactive trash television, digital videographics, and tabloid reality programming—show clearly what the new practitioners can only hype: that embellishment promises to intensify viewing and animate audience in profitable ways. As each of the five chapters that follow suggest, the aesthetic modes of exhibitionist television can also be understood as merchandising strategies: as boutique television, as loss-leader programming, as franchise packaging, as thrift-store video, and as ontological strip-mall. The oft-noted idea that ours is a culture of consumption means that marketing and packaging have become preeminent aesthetic rationales.

Part II

The Aesthetic Economy
of Televisuality

4   Boutique
Designer Television /
Auteurist Spin Doctoring

Above all, television is anonymous.
—Horace Newcomb and Robert Alley, *The Producer's Medium*[1]

Aaron Spelling is playing pinball on a customized machine. The machine, a testament to his power as a star maker, is decorated with images of Farah Fawcett, Kate Jackson, William Shatner, David Soul, Joan Collins, and others who have prospered from their associations with him.
—Mary Murphy, *TV Guide*[2]

Anonymous? Television has rarely been anonymous, although many people behind the camera used to be. Booming audio references to Quinn Martin and Mark Goodson–Bill Todson never left any doubt in the viewer's mind about where their shows came from in the 1960s and 1970s, but by the 1980s creative producers like Stephen J. Cannell actually appeared in dramatized filmed I.D.'s tagged on to each of their episodes. Consciousness of television directing now pervades promotional hype. Even an untrained twentysomething teen heartthrob like Jason Priestly was discussed extensively by the popular press for the creative "risk" he took when he moved behind the camera as director for an episode of Fox's *Beverly Hills 90210*.[3] Directorial and producerly distinction is now so much a part of the popular discourses surrounding primetime television that any useful analysis of the medium must reconsider the programming *function* of televisual authorship.

The whole point, however, is not to argue over which creative craft dominates television, as some theorists have done—the producer, the director, the writer, or the "performers [who] are the key creative figures of the medium"—but to examine how producerly "intention" is fabricated and wielded; both within individual programs and among programs in the flow.[4] Producers and authors are conceptual mythologies manufactured by production entities and broadcast corporations alike. Amassing and flaunting producerly distinction is one way for media corporations to cut through the televised clutter. Television's recent penchant for subjectivizing the origins of its mass-produced fare frequently takes place in boutique genres and formats. The all-pervasive flood of television programming twenty-four-hours per day has tended to monotonize the experience of viewing. Against this backdrop of mass-market retailing and monotony, producers were quick to learn the value of counter-programmed sensitivity. Boutique programming constructs for itself an air of selectivity, refinement, uniqueness, and privilege. The televisual excess operative in boutique programming then, has less to do with an overload of visual form than with two other products: *excessive intentionality* and *sensitivity*. Programming's cult of sensitivity may involve the kind of spectacular cinematic spectacle more typical of loss-leader event programming, but it also may involve more restrained forms of drama, writing, and cinematog-

"Viewer discretion" for Bochco's *NYPD Blue* also meant viewer "distinction"—adult themes, sensitive producerly expression, and painterly images. (ABC)

raphy—subtle orchestrations of televisual form that create the defining illusion of a *personal touch*.

This personal touch is pitched and programmed on a regular basis. During breaks in the NBA finals games between the Chicago Bulls and the Phoenix Suns in 1993, for example, NBC oozed with insider familiarity: "You remember *Miami Vice* . . . now watch *South Beach*." The picturesque *Vice*like images that accompanied this challenge to viewers indicated that the stylistic refinements of designer televisuality were being showcased, not character clones of Crockett and Tubbs. Here, apparently, NBC's boutique was open even to sports fans; at least to those who could discern the promos' stylized designer references. The same sort of designer challenge formed the center of *Midnight Caller*'s hype to potential broadcast and affiliate buyers in 1988. In extended marketing promotionals and shorter televised previews during primetime, the show was pitched to insiders and audiences alike with the mantra: "Produced by the same people that brought you *St. Elsewhere*."[5] Since absolutely nothing about the premise, the characters, the plots or the visual style of the two shows was the same, something else was being hyped. *Midnight Caller*, like *St. Elsewhere* before it, would seep with sensitivity and refinement. Fourth network Fox also learned the benefits of the boutique. When corporate needs and demographic incentives encouraged the network to upgrade its adolescent image—that is, to transcend the blandly framed pubescence of *Melrose Place* and *Beverly Hills 90210*—Fox signed Robert DeNiro's creative collective to produce sensitive artistic showcases in a series named after his New

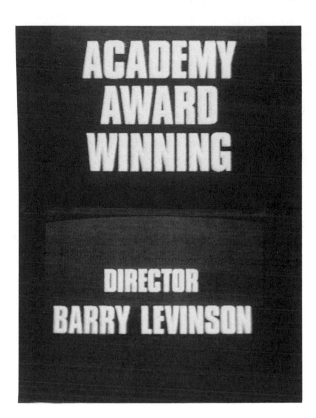

Auteur Barry Levinson's *Homicide* "brings 'em into the network tent" at NBC in 1993–1994. Austere white-on-black graphics tease the drunken Super Bowl audience about the aesthetic severity and importance of the event to come. (NBC)

York neighborhood: *Tribeca*. Although critics by and large supported DeNiro's risk taking, low audience numbers eventually forced the show's cancellation.

Major television corporations no longer relegate boutique intentionality to individual series ghettos either. Universal-MCA for example, in a historic alliance with television almost from the start, now bears none of the ambivalence toward television that characterized the early telefilm days in the 1950s. Full-page color ads by MCA addressed to the industry in June 1993 proudly announced to potential buyers the availability of "original" boutique programming: "Six Major Theatrical Moviemakers Creating Original Motion Pictures. All for Television. All first-run."[6] Trying to entice cable systems to come on board with regular MCA weekly series, executives bragged about MCA's "ability to line up several marquee movie producers, who are often contracted under the Universal Pictures banner": John Landis, Sam Raimi, Hal Needham, and Rob Cohen.[7] By producing big-budget "original motion pictures" in 1992 and 1993, like *Stalin* starring Robert Duvall, Ted Turner's TNT also thumbed its refined nose at the networks, theatrical producers, and noncabled homes alike. The boutique model of celebrated sensitivity was not just limited to expensive network risks like George Lucas's *Young Indiana Jones Chronicles*. Pay cable, basic cable, and syndicated programmers had all learned the benefits of showcasing artisans in discriminating boutiques and for an obstensibly limited clientele.

The way boutique television manipulates "excessive personality" becomes clearer once we set aside the interpretive burdens that come with notions of singular authorship. Although television usually provides a name and a face behind signature television, it also regularly rips intentionality from its human moorings, and hypes it as a guarantor and label of product relevance. We have already seen how relevance programming was an important part of the Lear-MTM era, but this designation usually referred to a show's focus on explicit contemporary topical issues or themes. In the 1980s and 1990s, authorial personality has become a more textual, rather than topical, force. Several examples show how intentionality permeates primetime shows in ways that contradict postmodernist explanations. An episode of *thirtysomething* entitled the "Mike Van Dyke Show" became a paean to the sensitive and long-suffering artist. On the surface the show's extended yuppie family has a love-hate relationship with its past: with 1960s television, music and mass consumer values. Underneath these references, however, the episode was an overripened vessel brimming with seriousness and sensitivity. Faith is critically injured in a car crash at Christmas, after which a burning bush calls Michael back to God and to the Jewish faith. In a heavy and highly cinematic scene, Mike takes his cousin Melissa to synagogue in an old Volvo with a Christmas tree tied to the roof. As the snow falls Melissa explains Michael's interfaith burden—of clashing symbols and familial angst—as "God's punishment." Although the show was rife with cute intertextual references to Dick Van Dyke, the episode was far from the blank or ironic pastiche that postmodernism theorizes. This highly stylized series was, in fact, *very serious business.* Michael's traumas and intertextual daydreams, in fact, formed the basis for a religious conversion. No wonder the academic left hated the show. Its nostalgia was based on belief that suburban living and homeowning are meaningful contexts for emotional struggle and growth. There was certainly no oppositional postmodernism here either. Although the show was celebrated for having complex interrelationships between characters and sophisticated plot lines, many elements reinforced the rather traditional notion of a sensitive and subjective center within the show's narrative world. The character Michael was, after all, a creative director at an ad agency, a persona capable of on-screen televisual flights of fantasy, and also an easily recognizable surrogate for the show's creative producers Zwick and Herskovitz.[8] This overdetermined layering of sensitive points of view—where the dramatic and producing worlds aligned as a composite, and very visible, emoting source—gave any topic the show dealt with an aura of seriousness. Nothing was *ever* blank.

This kind of traditional subjective centering, within an apparently decentered postmodernist series, showed up in other boutique programs as well. Far from a blank use of empty style, the first episode of *Beauty and the Beast* actually made rigid stylistic distinctions between the world of the city above ground (cold, blue-filtered, hard-lensed, and angular), and the world underground (coral-filtered, diffused, baroque-lit, curvilinear, and soft-textured). Although all the elements of postmodernist collage were evident in the detritus underground, when the characters read "literature" aloud they meant it. The show was driven by classical dramatic appeal, pathos and empathy, and self-consciously placed the ritual of victimization at its center. The physical deformity of the beast and of Catherine's slashed Caucasian face, for example, were described in the dialogue of the very first episode as a model for all minorites—for victims who have been shunted

thirtysomething was signature TV par excellence. A burning bush calls show's producer-clone and emotive-artistic center Michael back to Jewish faith and away from postmodern disinterest. Final leap in *Quantum Leap* is orchestrated in press as evocative homage to simple origins of creative source Bellasario. (ABC, NBC)

off to culture's margins. Catherine's visual disfigurement, then, became both a key to the plot and a premise for the show. *Beauty and the Beast*, for all its cutting-edge postmodernist looks, was a highly subjective and narratively centered show. The extreme flux and visual complexity of the show was inevitably motivated and intensified by victim Catherine's extreme sensitivity.

One final example suggests the textual and programming power that excessive intentionality played. During the last episode of *Quantum Leap* in 1993, Sam jumped into his hometown on the day of his birth. Choreographed by network press-releases, the media ran with the story that this final destination was actually the childhood hometown of series producer Donald Bellasario. In the smoke-filled Caravaggio-lit tavern that Sam leaps into—set deep within a rural, coal-mining Pennsylvania town—viewers are *taught* not about Sam, but about the origins of *Quantum Leap*'s producer and creative source. What possible function could the exhibition of Bellasario's authorial backstory fulfill in this primetime science-fiction? As *Quantum Leap* headed for syndication, this episode's display of authorial intentionality gave the series package a very lucrative spin—one that aimed to motivate interest in the show's afterlife.

All three signature shows then—*thirtysomething, Beauty and the Beast*, and *Quantum Leap*—created overdetermined, sensitive, emoting centers from which their complicated visual worlds were seen. This shared trait gave any experiential journey within an episode's plot—no matter how excessive—ample motivation. Although filled with hallmark forms of parody and intertext, none of these shows shared two of postmodernism's most important ideologies: blank disinterest and decentered subjectivity. Far from it. Sensitive relevance was lurking everywhere. The conflation of highly styled, heightened relevance and sensitivity went hand in hand. Once authorial intentionality was disembodied, it could be used as a textual force that allowed and justified extreme forms of presentation: time travel, fantasy, daydream, parody. Boutique television, then, became a selective, signature world where artistic sensitivity went hand in hand with social relevance and viewer discrimination. The degree of excessive intentionality in these shows was directly proportional to the amount of formal and visual excess that each series needed to restrain. Expressive intent, then, both exploits and channels the flooded televisual image. *The boutique signature is used to discipline excessive style.*

Boutique intentionality, however, does not simply work to constrain the complicated visual world inside of television programs. It also plays a regular role outside of and among individual programs. Programmers in both cable and broadcasting, for example, regularly *strip-off* intentionality for their own institutional and political ends. Intentionality is not then just a generic or aesthetic force, but provides fuel for many of television's nightly economic, journalistic, and cultural rituals. The case study that follows analyzes one such situation where programmers appropriated auteurist intentionality in order to structure and manage an extremely volatile political crisis. Based on coverage of the U.S. invasion of Panama in December 1989 and January 1990—and the dramatic and fictional programming aired during the period—the analysis targets a symptomatic process that happens nightly on thousands of channels across the country: the televisual hybridization and repackaging of a feature film for broadcast.

Theorists, still bent on pursuing the medium's essential qualities, tend to overlook the fact that television includes a great deal that comes from elsewhere. Television's "feature presentation" of films, in particular, is a widespread ritual that needs to be acknowledged and addressed on a more systematic basis. Historically, televised films have been as much a defining form of television as the sitcom or any other genre, although they are seldom recognized as such. Broadcast features also regularly infuse television with heavy doses of designer intent. The complex ways that signature television both fuels and feeds off of the broadcast hybridization of films demands a much closer, and more patient, presentational analysis. Such an analysis, for instance, shows immediately that televisual modes do not exist in isolation. In practice, the broadcast flow frequently enmeshes boutique sensitivity with cinematic spectacle and videographic embellishment. The extensiveness of the film-to-television hybridizing process also demonstrates that televisuality is clearly not just an aesthetic issue. Signature and marquee television also have sharp political teeth. In some ways, then, the boutique is just a front—a programming space available for lease to any and all ideological comers.

Case Study: *Salvador,* Noriega, Stone
Video Couture: Convulsions of Topicality

War makes good television. Especially if by *good* one means spectacular, visual, and all-encompassing. But if the Gulf War is any indication, there is more to the politics of war on television than objectivity, censorship, and "coverage." Even though many accounts were rightly critical of the way television covered the war and the way the government managed information about it, fewer dealt with the effect the conflict had on programming in general.[9] Wars have become special events that send ripples throughout programming. Broadcasters (not just Pentagon briefers) have become spin doctors for such events, and learn to exploit unfolding conflict in the name of viewership and ratings. Not only have wars become media texts, they have become contexts that transform other texts in substantive ways.

The study that follows examines the formal ways that television appropriates, resuscitates, and redefines dated films in response to fast-breaking historical and political events. Television does not just gather and re-present selected films around contemporary events, however. It also encroaches upon and stylistically reworks earlier films—frequently by exploiting the signature potential of those works—into fundamentally new and hybrid forms. It is important to better understand the ideological effects of this hybridizing process, since features are broadcast nightly to mass audiences and are extensively interspersed with late-breaking "reality" material in the process. As the evolution of videographic and electronic production equipment provides more and more stylistic options for programmers, the actual visual and narrative presentation of features also becomes increasingly complex. A closer look at these stylistic operations suggests that more is at stake politically than television's avowed interest in a fast-breaking present. Urgency is merely one explicit interest that broadcasters have in the transforming process. The televisual soup that results from broadcast hybridization involves a play and conflation of cultural signs that has far wider implications.

Mass culture in the past decade has been particularly fond of turning history (war) into fiction (film and television) and fiction into history. This late–cold war period has also witnessed a string of imperial practice wars—the Falkland Islands, Lebanon, Grenada, Panama, and the Persian Gulf. In the absence of anything larger, more nuclear, or more apocalyptic, first world governments have repeatedly used provisional conflicts to test out new, expensive and otherwise unused high-tech weapons, military tactics, and global political strategies. There are clearly practical and economic benefits from these neocolonial practice wars. Military budgets (along with hardware) are justified, and the resulting globalism legitimizes continued growth. But broadcasters who cover or even refer to such events also benefit. The analysis that follows shows how one recent practice war, Panama, honed the programming and aesthetic skills of broadcasters by economically resuscitating dead media texts, that is, by reviving old films that still housed surplus intentionality. Spin doctoring and impression management are stylistic and aesthetic processes—as well as political tools—when these texts are tapped.

Consider, for example, the following three incidents. A Los Angeles news anchor appears in a newsbreak during the broadcast of a film and asks viewers to "stay tuned following our movie" for a major exposé on the mob in Los Angeles. The film interrupted by the newsbreak is *The Godfather II*.[10] After over three years in captivity, a second American hostage is released by Islamic terrorists. KTLA television programs and broadcasts its miniseries *Voyage of Terror: The Achille Lauro Affair* the same week. The miniseries is advertised as "a remarkable true story" of Islamic terrorists, with "new facts never before released."[11] Along with

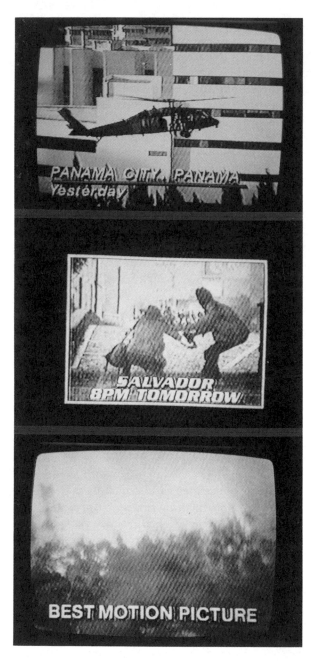

George Bush's 1989–1990 invasion of Panama is televisually hybridized into topical amalgam with 1983 film *Salvador* and Oliver Stone's 1989–1990 "award-winning" *Born on the Fourth of July.* (KCOP)

these proclamations of the real, less visible disclaimers within the program suggest that portions of the show are dramatized fictions. Finally, for weeks preceding the premiere of the miniseries *Drug Wars: The Camarena Story* NBC heavily promotes the broadcast event as having broken "Long before Noriega."[12] While the miniseries itself dramatizes covert American DEA operations in *Mexico*, advertisements and promotions for the show refer instead to the unfolding U.S. invasion of *Panama*—an operation launched to depose accused drug lord, President-General Manuel Noriega.

All three incidents suggest the varying degrees to which television is fueled by textual shifts between fiction and reality. In the first instance, a fiction film is used to contextualize a factual expose on the mob. In the second, the historical climate of contemporary terrorism is infiltrated by docufiction in a way that promotes both the station's news and programming departments. In the third case, explicit reference is made to current military events and the history of Noriega as an interpretive allegory for understanding NBC's previously produced and reality-based fiction on Camarena. It is curious that in each case, television does not try to hide its ontological and textual distinctions. Rather, it flaunts and exploits the distinctions: news versus film, history versus entertainment, and reality versus fiction.[13] The terms in each of the polarities are wielded by stations as part of a process of self-interpretation and self-valuation. Such distinctions are in fact explicitly part of the language of programming and marketing.

This phenomenon suggests perhaps that good viewers are in part "good" because they know and can value such distinctions. At the same time, the act of *assigning* historical and fictional status by television, is volatile, shifting, and prone to reversal. Late-breaking historical events can rapidly change the value and currency of preexisting fictional texts. Good broadcasting in this respect, and in an economic sense, can be seen as analogous to Lévi-Strauss' "bricolage"—it is able to fabricate new value out of existing and outdated textual material. In programming bricolage, history and reality are frequently marshaled as the material stuff of refabrication. The process of assigning fictional and historical status in television, and the formal ways that texts are transformed, deserve to be looked at more closely.

I have chosen to look at the broadcast premiere of Oliver Stone's film *Salvador* because of the important role that televisual embellishment played in its adaptation. The event is worth considering more closely, for both the film and the broadcast made explicit reference to recent historical events. The feature was also clearly programmed in and around the context of "current" American involvement in Central America and was specifically tied to the American invasion of Panama. KCOP-Channel 13 in Los Angeles aired and adapted the film, in fact, on the very day Noriega was brought to arraignment in Miami. Along with numerous references to events occurring in Panama, the broadcast was also used as a mechanism to set up and promote other programming on channel 13.[14] As we will see, national distinctions of the sort that separate the countries of Central America seemed, in this case, to have little value to American television programmers. El Salvador, *Salvador*, Nicaragua, and Panama, appeared as one and the same phenomenon in this week's multitextual programming soup.

The broadcast of *Salvador* was not the only example of fictional programming presented around the theme of U.S.–Latin American relations. The week of January 1–8, 1990, was in fact loaded with fictional broadcasts related to this theme. This context, the first week of January and the second week of the invasion, included programming rife with references to Latin American relations, politics, and mythologies. Within the four-day period in which *Salvador* was broadcast twice, the heavily advertised and promoted NBC miniseries *Drug Wars: The*

Camarena Story—produced by *Miami Vice*'s fashion-setter Michael Mann—premiered, on January 7.[15] In the face of the same recent historical events, ABC came up with its own timely offering and fictional allegory, the feature film *Heartbreak Ridge*. Clint Eastwood's version of the 1983 Grenada invasion offered to programmers an explicit and accessible foreign policy map of the Panama action. Both as a generic update of the war picture and as Hollywood's vision of gunboat diplomacy in the Reagan era, the film helped naturalize the more recent invasion. Grenada, El Salvador, Panama were merely interchangeable names in the reality-based archetypal programming formula operating that week. Other films also kept Hispanic and Latin-American issues on the fictional agenda. The films *La Bamba* and *The Border* were broadcast at that time. In addition, the news discourse during the week constantly referred to, and tied together, the fiction-film and historical realms. *Salvador* aired the day Noriega was introduced to American justice and was repeatedly interrupted by news footage of Noriega being taken from Panama. Television as an institution was clearly pondering recent media and political events. It did so by resuscitating and working over old fictions.

What Television Adds

Televisual adaptation is not a subtractive process, even though many assume that the broadcast of a film is, by nature, reductive. While the scenes that are deleted for broadcast elicit protests on the part of many viewers and fans of films, a much more additive process is also at work. The analysis that follows is directed at this additive, *hybridizing* process—a recurrent practice that characterizes and defines broadcast adaptation. Even subtractive operations (for example, those involving the elision and deletion of source material), create new structures and relationships within adapted texts, but those practices—including censorship and automatic dialogue replacement—are beyond the scope of this study.[16] Even apart from those devices, however, a close analysis of the televised version of *Salvador* reveals a rich array of stylistic operations whereby the original film was expanded and recombined with other elements. Such operations reconfigure adapted works for viewers, even as they offer to prepackage interpretations.

Promos, pretext. Marketing and promotion are at work both before and within film adaptation.[17] Consider, for example, how the announcer sets up the broadcast in the following station promo: "Tomorrow at 8:00, James Wood stars in the Oliver Stone production offering a scathing look at a torn country's social injustice—as seen through the eyes of a news photographer—in *Salvador*." The emphasis here was not just on time-place information but also, through narrative synopsis, on how the viewer should interpret the yet-to-be-aired film. The interpretive pitch, for example, clearly worked to personalize the political dimension of the film and to use the stardom of James Wood to achieve this end. Then, in short order, a hierarchy of personal points of view is created whereby the emphasis is not on the country of El Salvador, but on a second subjectivity: the distanced and personal ("through the eyes") perspective of the fictional journalist. Finally, the promo gives a tertiary personal spin to the event by making reference to it as an "Oliver Stone production." This tactic clearly linked the televised film with the notoriety that Stone was receiving both on- and off-screen for his award-

Censored image of blood-splattering execution cut from film is gleaned and reused in hybridized promo for broadcast to home viewers. (KCOP)

winning film *Born on the Fourth of July* during its release in December and January. By prefiguring and framing *Salvador* as both a directorial statement and a star vehicle, the experience of the film *before it ever started* was heavily subjectivized and mediated by frameworks that distanced the viewer from any of the messier realities of Central America. The transformation of an overtly political subject into an aesthetic framework—of the broadcast event into a showcase for expression—became a master code for much in the program that followed. The aesthetic pose of the film, its pretext and pretense, was then twofold. The event's prepackaged aesthetic status lured viewers with heavy doses both of acute sensitivity and vision-driven spectacle. Such traits—formal excess and aesthetic consciousness—are now vernacular parts of feature presentations on television.

Cut scenes as video stills. This televisual adaptation, like others, included a hybridizing operation that one might not expect to find in the reduction of a film for television. KCOP first removed a scene from the film and then manipulated it as a freeze frame for other purposes within and for the broadcast. The reincarnated image in this censored-then-reused sequence is particularly curious, since it headlined a promo repeatedly used by the station to advertise the broadcast of *Salvador*. The practice suggests that television feeds on the iconic and visual power of certain images, even if they are too hot for the standards-

Decisive-moment photographic code—à la Capa's Spanish Civil War icon—mentors film's visual and directorial style.

and-practices censor. Of all the other images in the film, KCOP used this image to advertise the broadcast: a scene that depicts a government soldier being executed by a leftist guerrilla; a scene apparently either too graphic or too politically partisan for station censors/revisionists. Why would a station advertise its broadcast with an image that it also censored from the film?

The formal properties of the shot give some indication of the privileged status that it had in the original film. There is a curious and violent power to the shot, taken as it is with a wide-angle lens, low, and in close-up. In fact, it is this shot, along with one other, that suggests an important subtext for the characters early in the film. Boyle's buddy Cassidy, a more successful professional photographer, asserts that his ultimate objective is to get shots equivalent to Robert Capa's powerful images taken during the Spanish Civil War. As Cassidy states, those were shots taken "while looking into the face of death." Boyle underscores this ideal by acting out the gesture of one of Capa's victim's caught at the moment of bullet's impact. Cassidy eventually gets what he wants. He aims his final exposure head-on at a strafing fighter plummeting toward him. The photograph costs the photographer his life. Cassidy photographs the censored-then-reused execution image shortly before this scene. In order to photograph the execution, he discreetly places his 35mm rangefinder Leica at waistlevel and fires. The feature film places heavy directorial emphasis on these two sequences—the face of death scene and the co-vert waist-level execution shot—as *the* Capa-like images aspired to earlier. Oliver Stone sets up the last incident with an ecstatic interchange of shot-reverse-shots between Boyle and Cassidy; and sets up the earlier image by dramatizing how violent leftists prevent main character Boyle from witnessing the execution. Stone belabors Cassidy's suicidal impulse—"to get the ultimate shot"—in order to ex-ploit the film's Capaesque decisive moment. This visually induced death is sealed with an extreme close-up reaction shot of Cassidy secretly firing off his shutter.

Why is this scene too hot for the film's broadcast, but useable in broadcast promos for the film? The answer may lie in the nature of the image as a "deci-sive moment" shot—a photographic aesthetic that privileges the smallest slice of time as the most profound, and the most visual and dynamic instants as the tru-est.[18] What is characteristically lost in this aesthetic, however, is any sense of con-text, background, and political understanding. The decisive-moment aesthetic privileges accidental compositions, extreme visuality, and spectacle. In a phenom-enological sense, then, decisive-moment images are typically open to diverse ideo-logical readings and interpretations. As an existential aesthetic view, the theory dominated mid-twentieth century photojournalistic practice, but has since been critiqued by revisionists as naive and apolitical. Significantly, this visually induced ideological naivete and apoliticism may make decisive-moment imagery particu-larly attractive for televisual exhibitionists. The censored-then-reused image, more than others in the film, is open to appropriation precisely because it is visually fragmented and stripped of any context. Its violence is presented as a frozen and emotional *composition*, rather than as a political *act*. A fragmenting and freezing operation like this is a paramount televisual strategy. By setting up Capa's deci-sive moment as part of the film's overall design strategy, and then dramatizing that aesthetic at key moments, the film's cinematography begs for appropriation by the televisual text. In digital video graphics terms (the subject of the next chap-ter) the film's frames dare broadcasters to grab them, regardless of their ultimate use in or deletion from the final televised scene. Televisual adaptation is clearly not subordinated to broadcasting's standards and practices editor. Censorship is a mute issue in cases like this one, since televisuality finds alternative ways to hy-bridize and reuse questionable, censored material.

Graphic station I.D.s and intertitles. Like many televised movies, *Salvador* is graphically stapled to its programming slot with various devices that signify own-ership. Chief among these graphic displays are the station I.D.s keyed or burned into the film's imagery in the lower third portion of the picture. One cannot imag-ine this sort of stamp of ownership being allowed in other artforms—the signa-ture of the purchaser rather than the maker stamped directly into the artform—yet this is an almost universal practice in both broadcast and cable television. Televi-sion mise-en-scène is far from sacred or inviolable ground. In practice, appropri-ated films offer broadcasters an open, visual terrain, visual turf upon which stations erect graphic promotional signs. While the situation in El Salvador may have been confusing, and the social implications of both the movie and recent events threat-ening, this graphic insemination of the spectacle worked to mark, identify, and distance any horror that might reside in the original film. Graphic stapling sug-gests two things about televisuality: first, that the unfolding spectacle is known, and second, that it is also apparently owned by KCOP. The ideological packaging of film, then, is not just a process of *extratextual* promotion. The process also occurs visually within the frame. Graphic packaging further reduces the televisual spectacle to a known and owned status, to that of object and commodity.

The most repetitious graphic hybrid in this adaptation is surely the "8 o'clock movie" identification. This icon precedes each program break, as well as each of the film's televised segments. The shift from dramatic scene to advertisement is not then a direct one. The segue from program to break and back only occurs through the agency of a program identification graphic. Whereas in the earlier example keyed graphics directly intervened in the film imagery, here connection with the film is enacted through external visual and verbal reference. The station announcer repetitiously intones in each 8 o'clock movie identification that "we will return to the KCOP presentation of *Salvador* after these messages." By changing the audio reference each evening to a new and different film title, even the repeated visual I.D. is narratively redefined on a nightly basis. As a result of this nightly process of renaming, numerous films are lifted and verbally framed over the singular and unchanging station graphic. The infiltration of texts however is bi-directional. While keyed I.D.s over dramatic scenes redefine the fiction world, audio references to the fiction world laid over graphics redefine the real world of broadcast. The process then is twofold and complementary. The seeming redundancy of this graphic-sound liturgy in station-program I.D.s suggests that the industry is skeptical of the viewer's short-term memory. The tactic also clearly demonstrates however that the supertext is not simply the result of programming proximity or the juxtaposition of diverse texts within the broadcast flow.[19] The supertext involves an active and aggressive process of formal and textual infiltration.

Other graphic operations work to secure and orient the spectacle. The late-night news show in Los Angeles, *News 13,* follows the film and immediately reacts to it with a graphic-verbal sequence thick in image and sound. The first shot in the sequence is a freeze-frame graphic of Noriega taken during his arrest and incarceration. The image is presented to viewers as a mug shot. Colored video borders are added by the broadcasters to either side of the televised "photograph." The title "TV-13" is visually keyed into the mug shot in the lower left corner of the frame. By doing this, channel 13, in effect, takes visual possession of the real Noriega for its programming purposes, much in the same way that it earlier appropriated the fictional *Salvador*. The pictorial convention and code of the police mug shot carries with it a heightened impression of the real. The visual convention of the date, frame, and wall suggest that Noriega is a hunting trophy. In these ways, graphic appropriation bestows upon KCOP more authority, skill, and ownership.

While critics of broadcast journalism frequently attack television's penchant for decontextualizing interviews and simplifying complex stories into sound bites, few have tried to explain the manipulative importance of visual bites in the news. For example, what does it symbolically mean to burn the *local* station's I.D., over an *international* figure like Noriega? Appropriating and fabricating a sign in this way does several things. It suggests that the station now both owns the story and also controls the figure's persona for its own ends. Both effects help legitimize

Noriega's Miami mug shot heard round the world is tastefully framed by KCOP on its embossed, blue digital wall. Channel 13 staples ownership *inside* of diegetic mise-en-scène. (KCOP)

the station's angle and coverage of the event. Certainly this kind of keying, framing, and bordering is a form of *claim-making*. Televisual claim-making devices like these announce to viewers that the relationship between the station and the story is indeed a special one.

Blurring borders: video-film-video. Television and film today frequently use visual images borrowed from each other. From music videos to commercial spots,

Replicant geographical shorthand used to bolster both 1983 feature film and 1990 broadcast.

filmed images of scan lines, video pixels, and videotape shuttle effects are frequently mixed with cinematic images. This tendency suggests that recent television and film viewers are, in effect, apparently familiar with media-specific production elements and technology-dependent production styles.[20] *Salvador* exploits this viewer consciousness of distinct film and video looks in its presentation as well. During this event, as in any broadcast, the station airs a newsbreak sandwiched between numerous ads. The voice of Madison Avenue is thus briefly set aside for that of history and reality. After the break, the station chooses to return to the film by fading up, not on an image of cinematic spectacle, but on a videotaped scene within Stone's original film. Both the placement of this video footage within the original film, and its selection and isolation by broadcasters *as a lead-in* to the next segment of the film are significant. The transition back into the film could have occurred at almost any scene in the film, but the broadcasters chose to link their video imagery with video footage from the film. The net result clearly disguised the transition between media.

This hybridizing strategy exemplifies a process by which feature film footage is strategically repositioned during adaptation. The function of this strategic placement is to blur and camouflage distinctions between program and program breaks; between filmed and televised material; between present and past. The television viewer has just seen a long sequence of station materials and ads, and then sees a televised head shot of President Reagan on videotape. The source of the footage is obvious. This is Stone's archival news footage of the "real" Reagan. It has undergone a double appropriation however. First, Stone's feature film claims and uses this reality video footage in order to contextualize and legitimize its fiction. Later, television isolates and transitions around the same scene further deteriorating the distinctions between television station and television text. The ontological status of the footage is ambiguous to viewers since: (1) it is television footage; (2) it has been re-filmed and printed on motion picture stock; (3) it is subsequently rebroadcast on television. The footage carries with it both the aura of presentness and direct address that one associates with video news. These liveness traits are also noticeably present in the anchor's broadcast newsbreak that directly precedes the (cinematic) news scene. As a result of this collision, the local broadcast newsbreak also manages to take on and to exude the epic proportions of the cinematic context from which it comes.

There is, then, a double appropriation. Stone strips the credibility and urgency signified by video for his own diegetic ends. KCOP in turn strips off the epic pretense of film in order to validate and bolster its own speaking position. The formal process goes through three stages, television to film to television. This double appropriation is a clear example of one of the ways that the televisual text spreads out—both program into program break and break into program. The textual ambiguity that results from this spreading out, helps create a context essential for effective broadcast adaptation. In short, by leveling categorical distinctions that exist in the source film, televisual adaptation is better able to reanimate and resuscitate the older text for its own ends. Textual spreading and ambiguity mean, in effect, that the source material is no longer the master of its signified. It is worth noting as well that this leveling and reanimation process uses televised political footage to realize its ends—footage that the viewer associates with history. The actual verbal content of the doubly transformed footage—Ronald Reagan in a xenophobic diatribe about communist hordes "on the banks of the Rio Grande"—is an issue to which I will return later for it implicates another important televisual operation.

Splitting tracks in abridgement. Another less obvious but pervasive form of stylistic revision involves the electronic splitting of audio and video tracks, a postproduction method typically used to conceal elision. Film-to-video broadcasts inevitably mean that certain scenes are cut down in duration. The formal operation of splitting—that is, the electronic separation of sync audio from picture—does not just conceal cuts; it also re-creates. Several of *Salvador*'s source scenes, for example, had major sections deleted when broadcast, yet splitting was used to reconstruct a sense of continuity. Given the fact that a great deal of dialogue was lost in these sections, the way that continuity was maintained is worth noting. In one scene Boyle seeks identification papers for his Salvadoran girlfriend by offering American advisors photographs that he has taken of leftist weapons. Most of the original scene is deleted for broadcast, but splitting is used on the last line of dialogue that is common to both the film and televised version. When actually broadcast, the picture corresponding to the original sync statement disappears,

Stone/Boyle's long, angry political criticism of American foreign policy in film becomes, through splitting on television, a psychologically sensitive reaction-shot about lost love.

while the original line of dialogue is kept as a bridge for use at the end of the revised and greatly shortened scene. The split-off line of dialogue in this scene is then laid over a completely different reaction shot of Boyle at the table. As a result of this operation, the image of Boyle—ripped from its initial shot-reverse-shot sequence—no longer forms part of an interchange designed by Stone to showcase heated political dialogue. The new image becomes instead an emotive reaction shot, and so is redefined in an explicitly psychological, rather than political, way. The scene, in its split and abridged form, is now singularly and primarily about Boyle and the loss of his girlfriend.

The simple operation of splitting off and later reusing existing dialogue as a voice-over bridge redefines existing visual images, but not in a syntagmatic or Kuleshovian way.[21] It redefines instead by using new and nonsequential sound-image relationships. Unlike censorship and automatic dialogue replacement—operations that typically flag themselves to by mismatches in audio presence—the technique of splitting tracks is even less obvious to motivated viewers. Splitting and bridging maintain the original audio presence or ambience in a scene, even though the same images and sounds have reconfigured relationships. The stylistic potential of the operation—driven by the mandate to squeeze a film into regimented slots between commercial breaks—make any original meanings in a film

The world of television lurks underneath the film world. It enters the films on its own terms when the digital door swings open.

volatile and prone to change. When KCOP's splitting and elision in postproduction work over *Salvador*, the film's overt and angry rhetoric about Latin American military involvement evaporates. The very same scene in the film becomes, for television, a portrayal of lost love.

Boxed endtitles with live anchor. While the source film appropriated historical footage from television as a way of injecting currency, a counteracting process was also at work in broadcast adaptation. Blurred media boundaries and textual ambiguities create a general context in which hybridization takes place, but other televisual operations are less passive in the way they work over source material. For example, broadcast news forms constantly work to mirror or infiltrate the fiction of the film, either by narrative analogy, specific reference, or through the use of hybrid televisual forms. Notable among these hybrid visual forms is the use of a picture album–style graphic box, typically inserted over the film's endtitles. This device, of course, brings the broadcasting station "into" the film even before the movie is completed. As the endtitles roll and the theme song plays, a female news anchor from the upcoming 10 o'clock news previews the programming that follows in a three-point outline. From a graphic box on the right half of the image, set back from the screen at a 45-degree angle, she comments: "Up next on News 13: We'll tell you how a suicide and a new suspect may help solve a headline-making murder. Consumer reporter Ken Daly tells you about the best buys in tiny TV's. Also . . . a surprising change in . . . child support."

After these three stories are previewed, the graphic picture box rotates back toward the picture plane, and the endtitles and theme music continue without further infiltration. Apart from the general sense that television has encroached upon the fictional world, this operation lays bare an important form of exchange. First, the anchor is placed over the graphic field while the swelling and tragic orchestral score continues underneath. The anchor's place and import are—as a result of this fluid operation—heightened to epic and tragic proportions. She appropriates the film's rich connotations of history, tragedy, and passion for her own presentational ends. Her news discourse, then, gives to itself the same kind of earth-shattering import that viewers confronted during the two previous cinematic hours. Although music is conventionally seen as an emotive or editorial device that comments on the film-video program, in this case it is reality-news that seems to editorialize and comment on the music. The anchor's presence is literally imaged- and voiced-over the film's footage and soundtrack. By forcibly infiltrating and unseating the signified of the film-music in this way, the anchor's presence makes itself the new target and referent of musical connotation. Formal devices within the film then, like music and titles, are fair game for appropriation by television in its repertoire of adaptation.

News previews. Previews of the "upcoming news" are among the most persistent stylistic infiltrators of films when broadcast. Their nightly presence keeps the shadow of historical consequence always on the viewing agenda. Again and again, the fictional world is halted, set aside, and frozen, while television shifts its discourse to the "reality" of ads and news. The attitude evoked by newsbreaks within the televisual flow of a feature film is both catchy and exaggerated. With only a few seconds to hook viewers, newsbreaks typically start with an interview quote or sound bite presented out of context. This is followed, first, by the challenge or premise tackled by the news staff that evening, and then a just plain folks invitation from the anchor to watch. Prominently displayed in newsbreaks during the broadcast of *Salvador* was a graphic title that displayed the same ponderous statement that viewers heard on the soundtrack. In image and sound, this evening's newsbreak earnestly asked: "Can Noriega get a fair trial?"

News previews regularly exploit their temporal proximity and juxtaposition with other components in the flow. If we consider that segment one of the film—or act one in a classical dramaturgical sense—aims to set up a dramatic fiction by presenting the story's chief or underlying problem or conflict, then the newsbreak in this broadcast is significant in more ways than one.[22] In the case of *Salvador*, the violent and confusing first act of the film seems to ask, "What in the hell is going on in this Central American horror?" Television answers this hook and premise with a question, with a matter-of-fact news preview that asks whether Noriega can get a fair trial. It is a commonplace of narrative theory that narrative discourse depends upon cause-and-effect relationships. Elements in the diegesis, even fragmented and apparently discontinuous ones, are made causal by what Robert Scholes terms the viewer's "narrativitous countertendencies."[23] Here, through televisual segmentation and juxtaposition, an overarching dramatic premise or question in the film is narratively and falsely answered by the news.

Cause-and-effect logic:
If (slaughter in El
Salvador), *then* (can
Noriega get a fair trial)?

That is, *if* war in Central America *is* a nightmare, *then* can there be "a fair trial for Noriega?" Though geographical differences in Central America are evident in both the film about El Salvador and in news about Panama, the cause-and-effect relationship set up between El Salvador and Noriega is absolutely artificial.

Broadcasting uses this artificial, narrative causality to set up, heighten, and enhance the vantage point from which it sees; and the authority by which television speaks. To a fictional world in which there is no answer, television promises one. Though the issue of violence and war is tough and pressing, the televisual response is polite, authoritative, and helpful. The anchor smiles to the viewer, and says: "hello." Other breaks continue the cause-and-effect pattern. In one transition, the specter of Noriega is raised again as *the issue* by which the tonight's cinematic musings on Central America will be solved. The fourth major station break leads again with a news preview that explicitly postures Noriega as the political problem for the news editor's narrative answers: a drive-by shooting in Los Angeles and "prostitutes and the spread of AIDS."[24] Repeated causal linkages like these create a de facto Central American viewing agenda that spreads out and claims other news items. A hail of bullets in Los Angeles, and the issue of prostitutes and AIDS—although both secondary stories—resonated explicitly with events in *Salvador*'s fiction. Mimicking Stone's heavy-handed cinematic blueprint blow for blow, the news threw urban gun battles, prostitution, sexually transmitted diseases, and Noriega's "many women"—overt ethnic stereotypes that recurred in both *Salvador* and in television coverage of Panama—into their videographic vision of Central America.[25] Urgency rubs off and permeates even secondary stories abducted in what nightly masquerades as a nonnarrative flow. Television awarded itself credibility on this particular evening by stitching together an abiding sense of narrative causality. The news knows well how to narratively style an urgency of international proportions.

Piggybacked studio promos. Little is gained by viewing television as the culprit responsible for the degeneracy of higher texts, or as a crude conspiracy that misrepresents original cinematic works. Broadcasters, clearly, are not the only ones interested in the appropriating potential offered by televisual adaptation. The process of textual appropriation, in fact, characterizes much in the mass media, and broadcast adaptation includes regular forays and investments by Hollywood's feature film industry. The ways that a film studio's product is drawn and quartered in previews and trailers on television is not just a contemporary issue tied to current film releases. The owners and marketers of older features have a clear stake in the kind and quantity of trailers aired on television as well. The promotional inertia provided by current trailers can, in fact, revalue dated films if an authorial link is established when the trailers are broadcast. One sequence aptly demonstrates how this kind of promotional inertia revalues and resuscitates dated films and dead texts. In a broadcast preview for the feature *Born on the Fourth of July*, the American flag frames the face of its star, Tom Cruise, and graphic subtitles boldly announce the prestigious film's current nomination for many awards including "Five Golden Globes" and a nomination for "Best Director, Oliver Stone."

This televised preview raises clearly the question of who is appropriating and exploiting whom. While the videocassette of the feature *Salvador* was still in circulation at the time of its broadcast, its theatrical run had been over for many years. The studios could still cash in on the older film through a retroactive process of transference. What was being marketed here was not just a film, but also a director's reputation and body of works. For the price of a thirty-second spot on KCOP-13, the film studio gets two hours of viewer experience *defined as coming from* Oliver Stone. There is, the viewer is lead to believe, an Oliver Stone package of works that now includes the headlining star Tom Cruise. Whereas the programming of *Salvador* was clearly related to current events in Central America, it was also related to the release of Stone's new film, *Born on the Fourth of July*, a film that had itself become a celebrated media event by January 1990. Images or sounds of Stone, Cruise, Kovic, and clips of the film appeared in almost every venue of the mass media that covered the entertainment industry.[26]

Reminiscent of the way that Universal tried to shield itself from vocal critics of the *Last Temptation of Christ*—that is, by marketing Martin Scorsese's film as "personal artistic expression"—*Born on the Fourth of July* was pitched to the public as a "true and personal" story from Stone and Kovic. The highly intense eyes of the composite Stone-Kovic-Cruise, heightened through extreme close-up in the preview, provided a sensitive, composite eyewitness to the war's horror. The broadcasting event in January of the film *Salvador* clearly paled beside the simultaneous national promotion and release of *Born on the Fourth of July*. Nevertheless,

Previews create composite persona of Stone-Kovic-Cruise—the experienced veteran, politico, and sensitive artist capable of grasping the spectacular terror and carnage that approaches the "banks of the Rio Grande." The visionary and tortured composite artist becomes an eyepiece for Reagan-era xenophobia.

the broadcast adaptation of the earlier film afforded the producers a timely opportunity and site for marketing the later film.

The connection and similarity between the two films did not only exist at the level of plot, although both works—like almost all of Stone's films, from *Midnight Express* to *Platoon*—were premised on the idea of "one man's tortured journey through a violent hell and back." The loudly stated ideologies and political commitments of the two films were also similar. Promotional and on-air quotes by the marketeers and promoters of the later film went to disingenuous extremes to contextualize its fundamentally antigovernment stance as actually pro-American. This *repatriating* tactic did not just depend upon having the characters of James Woods and Tom Cruise assert their patriotism within each film, although both do. Instead, televisual repatriation was also a matter of politics outside of the film or broadcast. Ron Kovic, by January, was being touted as the next important candidate of the Democratic party in Southern California. Meanwhile the Republican bastion that dominated Orange County attacked both Kovic and Stone's left-leaning liberalism. Published statements ripped Stone and his liberal boys from Hollywood. Given the horrifying public speculations that *Born on the Fourth of July* film might propel Kovic or Stone, or both, into office, arch-conservative U.S. Representative Robert Dornan dared Kovic and Stone to take on the Republicans:

"Kovic is going to have to defend the Hollywood far left. If he thinks he's going to recruit Oliver Stone and *Top-Gun*-turned malcontent Tom Cruise, and bring the whole Jane Fonda team down here to Orange County, I welcome it. Let's go." [27]

The mass marketing of American films pretends to and depends upon an ideology of universal appeal. Probably because of the industry's burden and mythology of mainstream populism, the obvious anti–status quo stance in both of Stone's films on this evening was effaced and neutered in marketing rhetoric. Both of Stone's film characters, Boyle and Kovic, are praised for having loved and sacrificed for their countries. Both films were, in addition, presented as fundamentally autobiographical and personal. Consider the strange and striking similarities between the the following three "patriotic" public statements, all collaged together in this same broadcast event: "Well, I *love* America"(Tom Cruise, in *Born on the Fourth of July* trailer);[28] "I *believed* in America" (Dialogue from Stone's *Salvador*);[29] "Many *Thanks* America" (Placard held by grateful Panamanian citizen thanking the American military for invading her country).[30] From such combinations, one might suspect that the Stone-Kovic-Cruise-Boyle persona had snowballed to hemispheric proportions. This composite subjectivity—at least as it appeared on television—was now shared as well by the entire Panamanian citizenry! Everyone now, apparently, is swept away by ecstatic devotion to the United States—including those who were having their neighborhoods violently razed by the U.S. Army Special Forces. In addition to selling tickets, then, the plug for the more recent movie during the hybridized broadcast of *Salvador* actually worked to stress not the fundamental *antiheroism* of both films, but just the opposite: to laud the corporate *ideology of weathered and sacrificial patriotism*.

While the main thrust of programming related *Salvador* to Noriega and Panama, this secondary connection invoked a very different kind of history—a biographical history. Personal histories have proven effective ways of marketing the present by reference to the past, as *Born on the Fourth of July* did by reference to *Salvador*; as Stone did by reference to Kovic; as the studios did by reference to Stone. The history of director Oliver Stone is invoked to establish the validity of both film accounts. Whereas television news appropriated the fiction to insure its urgency and import, Hollywood appropriated and infiltrated the same fiction to legitimize its current aesthetic and (very apolitical) economic program. This broadcast became a consensual site for multiple appropriations—a kind of win-win rental text.

Previews for *other* broadcast movies. In addition to promos for the broadcast, and to previews for Stone's other films, the televised *Salvador* also included station promos for other films slated for broadcast. The short promotional spot for *The Border,* for example, scheduled to air two days later on KCOP-13, is particularly instructive. Short and to the point, the promo occurs as the ninth nonprogram item, exactly halfway through the second television break. The ad throws a pressing voice-over appeal to the viewer: "Saturday night at five, personal problems force a border patrol officer to risk his career—and his life—in the exploitation of illegal Mexican immigrants. Jack Nicholson stars in the action-adventure *The Border.*" If one were to substitute the following words, the synopsis would apply as easily to the televised presentation of *Salvador*, in which *The Border* promo was broadcast. "Personal problems force a *photographer* to risk his career—and his life—in the exploitation of illegal *Salvadoran* immigrants." The border, exploitation, and immigration are all celebrated themes in *Salvador*. This promotional ploy forced programming congruence based on perceived similarities of plot and point of view. The same promo, significantly, also hyped the film as an action-adventure genre picture. Both the *personalization of the story* and the self-conscious *attribution of genre* are categorical methods by which television distances and mediates films. *Salvador*, even more so than *The Border*, is characterized by aggressive political critique, but the promotional language of the televisual text however effaces this fact. It invites viewers to see both films as a dramatization of two recognizable forms of pathos: male alienation and male generic action.

Breaks: segmented continuities. If promos knock the viewer in the head with generic and gender-loaded interpetive frames, other additive elements are less self-conscious in their connection to the televised film. Chief among these institutional adhesions are the large number of ads strapped to any feature film presentation on television. While Raymond Williams referred to this process as part of the flow of programming, later theorists like Nick Browne and others have embraced the idea that this infusion and expansion of the text creates a kind of supertext.[31] This academic vision of television as an expansive and unifying text contrasts with the industry's own explicit view of such ads as part of program breaks, that is, the material inserted between segments of the show. Although critical theorists like Tania Modleski have clearly shown that a symbiotic relationship exists between ads and the programs during which they air, *industry writers* still conceptualize their task as basically intratextual; as concerned with the narrative as a unit.[32] An analysis of the first programming break in *Salvador* reveals some provocative and overt linkages between program and nonprogram material.

Of the commercial ads sold here the majority focused on domestic and hygienic goods and products.[33] Of eleven ads, the four that appear during the first half of the break are broad in their address, and appeal both to female and male consumers with a variety of goods (food, audio tape, a charge card, snack cakes, and Italian sausage). During the second half of the break, the audience is also idealized as broad based (underwear, food products, cold medicine), although more of these ads clearly appeal to female consumers (feminine douche, finesse hairspray). This, then, is clearly not the male focus one might expect to find supporting a violent boy-action picture like *Salvador*. This emphasis on the familial, domestic, and hygienic (middle-class) goods and services may be a key to the kinds of revisions that go on in the televised version of the film. *Salvador* does not just focus on other values. It dramatically attacks middle-class values: industriousness, decorum, manners, restraint, morality, cleanliness, and women are systematically denigrated by the film. At the same time, however, ads in the televised version suggest that the film is programmed for an audience defined by those very traits. As one might guess from this blatant contradiction, television elides those scenes most repugnant to the sanitized world idealized by its ads. When Doc falls to his physical lowest—a victim of amoebic dysentery, sexually transmitted diseases, and flesh-covering grime—he steals food from a street person's plate and finally explodes at Boyle: "I'm stuck in the middle of this fucking country. I can't speak the language. I got the shits, and I only got three fucking dollars . . . I gotta get outta here, man." The downward trajectory of these characters toward amoral anarchy is clearly at odds with the upbeat world of bodily and familial maintenance suggested by the ads. Modleski and Flitterman have argued that the complementary relationship of ads and program result from a need to address and exploit the viewer's daily schedule and viewing habits.[34] In the case of daytime soaps, the continual lack of textual closure and tension is solved in the falsely resolved world of the ads. A different kind of viewing habit is implied here, however. The net effect of the *film* text is less one of irresolution than of perpetual *estrangement*. *Salvador* works by progressively alienating its characters from their increasingly bizarre surroundings. To the viewer of the film, the experience is no less bizarre, since the Central American physical and bodily horrors depicted are absolutely alien to the North American suburban culture associated with mass market television. The relationship between soaps and the daytime viewer-homemaker, however, is posed artificially as congruent, since their ads work together to connect the viewer's world to the televised world. The cinematic fiction here, by contrast, *explicitly* signals its extreme dissimilarity to, even loathing for, the world of the television viewer.

Broadcast adaptation, in this case, addresses the film's strategy of estrangement head-on. Whereas the original *film* exploits the anxiety it creates in viewers through extreme dissimilarity and dissonance, the interjected *television* station material does just the opposite. Material added for broadcast works by constant *analogy* to the film spectacle, and so tends to fill the vacuum caused by the film's acts of estrangement. It soothes and mediates the dissonance between the viewer's world and the world of the filmic spectacle. Broadcasting offers itself as the known and knowing agent; as an institution capable of contextualizing and accommodating cinema's political and existential abyss. Connections and analogies to other programming forms are continually broached and bartered in a process that directly promotes the station's other offerings. This process is logical, given the fact that the site of televisual adaptation is, in essence, a marketplace. The televised film is a product that is stylistically drawn, quartered, and packaged under many different labels. A close analysis of *Salvador* suggests that the stylistic repackaging of the film works to make its cinematic spectacle more palatable and more universal. Feature films, it seems, are forever open to redefinition as historical events change the viewer's agenda and expectations. In this particular case, the visual spectacle of *estrangement* in the *film Salvador* actually intensifies the sense of *congruence* between viewers and broadcasters, in this user-friendly televisual spectacle.

Televisual geography: the new world order. Although my primary aim has been to examine formal encroachments and additions to the adapted film text, I want to add one example of the way even subtractive or elided material reconstructs a new text for broadcast. Along with a general strategy of deleting personal, biographical, and causal motivations from character, the broadcast of *Salvador* also persistently deleted all transit sequences. With their absence vanished any clear sense of geography for the viewer. One of the earliest and most obvious elisions of film material depicting travel and geography is the sequence dramatizing Boyle and Doc's down-and-out drive from California to Central America. In the original film version, there are a total of fourteen shots in the montage sequence depicting the characters on their way to El Salvador. In the television version the

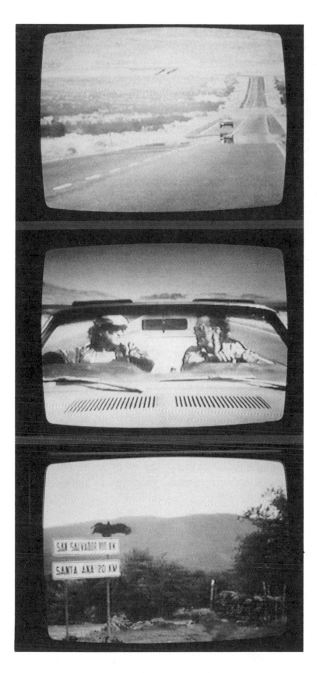

Driving the death mobile, the alienated boys flee their insensitive "bitches." In television's massive elision, however, the transit montage across the hemisphere is reduced to three shots. Within seconds of a recognizably Southern Californian desert, they now pull into vulture-infested El Salvador—somewhere just outside of Los Angeles—destined to pay for their lifestyle excesses.

2400-plus mile journey is made in a total of three shots that last a matter of seconds. The car in each shot, nevertheless, suggests a continuity of action even though time and space are filmically and massively condensed. Dialogue is done entirely in voice-over in the new scene, the soundtrack is pop, and the few surviving shots are stitched together with a series of video dissolves. A close analysis of this problematic transit scene is instructive. The massive geographical ellipsis

that results from the editing of the scene typifies the narrative condensation pervasive in the first half of the film. Seconds after a couple of traveling shots in
the opening of the montage—clearly filmed in the deserts of Southern California—the viewer immediately confronts a mileage sign in El Salvador itself. One
wonders whether this televisual ellipsis (rendered in video dissolves) implies that
the men are traveling to Santa Ana in suburban Los Angeles, rather than to a city
in Central America. In any case, the net effect strongly implies that the land where
"they kill people" is actually just on our doorstep.

The geographical threat and xenophobia visually implied by this hybrid
televisual montage is reduplicated in explicit terms two important times later. First,
news footage in the film of Ronald Reagan warns that the spread of communism
in Latin America will soon threaten North America itself. Later, a U.S. embassy
military advisor baits Boyle by suggesting the specter of Cuban tanks on the Rio
Grande. This isolationist model of fortress America, threatened by communist conspiracy and the influx of illegal aliens, was in fact a common theme in American
mass media during the 1980s. Here, perhaps in the interest of narrative expediency, the mythology is perpetuated through televisual shorthand. Whereas the narrative cause of their trip is shown through a series of dramatized personal
rejections, the narrative effect is to step next door into the kind of hell the media
know as El Salvador.

Personal rejection, a car, a border mileage sign to San Salvador and Santa Ana;
this is the symbolic route that television viewers travel in order to *play* the new
narrative. Unlike the film, the televisual route is efficient, visual, depersonalized.
Boyle's personal history is completely removed as a justification for the sequence.
The textual vacuum created by this elision is filled by a mere transition. Simple
video dissolves subvert the original narrative flow and create an artificial causality between the visual and textual elements that survive. Although broadcast adaptation may be driven by the fundamental imperatives of programming time,
hybrid forms of space are left in its wake; televisual geographies scaled to tortured political maps.

Programmed Confusion: The Spectacular Other

This analysis began by reference to several broadcast incidents that betray
television's appetite for shifts between history and fiction. It will end by citing
two additional examples that suggest the *ideological* stakes involved in the
televisual process of encroachment and resuscitation. A KNBC television correspondent covered the inauguration of the newly elected government of Chile on
January 12, 1990, and described Vice President Quayle's participation in the event.
Not once, but twice, the reporter mistakenly referred to Quayle as meeting with
Noriega (the deposed president of Panama) rather than with his actual contact,
Daniel Ortega of Nicaragua. Although the political positions of the two men could
not have been more different—one from the right, one from the left—the names
were apparently interchangeable to the news staff and editors. No retraction was
forthcoming in the news show that followed.[35] Two months later, as the new government of Violetta Chammoro was being installed in Nicaragua, American stations (and even local affiliates) sent their various headlining correspondents to

cover the occasion and boost ratings. An on-air phone-in report from Nicaragua by ABC commentator Bruce Herschenson detailed the changing situation in Managua. After an extended discussion with Herchenson, perplexed studio anchor Paul Moyer asked, "Are you in El Salvador?" Herschenson responded, "No, Nicaragua." Only slightly less ambivalent, Moyers concluded, "You're in Nicaragua?"[36]

Countries, nationalities, leaders, political affiliation—all are apparently interchangeable when it comes to the worldly spectacle "out there." By neutering particularity and partisanship, televisuality makes the global spectacle open to infinite appropriation. I am suggesting that the process of appropriation and resuscitation is not just a transformation that targets and revalues dated Hollywood products like *Salvador*. The constant unfolding and decontextualization of current historical events on television provides an ample resource, an excuse, for resuscitating any dead cultural text. Through "signature intentionality" and stylistic encroachment, televisuality adapts and hybridizes existing texts in a way that makes them acutely open to appropriation. Televisual auteurism—typically driven by a surplus of concerned sensitivity and social topicality—is, then, both a pretext for economic intervention and a programming tool used to flaunt and throw around ontological distinctions: history/text, news/film, reality/fiction. In broadcast adaptation, these dialectical terms are used as currency in the televisual system of exchange. Signature television—whether homegrown or cinematically imported—frequently acts as the system's broker. Yes, war makes good television, even as it remakes dead fictions.[37] Televisual authorship is, then, both an index that hawks excess intention and an instrument that programs real-world authority.

5 Franchiser
Digital Packaging / Industrial-Strength Semiotics

Metaballs . . . generate smooth holes, cavities with soft edges;
attach blobs or combine blending groups to create
melting objects, moving molecules.
—Thompson Digital Image Corporation[1]

What ever happened to the good old-fashioned dissolve?
—Rob Wyatt, Telezign, New York[2]

Lurking behind the counter of the televisual boutique, and on each facade of the televisual marquee, lies a much more pervasive and vernacular form of stylistic exhibitionism. Digital videographics so dominate mass-market television that they have become an obligatory—even if unremarkable—part of the cutting-edge package. Seldom betraying a prestige signature maker, digital videographics mass produce acute stylistic looks for both primetime and off-primetime. Between shows, in title sequences, in newsbreaks, in previews, and throughout many live genres, videographic televisuality has become a requisite form of packaging, and an indicator, *not of quality, but of quality control.* If the cinematic boutique overflowed with excessive intentionality and refined sensitivity, videographic televisuality evidences no similar illusions of restraint. In fact, by contrast to signature television's fantasy of the human touch, videographic televisuality is really more of a franchising operation—a way to clone, mass produce, and distribute stylishness. Like gilt-plastic rococo frames, warehoused in discount department stores, videographic televisuality can be downloaded into user-friendly regional technologies. Stored in postproduction electronic banks, these technological flourishes can be punched up, on command, in the monotonous workaday world of industrial video and primetime alike.

The eye-popping nuances of "metaballs" and other techno breakthroughs are part of a more widespread industrial practice, one whose very mission is to intensify and complicate the the television image. Consider, again, how this videographic form of televisuality undercuts high theory's definition of television's visually distracted audience: "The broadcast TV viewer is not engaged by TV representation to any great degree";[3] "There is a specific way that television is watched";[4] "We turn on the set casually; we rarely attend to it with full concentration."[5] According to this litany, the medium, on the one hand, is textually boundless but amorphous and stylistically crude. The viewer, on the other hand, is distracted by other activities and is fundamentally inattentive. Outside of sanctioned academic discourse, however, industrial players also theorize about the ideological and aesthetic effects of the basic televisual apparatus, but they do so in ways very alien to high theory.

High-tech Orkin Man as
cyborg-robocop. (PDI)
ABC's computerized
optical spectacle. (ABC)

The commercial is like a dragster. You jump on it and you've got a quarter-mile to break the record. Commercials push the envelope, and that offers a kind of excitement you don't find anywhere else.[6]

Citing a trend, preparing for it, and riding its popularity can spell success for the agency and its client. Surfers talk about a swell, he says. When you're sitting out past the waves you see a swell in the horizon and three waves later you know that'll be the big one. We're like that.[7]

The driving force in the broadcast graphic environment today is the relationship between the artistic demands of the designer and the advancement of technology. Each challenges the other to "push the edge of the envelope."[8]

Discourses in the production world, then, argue that television is like drag-racing and surfing and, in a reference to both racing and test pilot aviation, various promos imply that good commercial television "pushes the envelope."

Given these declarations, one might surmise that *aesthetics* is perhaps the wrong disciplinary arena for television scholarship. Perhaps *athletics* and *spectator sport* are better frameworks for analyzing the kinds of production practice currently favored in the industry. Such a shift would, after all, enable frustrated critical theorists to become play-by-play analysts of popular culture and so earn the respect so long denied them by mass culture. Media practitioners now explicitly describe their art in physical, active, and bodily terms. This shift to bodily and athletic discourse in the industry is all the more striking given recent developments in technology. The ever more complex technology of television, in contrast to the relatively static development of film technology, has come to overshadow its users. But whereas the engineering of television equipment has become increasingly complex, the method of controlling that equipment has become more user-friendly, refined, and conceptual. In fact, the actual *physical* interface with television equipment has become extremely cerebral and nonphysical. Control of the new televisual equipment is driven by the stylus, icon, and palette—not by potentiometers, switches, and patch-bays.

Given the industry's ongoing discourse about meanings and modes in television, and recent developments in technology, it is worth reexamining the role that critical theory can play in an analysis of the new television. Depending on the intellectual models used in television study, the practice of figuration is very much a problematic issue. Television scholarship that comes out of the tradition of film and literary theory generally shares contemporary theory's denigration of the naive metaphors and figures favored by classical film theory.[9] In classical theory, the conclusions one made about film greatly depended upon the imported figures that one first brought to cinema from other art forms, life forms, or human activities. Yet even later revisionist theorists, who ostensibly shared in a materialist and modernist penchant for media specificity and hermeneutic rigor, continued to freely invent and import their own master paradigms from other fields—medical suturing, the mirror stage and voyeurism. Some of the most recent and provocative work on television, argues persuasively that television is like "the freeway, the mall, and the automobile."[10] Accounts like this do much to suggest how television functions within a broader cultural dynamic. More traditional accounts of broadcasting foreground the framing paradigms for American television, but do so with economic and political categories so broad and extratextual (for example, nonpaternalism, competition) that they describe what television is not, rather than what it is or can be shown to be through critical analysis.[11]

Yet even a cursory look at contemporary television shows that the paradigming process is not the exclusive domain of academic discourse. Rather, figuration and the importation of paradigms are fundamental conceptual processes that define the media industry as well. In fact, rapid theoretical figuration so pervades the contemporary television industry that high theory pales by comparison. The question is not, then, whether paradigming and figuration are appropriate methods for television critical theory, but why high theory pays so little attention to the centrality and self-consciousness of industrial figuration.

This chapter, through a case survey of videographic exhibitionism, engages fundamental issues and categories evident in low theory's process of industrial figuration and self-conscious aestheticization.[12] Despite their centrality, both processes have generally been overlooked in television high theory. Some recently published works on television stand as notable exceptions to this disregard and demonstrate the importance of considering material practices and popular discourses in historical and critical analysis. Lynn Spigel utilizes a rich array of popular sources (advertisements, fan magazines, television marketing literature) to demonstrate television's role in the construction of domesticity, the home, and the feminine in the postwar era.[13] William Boddy incorporates an extensive group of primary sources dealing with regulation, management, and programming in order to account for the negotiation liveness in 1950s television.[14] Both projects differ from the general tendency of high theory (in both its poststructuralist and social-scientific variants) to impose and *overdetermine* external schemas in critical and historical accounts of television. This categorical overdetermination invariably reconstitutes television as an object of study in significant ways. Even if such a reconstitution is justified by the abolition of subject-object distinctions under poststructuralism, the reconstitution comes at a practical cost for theorists of the mass media. The process cuts off consideration of a number of important mainstream practices that have political consequences for viewers. The industry's performance of digital style, my focus here, is one of these important but banished discourses and absent categories in high theory.

Arguments over the relative merits of insider versus outsider perspectives for interpreting cultural phenomenon are common in linguistics, ethnomusicology, and anthropology. Film and television critical theory has frequently utilized what some linguists would describe as an -etic approach in analyses and taxonomies of media-specific languages.[15] This tendency is due in part to the iconoclastic legacies of structuralism and psychoanalysis, trends that presupposed fundamental suspicions of surface structures, meanings, and interpretations.[16] Since I am arguing that televisuality is a *self-conscious* cultural practice, however, it seems important to shift the analysis from the analyst's theor*(etic)*al culture to the producing culture itself—that is, to an -emic analysis of television's sign system. Such an approach would take seriously the industry's attribution of meaning and its own semiotic analysis.[17] It does not, however, necessarily buy into the industry's theorization, but rather considers self-theorization as a key signpost to the cultural significance of visual codes and systems. As anthropologists Malinowski and Radcliffe-Brown suggested, even "incorrect" answers by informants are not without value, but are socially significant facts in their own right.[18] The industry is far from naive about its visual stylistics and image-making technology. Indeed, it is highly self-conscious about its visual codes and expends great effort codifying and elaborating the meanings of those codes. Because televisuality has emerged from a market economy (rather than a centrally controlled economy), marketing and promotion have played an important role in the proliferation and official interpretation of the new sign systems. By invoking television production as drag-racing and television style as surfing, practitioners are utilizing aesthetic means to differentiate their products and services from competitors. In the television industry, the classic strategies of product differentiation and marketing use explicit aesthetic and semiotic methods to achieve their ends—even as industry practice itself is a product of much wider cultural trends and ideological investments.

Primetime programs in the late 1980s clearly evidenced an increasing preoccupation with visual style, even in many serious genres identified by realistic narrative forms. Other genres and programming categories demonstrated an even more extreme aesthetic evolution, especially those program forms that utilized and depended upon a high degree of electronic postproduction. These *post*-dependent genres include the entire field of commercial television advertising, many studio-originated programs, and a wide variety of nonprogram materials that are

The "Two at Noon" news opening. (KTVU-TV/One Pass)

broadcast, including promotionals, and pre- and post-title sequences. One cannot easily overlook the sheer semiotic density of visual signs in these programming categories. In fact, each year the semiotic density of the postdependent genres and forms increases and threatens to encroach even more upon the diegetic world of featured programs. This trend suggests that the end-state idealized for television by television might be one characterized by complete videographic manipulation and stylization. Given this overall trend toward televisual stylization, this study will examine the specific ways that users in the industry theorize and enact the videographic art of television.

My goal here is (1) to articulate what several of the favored televisual semiotic modes consist of from the industry's perspective; (2) to suggest how these modes are organized and perpetuated as semiotic systems; and (3) to briefly summarize how these trends make problematic several privileged notions operative in high theory. The televisual industry *postures* as a self-perpetuating, and self-permutating semiotic machine. Television now evolves and explains new stylistic codes on a regular and on-going basis. The industry clearly has its own semiotic "theory of video practice," and it can be found, ironically, in the hype and verbiage of marketing brochures, corporate ads, and equipment data sheets.

Case Study: Videographics
New Modes, New Codes

Directors and designers who work in the new television do not just speak with and utilize a different formal language, they also freely interpret and discuss the meanings of televisual form. Consider the imagery from a television news program and its verbal description. "At One Pass, editor Mike Dennis and KTVU-TV's art director Jim Rodgers developed the KTVU news open Two At Noon entirely online from a Paintbox storyboard. The piece is a soft-edged combination of transparent layers that gracefully merge to show the urgency and professionalism of news."[19] Although the reference to "soft-edged" technique and transparency is interesting, it is the ease by which the transition from signifier to signifieds is made that merits attention. In repackaging the production company's press release, the trade magazine makes this semiotic jump from the signifier (the graphic devices) to its referent (what the graphics express) without hesitation, as if the sign system is an accepted and logical one. Consider the link constructed between these two sides of the sign. Since when and by what logic do soft-edges equal "urgency," or transparent layers equal "professionalism"? At one point in film history, graininess and awkward camerawork meant urgency. To say, then, that the use of soft-edges hints at a paradigm shift is an understatement. Clearly, these signs are recent inventions of the users. Yet the fact that an industry trade magazine presents such things as fact betrays the system's desire to transform visual forms and the connotations of those forms into more concrete denotations. The apparent desire is twofold: to find a logic for the use of televisual forms and to publicly establish and codify that logic. Through this operation, *industrial semioticians* attempt both to create and to own signification.[20]

Before attempting to explain the ways that self-interpretation takes place in the industry, it is important to examine more closely the language and categories that are being used to describe the new visual forms. That is, before dealing with the broader issue of how signifieds are controlled and performed, it is important to get a fuller sense of the texture and demeanor of the new televisual signifiers themselves. When this is done, it becomes clear that the industry no longer necessarily conceptualizes its task around traditional notions of narrative or sequence, nor solely around television's privileged myths of liveness and immediacy.[21] After surveying the televisual modes, and the models by which those modes are organized and performed, I will conclude by summarizing how these practices challenge a number of important tenets of critical theory.

Among the many different effects that proliferate in videographic practice, four popular -emic modes coalesce as organizing impulses in the new televisual syntax: (1) the painterly, (2) the plastic, (3) the transparent, and (4) intermedia. These are not the only ways that the new television manipulates and constructs images, but they are pervasive and general tendencies. Each of these four stylistic modes actually includes many discrete video effects, since it is clear by looking at almost any high-end video postproduction device that the current array of specific effects is extensive. Yet, a number of common patterns and graphic inclinations can be isolated within the current range of available video looks. Since the quantity and types of video effects seem to increase monthly in the industry, it is important to try to clarify the differences, affinities, and similarities that these proliferating effects have among themselves. Hence, the following four modes suggest *families* of stylistic devices and also some shared ideological problems.

The painterly. References to many current imaging effects as painterly proliferate in the industry. *Painterly*, when used by manufacturers, directors, and designers, describes images that appear worked over graphically, as if by hand. The overt reference here, of course, is to painting with a brush, but since no brushes are actually used in an electronic post-production environment (styluses, track balls, and digitizers are), the term has taken on a broader meaning. Quantel, a major manufacturer of digital and analog graphic devices, advertises the capabilities of its Paintbox, by extending the painterly reference: "Every type of artist's medium is provided by the menu and stylus: paint, chalk, airbrush, wash, or shade. Pictures can also be magnified, blurred, crisped, or smoothed to achieve just the desired effect. Stamp, smear, smudge."[22]

Painterly deformations
and the handmade look.
(Celluloid)

Stamping, smearing, and smudging are anything but related to the ethereal, luminescent experience most associate with the televised medium. Television is typically thought of as a medium of electronics and light, as an end-product of high technology. Makers and producers, on the other hand, situate the medium in practice, as a *surface* to be worked over, marked on, blurred, and blended. In short, users import artistic methods that are alien to the immaterial and tenuous world of electromagnetic wave communication.

Standard now on even low-end video post equipment is an array of painterly effects. Cross-point Latch corporation touts its switcher's ability to include "mosaic" and "posterization" effects along with colorization.[23] Marketing images for this device depict the transformation of footage of animals into mosaic grids. A new generation of scene-by-scene color correctors were also developed by the late 1980s. The DaVinci was designed, in its method and layout, and marketed, by its name, with direct reference to classic painterly control. It included smart, interactive interfaces for use with the Rank-Cintel, Marconi, and Bosch film-to-video transfer devices. Promotional illustrations for the DaVinci used a photographic triptych of *Mona Lisa* paintings as surfaces for precision electronic reworkings. The ad promises "a geographical area isolation technique that allows the ability to independently correct identically colored objects in the same scene."[24] No longer is the producer dependent upon footage whose quality is predetermined in the field. Instead of transfering scenes correctable only as whole *takes* or *scenes*, new devices such as these allow for extreme precision and manipulation *inside* the *visual fields*. Field footage and unfolding takes are no longer considered the given, or basic, unit. Pixels in a videographic landscape are considered the basic building blocks. The artist's control now is seen as wide-ranging and residing over the visual geography of the frame. Viewing each frame as terrain surely is a far cry from the academic view of television as a crude and low-resolution medium. The new electronic postdependent TV no longer works under the burden that programs are predetermined during production. A qualitative change in scale of artistic control is evident in the medium.

Within the painterly mode, effects are stacked, ordered, and marketed to users

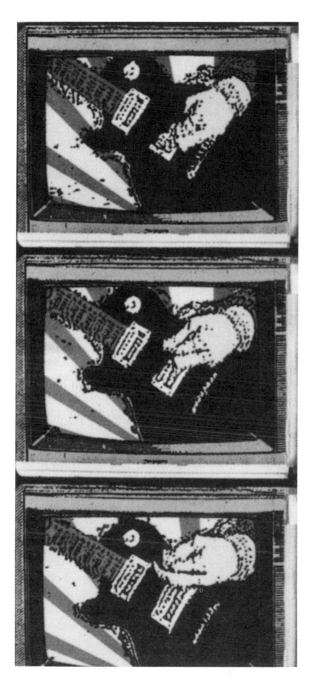

Video-povera and the physical mutilation of the image. (Klasky/Csupo)

in a systematic way. In fact, video effects are frequently marketed in an *art historical* way. Ads for the Encore graphic device promise "mosaic, posterization, borders, crop, replicate, shading, pop, mattes, and many more—Encore's range of *classic picture effects* is the most comprehensive available."[25] Granted, this canon of art history does not include Grunewald or Van Eyck, but it does include what the corporation calls "classic picture effects" like pop and posterization. One might conclude from this that Peter Max and Warhol are the great artistic patriarchs of the new Western civilization. Apart from this kind of art historical revisionism, it is clear that style now is merely a collection of effects to be served up on command to directors and producers.

The practical *appeal* of "painterly deformations" to the televised image is worth considering, especially since this mode flies in the face of the current appetite for, and developments toward, high-resolution television. Painterly effects achieve just the opposite effect of high-resolution; they are low-tech and low-definition. The Celluloid production company, for instance, was covered in the trades for developing a "hand-made" look in their commercial work. In spots for Kelloggs and others, director Wahlberg of Celluloid creates looks associated with hand-coloring techniques, collage, and flat surfaces. The irony in this method for Wahlberg is the extent to which high-budget agencies will pursue this low-budget, hand-made look. "It's coming full-circle—almost anti-advertising," he remarks.[26] The significant point is that this pose of radicality, is pursued by developing an anti-illusionistic visual mode, by deforming the image. By smudging, hacking, and washing-over the video "surface" with colors, the agency pretends to innovative radicality by evoking artistic spontaneity and directness.

More aggressive forms of painterly deformation are evident elsewhere. A spot produced by Klasky/Csupo of Los Angeles moved from the realm of feigned artistic effects (ones simulated in a digital environment) to actual painterly violence. In the fifteen-second open for Radio Vision International, directors Csupo and Schwartz used cut-outs, photocopying, and even the scratching and painting of the film footage itself, before its transfer to video in post.[27] As if this Van Gogh–like violence was not enough, the producers added various electronic-visual deformations in post. Video snow, stretching, and random image repositioning all drew attention to the extreme materiality of the segment. Violent, painterly deformations of the image in this manner make the painterly mode more than a collection of discrete visual codes. The painterly, aggressive method itself becomes a sign—a feigned reference to a rebellious attitude and avant-gardism. As a consequence, the painterly mode also effaces the status of the image as mass-produced.

When considering the reasons for these tactics, it is useful to examine stylistic modes as *responses to inherent limitations* or obstacles in a given medium. The painterly mode is very unlike the way many theorists have thought of television. The video image is, for example, characterized by its linear construction, since it is comprised of a series of scan lines on a raster. Even the technical trait of fall-off creates, in video camera originated images, vertical and curvilinear outlines by intensifying high-contrast separations between lights and darks. In short, the video image involves linearity on several levels. Painterly deformation, on the other hand, attacks this look and quality. It does so by blurring visual distinctions, lowering resolution, and defeating linearity. Since scan lines are apparent even on relatively high-resolution imagery, painterly effects betray and conceal those lines. In their place the viewer is given images that seem physically marked and mutilated by human hands. To suggest physical marks and residues on images is to participate in an act of denial. The process covers over the very way that the tele-

The plastic surface and Picasso's electronic legacy. (Rank-Cintel, Ursa)

vised image is constructed around lines, contrasts, and patterns. This stylistic act of disavowal is perhaps a key to the other televisual modes as well.

Plasticity. If many of the stylistic devices in the new television work to deform and generalize the image, then another group works to make the image plastic and tangible. In fact an array of videographic devices are premised on the ideal of transforming the video image into a material surface or three-dimensional object. The conventional television set itself has given the video image a quality of great imprecision. By its physical presence (its convex screen, curvilinear rectangular mask, and the extreme variance that exists between any two individual sets), the TV image has historically reinforced the notion that it is only a crude approximation of reality. Arguably in response to this crudeness and imprecision of the TV set/image, videographic production technology has evolved and focused heavily on developing modeling techniques that create the illusion of precise surfaces and objects. It is within this tendency of what can be termed "plasticity" that many televisual motifs can be related.

Plasticity includes four popular and suggestive video styling strategies: (1) molding, (2) pictorialization, (3) surfacing, and (4) objecthood. Once video imaging in the 1980s achieved a tolerable level of resolution, and the ability to adequately render film-style lighting, an increasing emphasis was placed on molding, twisting, and torquing the picture surface itself. This molding tactic is somewhat odd given the fact that, unlike painting, television has no tangible or discrete picture surface. There is a glass screen but the light sensitive phosphors are aligned in depth underneath its curved surface.[28] Yet, videographic devices have evolved that deny the possibility of reading images as either nebulous and immaterial or as illusionistic and having "deep space." One case illustrating the current urge to mold and stretch images is found in Rank-Cintel's new Ursa. Press releases boast that the new image manipulation device "can meld optical scan effects with techniques formerly reserved for digital video—prior to committing footage to tape. Here telecine product manager Peter Swinson has bestowed a new set of Picassoesque features on an otherwise normal model."[29] The invention of the Rank-Cintel flying spot scanner was a fundamental factor in television's developing ability to render heightened visuality. The Rank technology answered the long-standing need for a film-to-video transfer process that could approximate the rich tonal values and resolution of film for television. With the advent of Ursa, the makers of the landmark device (originally premised on the notion of realistic illusionism), gave media users the ability digitally to stretch, deform, and meld the surface of the image in acute and bizarre ways. This basic add-on device dramatically flattens out deep space, makes the viewer aware of the image as a surface, and then pretends to pull this illusory surface out at the viewer. Stretching and molding are signs of acute artistic control and technical bravura. The representational illusionism of space gives way to an illusionism of plasticity and malleability.

Such tactics are part of a broader strategy of pictorialization, that is, many stylistic devices within the mode of plasticity work by making the viewer aware that they are seeing pictures, not images of real things. This is analogous to the museum viewer who sees images of landscapes as genre pictures, in gilt frames, rather than as representations of the natural world. Viewers of genre paintings are no more fooled by the depicted space than are viewers of the new television. Consider, for instance, advertisements for Microtime's Genesis Act 3. Promotional copy touts the machines wide-ranging abilities to provide page turns, 2X Expansion, mirror images, and nine rotation axis.[30] The resulting image is not just malleable. With these devices, it becomes instead a kind of two-dimensional projectile that is infinitely twistable and moveable in space. In addition, the ad promises to allay producer anxiety now that the page-turn effect is finally available for less than $20,000. The industry does indeed codify its signs. But unlike the cultural signs that Roland Barthes had to e(value)ate through critical analysis in *Mythologies*, the televisual industry consciously places price tags and financial values on its own signs as signs.[31] Industry copy like this assumes that the reader senses a semiotic bargain at $20,000 for page turns.

Digital Services Corporation positions its variants of pictorialization in even

The representational
image becomes pictorial
artifact. (Microtime,
Genesis Act 3)

The warp option.
(Abekas Corporation)

more complex terms. It offers two types of page turns, plus scrolling (where the turned page rolls in on itself), concave and convex curved effects, automatic cube building, and their own picture twist[32] —stills taken from the device show landscapes wrapped onto the surfaces of a spinning cube and a twisting plane. Abekas corporation designates these same abilities, in its A-53D, as warp options: "Warp is a collection of shapes which allows you to bend and curve the picture to create warped effects such as, ellipse, page-turn, twist, roll, wave, zig-zag, split, burst, flare, slide and many more."[33] "Warp . . . bend . . . twist." Competing companies manufacture distinctive names to market their own versions of plasticity. Regardless of their patented names, however, the underlying urge in all of the cases is to take realistic deep-looking video images and to pictorialize them into material objects. Consider NEC's rhetoric in touting their DVE System 10: "Curl it, roll it, fold it or peel it. With NEC's DVE System 10, video images are as flexible as a page in this magazine."[34] More so than the others, this ad betrays the impulse toward materiality that fuels the graphic mode of plasticity. That is, the various plastic effects aim to create for the viewer objects that seem touchable, to create a parade of pictures that are turned before the viewer. This pictorial device turns the image into objects, but not into 3-D objects. Rather, images become 2-D pictures on physical surfaces that are thrown and spun through an artificial Bazinian space.

Other videographic effects achieve plasticity by inventing, displaying, and manipulating surfaces. Microtime does not just describe its mosaic effect as mosaic, but as variable mosaic *tiles*.[35] The reference to ceramic tile construction is common to many other design motifs as well. A television graphics software survey describes how sophisticated surfaces became the driving force behind graphics development: "The new look of chrome and glass was tied intimately to technological advances."[36] If the illusion of chrome and glass surfaces was the chief challenge in the development of video graphics, then more subtle surfaces and textures preoccupied later software developers. Abekas, "a leader in digital innovation," boasted that its A72 Character Generator could render bevels, neon, and embossing, and "drop shadow softness."[37] Advertising stills for the Dubner DPS-1 showed embossed/textured graphic surfaces under floating and digitized images of women with roses clutched in their teeth.[38] Once, in the mid-1980s, the mathematical modeling of reflected light, shiny surfaces, and transparent glass was made possible, a wave of frenetic, high-tech imagery followed. Witness any network title opening in a movie of the week presentation during that time, complete with reflective chrome and and flying glass surfaces. Ad copy under a still from AT&T's postproduction tool, Topas, proclaims this facility: "Reflective spheres floating over a pool of water. Chrome logos flying through space. They're easy with the right software."[39] The copy fails to point out that the illustration of these effects takes place in a digital atrium, with Romanesque vaults and Renaissance frescoes overhead.

Subtle shadings and surface renderings became an important way of discriminating one's proprietary graphic capabilities from that of other companies. Video surfaces by the late 1980s showed off inlaid and multicolored faux marble and stonework. Graphics devices and even character generators were able to create the embossing effect one associates with precious documents, handmade paper,

The female body as
ultimate surface for
texture mapping. (Post)

and notarized texts. Sophisticated televisual rendering, then, aimed to show a rich variety of surfaces whose textures were infinitely distinguishable. In their brochure, Pastiche gave potential users a kind of artist's primer that lined up various examples of handmade renderings and textures possible on their version of "digital paper."[40] The industry term for this process is "texture mapping," a technique that amounts to a kind of electronic inlaying of surfaces. Regardless of the specific form this inlaying took, the underlying preoccupation was to create a televisual world of textures. Photographic realism and cinematic space gave way to the materiality and texture of vellum and papyrus in these examples.

The painterly, the plastic, and surfacing tactics all seek to rigidify the amorphous world of television imagery into rigid material-like artifacts. Many video effects make this urge toward artifact status even more overt. One of the common bells-and-whistles options offered in television postproduction equipment is the ability to produce cylindrical, spheroid, and cubic solids from pictorial surfaces. Once boolean algebra was applied to solve basic technical and rendering problems (for example, how to conceal the hidden backsides of modeled solids, and how to move those models through pictorial space), then the three-dimensional modeling abilities of television became pervasive and popular.

The practice of flaunting flashy graphic devices in production and broadcast now stands for state-of-the art quality—an ideological myth in its own right. The practice of graphic performance tends, however, to resist analysis as content, since it comes across as an autonomous process based on the potentially endless permutation style and form. In actuality, a long chain of signs exists between the specific graphic signifier and the idealized connotation of state-of-the-art. It is worth examining this intermediate chain of references in more detail. What does it semiotically mean in television production to: (1) grab a videotaped image, (2) pictorialize that image into a surface, (3) wrap the videographic surface around a cylinder or cube, in order to (4) spin the resulting object in space? On the most basic level, this process is about *the television image itself consuming television images.* Television, in effect, performs the act of consuming images, and it does so for the benefit of the spectator. In this sense, television graphics have a built-in stylistic appetite for images. Because of this graphic appetite, images are transformed from the world of illusionistic realism into a frenetic world of spinning surfaces. Television is not just a succession of images or shots. It is a machine that consumes images within its own images. If one looks for a key to the paradigm shifts occurring in videographic forms of televisuality, this consumption and graphic appetite for new images may be a central factor.

Transparency. If the painterly mode functions to deform televisual imagery with human marks, and the plastic mode functions to objectify and pictorialize that imagery into surfaces and artifacts, then transparency, the third stylistic mode, aims to deny the frontality and coarseness of the television set's image. In the place of this apparatus-specific frontality and coarseness, many practitioners and graphics devices aim to create a layered look of depth. Notice that I am not talking here about a kind of depth dependent upon Renaissance linear perspective. Rather these images typically work by layering transparent images on top of one another. As in the graphic example from One Pass/KTVU that I first analyzed (see photograph, p. 138), the layering mode is frequently used to visually approximate emotional states. In a psychological as well as formal sense for instance, compare the "deep" feeling one finds in layered transparent imagery, with the shallowness of objectlike and artifactual images. Objects are impenetrable; depth seems engaging and accessible. Because of the psychological feel suggested by the thickened layering of images, transparency is presented as the most emotive, sometimes even romantic, variant of the basic televisual modes, as many music videos on MTV and VH-1 well attest.

A promotional diagram by Cross-Point Latch shows how layering and transparency are built into the very architecture of standard production switchers.[41] The image shows six image planes hovering over each other like aircraft stacked in their approach to an airport; the copy boasts, "six levels of video in a single pass." The specific stylistic option here, of "all six layers," is hard-wired into the switcher control board itself. The mode of transparency, then, is clearly prefigured and predetermined in the very design of this technology.

Television is frequently written off by critics, detractors, and film production people because it is flat. This perceived shallowness results both from the industry's traditional tendency to use flatter and more high-key lighting for studio video productions, and also from the popular notion that television directors should not compose scenes with deep, filmlike space. The irony in the widespread use of transparent stylistic devices is that a deeper space is indeed achieved, but it is not achieved through geometric construction or organization around linear orthogonals. Transparent layering suggests a hierarchy of image levels relative to the distance the image components are from the viewer: deep layers seem farther, shallow ones seem closer. The tactic, then, creates a sense of depth less from illusionism than from aerial proximity and atmosphere. This distinction might be best explained by reference to a classic art historical dichotomy. Transparent layering works to create depth in the same way that aerial perspective and sfumato worked for the painters like Da Vinci. Depth through aerial, rather than linear, perspective, and the use of gradations of haze and smoke in the horizon, also characterized the Venetian school of painters of the Renaissance.[42] Whereas the second televisual mode of plasticity (of digitally creating three-dimensional solids) follows the tradition of mathematical, linear perspective, the mode of transparency depends upon a more generalized and atmospheric illusion of depth.

1989 fall season promo for ABC by the Post Group. Layered transparency equated with hypersubjectivity and the network family. (The Post Group)

By using the kind of atmospheric perspective found in layered transparency, video practitioners now challenge one of the commonplace assumptions about the medium's limitations. They overcome the inherent coarseness created by fields of electronic pixels, or scan lines, by creating images that seem to be layered in depth below the glass surface of the screen. The reasons that televisuality seeks to overcome the image's shallowness is of course a bigger question. A key to the question, however, may lie in the tendency of such images, because of their visual complexity, to suggest psychological depth or subjectivity. In fact, many title sequences and program openings that now use layering do so in a way that invokes their characters' personalities, feelings, and subjective states. In this way, transparency works to counter two clichés about television. One is the perceived coarseness of its imagery, and the other is the assumed shallowness of its characters and fictions. The mode of transparency is, then, more than just a state-of-the-art claim. It also frequently subjectivizes the experience of the flow and program for the viewer. This constructed subjectivity can also help personalize the otherwise anonymous nature and mass quality of the station or network.

Intermedia. Utilizing Julia Kristeva's term, various theorists have noted that intertextuality is favored in postmodernist practice. Since I have argued that television creates images that consume images, the importation of stylistic motifs and contexts from other art forms is a crucial issue in the study of television. Because the new television seems infatuated with other *visual* art forms (photography, film, modeling, performance art) rather than literary ones, I am choosing the term "intermedia" rather than intertext to describe this (nonliterary) mode. Television is rife with intermedia flourishes. The opening to the daytime soap *The Bold and the Beautiful*, for example, is shot and posed as fashion modeling photography. Beer ads on television show jumping film footage and sprocket holes along with the images inside the film gate. Dean Witter insurance ads fake grainy, high-contrast, scratched black-and-white footage to suggest turn-of-the-century values and commitments. Obsession perfume ads pose as restrained, statuesque, and classically Greek tableau. Television seems to want to be anything *but* television, to be anything but its own unique medium.

Editel of Chicago post-produced the commercial spots for Pepsi-Cola's new Miranda soft-drink. In addition to the obvious technical wizardry involved in

The camera-eye of Pepsi's spectator/consumer. Optical facticity anchoring the world of consumer fantasy. (Editel)

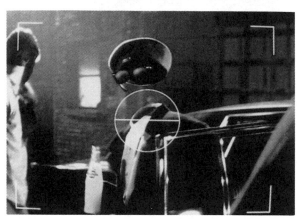

animating dramatic characters that had no visible bodies, the spots depended heavily on references to the photographic medium on several levels. One visual device used throughout the spot—SLR-like framing and focusing marks within the televised image—was pervasive in programming of the 1980s. Like *Max Headroom*, and *Robo-Cop* before it, Pepsico frames its vision of the new world of Miranda with a device linked to precision and factual documentation. A split-focusing prism dominates the center of the image. Cropping marks for "safe-area" fill the corners. In a commercial spot about *fantasy*, the producers use visual devices instead that signify great *facticity*. Optical and analytical marks help create the sense that the photographic medium is striving to document and contain this spectacle. A later shot reduces the spot's mise-en-scène to a Polaroid SX-70–like instant still. In the final sequence the visionary world of Miranda simply sits on the concrete pavement, a mere photographic record of the earlier spectacle. By appropriating optical devices from other media, television here shows itself struggling to get a bead on the visionary realm. The viewer is left with a thrown-away, but highly mediated, visual document of the past.

Commercial spots also appropriate highly visual and bodily performance modes from other media. Polycom productions of Burbank used "five layers of fire," the Harry digital effects device, and the Paintbox to approximate the steamy sexual atmosphere of *Dirty Dancing* in their spots for Coca-Cola.[43] In this example a multitude of televisual devices are used: painterly deformation, graphic surfacing, and transparent layering. All of these tactics are used to appropriate the sensibility and style of the earlier hit film and to fuse it with Coke's *own* sensibility

Table 5.1

Televisual Mode	Media Limitation to Mitigate	New Ontological Relationship
Painterly	Linear construction of image; scan lines; contrasty outlining; fall-off	The human hand, mark, and residue in the deformed image
Plasticity	The ambiguous image; the immaterial, nebulous phosphors	The pictorial artifact; the image as object, surface to be known and consumed
Transparency	Image coarseness; fields of discrete video pixels; frontal, convex screens	Psychological "depth" through layered imagery; hypersubjectivity
Intermedia	The isolated TV set; the picture box cut off from culture*	The expanding culture box; televisual imagery as surrogate that consumes images for us

* Obviously, no commodity in a consumer society can be cut off entirely from culture at large in either an economic or *ideological sense*. What I am referring to here is the pervasive sense that many viewers watch TV in apparent isolation, alone in the private sphere of the home. The sense of this *physical isolation* is constantly covered over and denied by the excessive flow of cultural imagery that issues from the TV—an image-eating apparatus that becomes a consumer surrogate, a mechanism of collective identity, and a means of virtual travel.

and pose. This hybrid televisual operation, in effect, creates a videographic soup in which the bodies of performers move, sing, and lust for each other. Again, the apparent aim is to create a kind of interpersonal subjectivity and emotive depth that one does not normally associate with mass-produced soft drinks or marketing.

One thing all of these examples have in common is that they work to extend television's arms out into popular culture at large. In this sense television, with its appetite for cultural images, is an expansive medium, rather than an inwardly preoccupied one. Televisual operations work to deny the very media-specificity that modernists idealized. Through intermedia and pictorialism, television becomes a boundaryless image machine, gobbling up any cultural visage that hesitates long enough to be abducted. If as I have argued, television favors images that are specifically about consuming images, then the intermedia mode is a key strategy that works to satisfy the medium's appetite for and consumption of imagery.

Summary of Basic Videographic Modes

Table 5.1 summarizes some of the key tendencies of the four televisual modes examined thus far.

In fundamental ways, all four of these televisual categories are formal *acts of denial*. All work to overcome a number of perceived limitations in the technical apparatus itself—linearity, immateriality, coarseness, and the physical isolation of the viewer. The stylistic and televisual process of denial operative here suggests that the industry works to retool its contextualizing frames, to overhaul and update the paradigms by which television is produced and consumed.

The economy in a mass consumer society is driven by the formation of need and perpetual desire. As a chief site of brokerage in consumer society, television analogizes and replicates desire. Like the ideal subject-consumer, television is not satisfied with its own unique qualities, aesthetic traditions, or technical limitations. Not only does television sell the new and the needed as products, it presents itself as constantly new and needed. In order to pursue this stance, practitioners stress innovation and stylistic distinction at all costs. All of the televisual modes examined here, and their many volatile manifestations as discrete video effects, constantly work against aesthetic continuity and stylistic stasis in the medium. Although televisuality may look merely like a performance of excessive style on one level, it also attempts to create visual analogues of feelings, products, surfaces, artifacts, and material pictures—representations of the stuff of mass culture itself. All of the televisual devices discussed here are alike in *showing off* their proficiency at *picture making*. The overt and explicit nature of this kind of imaging, then, challenges apparatus theory in both its televised and cinematic variants. Rather than the "window on the world" concept that was so important in the early years of television, contemporary televisuality flaunts "videographic art-objects" of the world; rather than the concept of the cinematic "fiction effect," a psychoanalytic notion premised on the viewer's need to deny the apparatus, televisuality flaunts the digital apparatus.[44] There is no attempt to deny the video picturing process in the new television. Rather, the objectification of the televisual apparatus is dramatically evident in its appetite for the pictorial artifact, surfaces, and images. The *new television does not depend upon the reality effect or the fiction effect, but upon the picture effect.*

Models That Order and Perform the Videographic Modes

Debates are waged in television study over the relative appropriateness of broadcasting versus cinema as models for the analysis of television. Academic institutions of media, such as the Broadcast Education Association and the Society for Cinema Studies, tend to be strongly polarized in their interests and methods for television analysis. It is clear when looking at current activities and discourses in the television industry itself, however, that programming is being theorized neither along the broadcast-communications-effects model of the BEA, nor on the cinematic-narrative-ideology model evident in much SCS work. Directors and designers, especially in the new postproduction environment, have set their sights on very different foci and concerns, specifically, on the play of visual style and excess. The basic contextualizing paradigms under which practitioners make television is being overhauled, in the world of digital video graphics.

The industrial acts that reframe television, however, are not all part of a singular or unified process. In industry marketing, for example, television might be conceptualized as surfing, racing, or test pilot aviation, while very different models are being introduced and utilized by television manufacturers and equipment operators. The above-the-line people (directors and producer types) tend to opt for

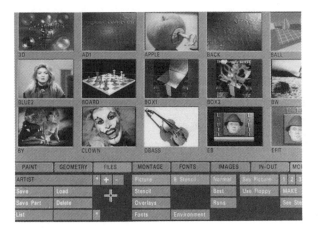

Production via video
library and visual card
catalogue. (Pastiche)

physical sport and the body as metaphors in promotion. This choice is evident in
the ways that independent production companies, postproduction houses, and blue
chip agency directors are marketed. Below-the-line staffers on the other hand (engineers, editors, and designers), go for more controlled models, with task- and
style-specific objectives. This difference recurs in equipment design and the industrial discourses surrounding new technology.

Such tendencies are evident in a recent ad for the Delta 1 Graphics Production
Tool. The ad copy boasts that "it's not just another video typewriter."[45] Set off
against an Edward Weston–like high-art photograph, the promotion suggests that
the Delta 1 is the "highest quality" ticket to creativity available. The historical
distinction and imperative is clear. Old television was bound to the verbal, linear
methods of the typewriter. Today's tasks, on the other hand, demand creativity,
refined tonal control, and great visual precision. Given these sorts of claims, one
might ask what new paradigms are being offered as substitutes for those models
bound to recording, broadcasting, and verbal typing? Certain devices, for example,
suggest that the pull from the film industry is still strong. A postproduction unit
named the Videola makes a direct reference to the Moviola—a traditional, upright, film-style editing machine favored in classical Hollywood. Other devices
and paradigms more quickly distance themselves from film. Consider, for example,
the Paintbox (posed as a high-tech art toy) or Harry (postured as a creative, non–
effects specific personality).

While the televisual modes I described earlier group specific videographic flourishes according to -emic categories, the paintbox, the Harry, and the Videola function as higher order models that contextualize and orchestrate those modes during
production. A brief survey of recent technical literature shows a number of contestants vying for dominance at this second order level. Consider, for example,
the design and layout of the controller for the Pastiche graphics device.[46] In both
its physical orientation and marketing the device presents itself as a library, or
video archive of electronic pictures. The controller, in essence, is more like a visual card catalogue than an editor's keyboard or controller. This design encourages producers to use Pastiche for digital archiving, image recall, and subsequent
graphic manipulation. The machine promises limitless variety and an expansive

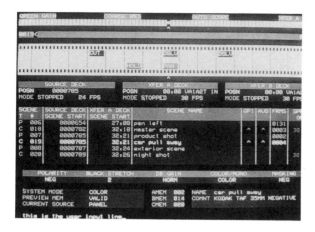

Film-based electronic
interfaces and the sync-
block paradigm.
(DaVinci)

digital *storage of pictures, not footage*. Again, television is fractured into pictures and indexed as stills.

At a basic level, each model of video equipment theorizes, since its design presupposes a specific kind of video task. Many devices demonstrate the burden that television still feels toward its prestigious media sibling, cinema. Take for example the DaVinci color graphics device. Its controller is laid out not as a library catalogue, but in the visual pattern one normally associates with film cutting rooms. In fact, its design is associated directly with the now-archaic film sync block. Even though electronically recorded images and commands are *stored* digitally in fields, they are nevertheless *displayed* analogically and syntagmatically, as if they were *spliced* together into linear footage. The visual image of film footage in television postproduction suggests the relative power that this particular model has for making sense out of the increasingly complicated electronic postproduction environment. The sync-block paradigm displayed on the surface here masks the complex interconnectedness of various devices throughout the postproduction suite. Unable to utilize the single, shared steel axle of the traditional film sync block during playback (an utterly simple device that rigidly ties film picture and sound production elements together in sequential time during viewing), this device and display works by electronically "slaving" various devices and their stored fields together, in order to simulate film footage in playback. The film-cutting illusion comes complete with simulated sprockets and edge numbers. By contrast, the high-end post-production controllers of the late 1970s and early 1980s were designed and engineered around lengthy lists of numbers and cues. The task of crunching lists of video time-code numbers in the other and older postproduction systems makes conceptualizing projects in terms of scenes and style difficult, since the user's aesthetic task has been relegated to the numeric realm. In stark contrast to the disorienting quantification required by older postproduction systems, devices like the DaVinci are attractive given the illusion of visual immediacy, simplicity, and directness that one associates with the visually simulated sync block. It is a look and design done more for conceptual clarity in the edit room than for improved recording quality. The technical quality of video equipment modeled after the sync block or film interfaces is not, obviously, necessarily any higher than that of other systems using numeric time-code lists. The promise, of the film-based paradigm, however, is that its design and interface is more user-friendly.

A third ordering and orchestration paradigm is influenced both by notions of artificial intelligence (AI) and therapeutic discourse. It is less tangible but no less popular than either the library-archive paradigm or the sync block–film footage model and is finding currency in technological discourse. Industry rhetoric now commonly presupposes an emoting and creating individual within the design of its technology. Although not as immediately apparent in machine architecture, the idea of an intuitive and emotive interface with technology is a notion driving many developments in the industry. Quantel corporation's Paintbox V Series is clearly pitched to users by appealing to intimacy. The machine comes equipped with an artist's stylus and a graphics pad rather than a keyboard. Incredibly, Colorgraphic Systems touts their machine's ability to provide a "programmable personality."[47] Not only do the new devices encourage creativity, but systems like Colorgraphic's actually claim that users can (by controlling technical parameters and defaults) assign design and stylistic personalities to the device itself! By its design and self-definition, the machine asks to be engaged and embraced by users as a *surrogate personality*. Clearly, then, the culture of simulation is not limited to Baudrillard's Disneyland. Engineers work toward a similar state-of-the-art ideal of simulation and surrogacy.

A recent trade account describes the new trend in the commercial television industry to team directors with editors at the start of commercial projects, rather than to segregate their work into separate production phases as was commonly done in the past. References are repeatedly made to the current need and demand for sensitive and intuitive artists who can work creatively together. This change is argued to be essential, given the fact that new and more radical genres have replaced traditional and less style-oriented approaches: "The proliferation of non-linear, contrapuntal image spots, which lessen the importance of the traditional storyboard, has created an agency warming trend allowing informal director/editor teams to develop with increasing frequency. . . . What makes for a blissful union between directors and editors is a mysterious blend of trust, shared sensibilities, and overlapping interpretations of the material."[48]

Such accounts make effective production seem more like sexual union or matrimony than work: "a mysterious blend of trust," and "shared sensibilities" create "blissful unions." This is not the rhetoric one associates with television engineering or preparation for the FCC exam. This is psychotherapy in the production suites. Informality and intuition are positioned as valuable production traits. Spontaneous video interpretation in edit suites is here favored over storyboards, precision, and planning. Directors are now marketed as *mysterious*, *creative*, and *emotive* individuals. Clients, on the other hand, have the opportunity to invest in personalities not plans; sensibilities not scripts; attitude and not content. This opportunity for self-actualization and therapy is offered to clients, ironically, along with significant but less visible costs. Rate sheets for postproduction suites typically and discreetly offer state-of-the-art services at between $350 and $500 per hour. This conceptual shift is a major change, one that reifies and constructs an emotive persona in the televisual apparatus. It is predicated on the psychological need for visual and aural change. The emoting-persona paradigm becomes a ghost

Table 5.2

New Ordering and Performing Models	Concept of Form/Focus of Transformation	Style-Implication
Video as library ex: "Pastiche" ex: "Harry" ex: still-stores	Views footage and imagery as *digital archives* (spatial)	Penchant for accumulation, recall, and highly dense manipulation of stored still imagery
As sync block ex: "Da Vinci" ex: "Animator Studio" ex: "MatroxStudio"	Forces imagery and time code data into look of *film footage* (temporal)	Visually stimulates linear footage out of video fields; thinks in scenes, not shots; program malleability
As emotive persona ex: "Paintbox V-Series/ Quantel"	Views postproduction workers as *visual artists*, not linear conformers (intuitive)	Transforms high-tech highly capitalized environment into impulse-driven and user-friendly surrogate emoting subject
As desktop ex: "Avid" ex: "Media 100"	Manages video-audio pictures like *cut-and-paste layout* process (entrepreneurial)	PC revolution boasts implosion of TV industry and personal mastery of segregated crafts. Teases cult of professionalism, but actually diverts massive film/TV equipment capital to digital storage manufacturers.

in the machine that permeates and motivates television's stylistic break toward excess and increased visuality.

These three recent and popular ordering and orchestration paradigms, along with "desktop" video—a marketing breakthrough that construes video post as a form of entrepreneurial, cut-and-paste graphic design work—can be summarized as follows. Thought of as *theorizations in practice,* they compete with other designs in popularity for the role of dominant model in the development of new television technology. The masters of ceremonies in the industrial contest are market share, sales, and profit margins.

These examples suggest that the industry manufactures syntagmatic and semiotic systems like it produces printed circuit boards. Each company, in essence, claims to have its own method of getting from image A to image B in a production sequence. The extent to which the corporation persuades end users of the uniqueness of its method is the extent to which the company succeeds financially. Formal and stylistic transitions are, in effect, patented alongside electronic controllers and interfaces. The *burden to interpret* that new equipment design bears means that semiotic and aesthetic systems are invented and marketed along with any new production tool. Effectively promoting and perpetuating these interpretive models increases the likelihood that the use of patented and proprietary effects will follow on a wider scale. Market permeation by a product (and its unique effects) insures acceptance of an allied semiotic system as well. Marketing and

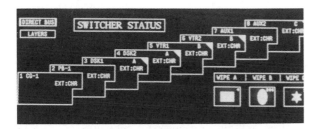

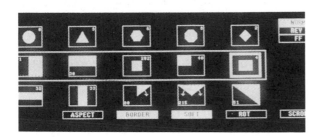

Video's hard-wired grand syntagmatique. (DSC/Chyron Corporation)

sales, therefore, cannot be separated from signification.

An analysis of television's ideational framing, its technical myths, and its practitioner discourse, shows that the medium is not just about programs and products. Television is a system of signification that is fueled by the manufacture of systems of signification. It is, as well, highly self-conscious about this skill and facility; about its ability to invent new systems of meaning. The televisual industry is quick to point out this semiotic capability in print and promotion. Television has become, in its stylistic pace and technical development, self-permutating. Its performance of style also endlessly transforms the meaning and social significance of signs. This cultural dimension of televisuality merits additional attention.

Rethinking the Televisual Apparatus

This study has examined only one of several favored categories of televisuality in American television—the videographic modes used in postdependent genres. Yet, the videographic modes discussed here operate in a substantial portion (15–20 percent) of each cable and broadcast hour and perform even within and around those program genres that use cinematic or minimal in-studio forms. The percentage is even higher when the programs themselves are videographic in nature. It is clear then, that televisuality challenges the *centrality* of several dominant mythologies in high theory, including glance theory and the ideology of liveness theory.[49] One of the most effective applications of the glance concept has been put forth by Rick Altman, who argues that the viewer's inattentive relationship to the low-resolution television image has helped position the use of sound as the defining aspect of television.[50] Unlike the high visual resolution found in cinema, for example, video's vague imagery forces broadcasters to fall back on an array of sophisticated audio tactics aimed at eliciting engagement with viewers. Altman describes these modes as labeling, italicizing, sound hermeneutic, internal audiences, the sound advance, and discursification.[51] While Altman shows that these tactics are logical extensions of what Raymond Williams terms the "flow" in broadcasting, he ignores the fact that these *six tactics are also operative within television's visual imagery* as well.[52] Extremely complicated visual appeals, contextualizations, *and interpretations* are made within images by images. The excessive manipulation of pictures, graphics, image-text combinations, and 3-D surfaces constantly codify and suggest to viewers that the imagery is produced just for them. Even at the time Ellis was writing, videographic televisuality had begun to perform and encroach upon programs and their diegetic space. Yet critical theory continues to ignore this practice in order to follow the logic of master paradigms flow and glance. Clearly, production practice shows that the image has not acquiesced to any inherent low-resolution nature of the sound-driven video image. Far from it. There is, in fact, an obsession with making images that spectacularize, dazzle, and elicit gazelike viewing.

Despite the specific differences among the televisual modes and ordering and presentation models, they all participate in one underlying dynamic: they self-consciously work to parade and perform visual style before viewers, and do so across a wide range of television programming, from network broadcasting to specialized cable channels. Why television does this, of course, is an important question. Faced with various theoretical determinisms of the past and reductivist effects models that constructed a passive audience, critical theory has reacted by trying to activate the audience in recent scholarly work.[53] We have, however, been so desperate to construe an active audience capable of counter-readings that we have tended in critical theory to forget or to overlook the capital-intensive power of the media, its production practice, and its technological base, all of which are important parts of television style and of the televisual apparatus. Has our disengagement with the industry in favor of studies of fandom and popular discourses come at a cost? Surely industrial and aesthetic concerns are not necessarily the most important considerations in media theory, but they are at least factors that affect how and what viewers consume on a daily basis.

An important part of the legacy of Jean-Louis Comolli and Jean-Louis Baudry was their demonstration that centuries-old ideologies of representation informed and inscribed the development of technologies of representation.[54] Clearly, the aestheticization of television that I broach here is also part of a broader trend in American culture, where artistic styles have been suburbanized, proliferated, and mass-marketed to middle-class consumers. The kind of distinction that manufacturers seek in designing and marketing trendsetting machines that flaunt style, mirrors the pervasive preoccupation with style in culture at large. Pierre Bourdieu has shown how the performance of stylistic competence is actually a mechanism for power and mobility throughout social groups and classes.[55] Yet, despite the clear relationship between ideological pressures and technology-style, Comolli overestimated the extent to which ideology could determine technological development. Among those criticizing his view, Brian Winston demonstrated the fundamental importance of other factors needed for technological, and therefore stylistic, innovation—specifically the issues of scientific competence, technical performance, and ideation.[56] In the late 1970s, for example, the first page-turn transitions in editing were made available. The function of the transitional device made little sense to many practitioners and editors, however, except as a flashy way to get from one image to the next. After all, by what logic and dominant ideology were electronic images construed as books? On the other hand, many of the televisual modes that I describe above fit perfectly with the kind of cultural capital that art-school grads brought to the production industry in the 1980s. Their aesthetic expectations and skills contrasted dramatically with those of their predecessors and colleagues from broadcasting or film school backgrounds. Although it is impossible to fully treat the question of determinations here, these examples suggest that excessive televisuality arose as a combination of internal and external factors—from fetishized technical experimentation to the mythologies of deregulation and consumerism that fueled the Reagan-Bush era to the cultural and educational capital brought to the industry by practitioners.

An extensive treatment of the politics of televisuality is beyond the scope of this chapter, yet it is important to note that the kind of stylistic performance and videographic packaging that I describe here was active, even frantic, in two highly televised social crises. Within minutes of origination in the Gulf War and the L.A. rebellion (the subject of the final chapter in this book) even raw, amateur videotape footage was crunched into highly stylized videographic configurations. Scenes of reality, chaos, and suffering were immediately rendered as pictures, reflective surfaces, and flying text-image projectiles. Social trauma and rebellion were turned into artifice. The face of the dangerous other was extruded into artifacts of the known. The response of critical theory should not simply be to revert to the iconoclasms of high theory, and to write off the televisual preoccupation with style as enemy. Rather, we should see and describe how power is wielded and represented through televisuality. Otherwise we simply revert to the safe havens of our own favored academic paradigms and discourses, narrative and realism. Televisual style is no more or less regressive than narrativity—both are *social constructions* of great complexity and both merit critical engagement. The centrality of digital videographics in contemporary television—a cost-effective, optical quality control that rotely packages political threats and programmed filler alike—means that ignoring the practice is also a kind of social disengagement.

6   Loss Leader
Event Status Programming / Exhibitionist History

We will plug the spend-spend-spend "black-hole" of the mega-miniseries.
—ABC/ Cap Cities Management, 1989

We're going toward more event status programming in addition to our regular series. The network is looking to capture one-time-only, once-a-year kinds of events . . . and plans to roll out the monthly blockbuster strategy in fall 1994. We want the network to have a sense of urgency to it. After a series has been on a few years, it's a little bit harder for the public relations mechanisms to be as excited about it. We want to keep the network invigorated.
—Paul Kaiser, executive, the Family Channel, 1993[1]

Even as relieved networks gladly proclaimed the death of the budget-busting epic miniseries in the late 1980s and early 1990s, many new competitors were trumpeting the very same programming approach that the miniseries had proven viable: event-status programming. Throughout the decade, the networks had made the miniseries a regular part of each new season's spectaculars and specials. Epic miniseries, like *Shogun, North and South, Peter the Great, Amerika, The Winds of War,* and *War and Remembrance*, comprised but one genre in what has come to be known as event-status programming.[2] Other variants include special movies of the week, or MOWs, like *The Day After, In the Line of Duty: Ambush at Waco, and Amy Fisher: My Story* (the last, but one of the MOWs simultaneously produced by CBS, NBC, and ABC on the infamous Long Island Lolita/Joey Buttafuco statutory rape–attempted murder case.)[3] Primetime soaps like *Dallas, Dynasty, Hotel,* and *Knots Landing* also regularly offered opportunities for special episodes that promised event status: year-end finales, cliff hangers, the introduction or expulsion of regular cast members from the series. Together with network coverage of international sporting events like the Olympics in 1984, 1988, and 1992, the World Series, and the yearly Super Bowl ritual—an event recognized as much for its initiation and showcasing of new advertising spots as for sports coverage—these shows all share a number of common attributes. They all require massive investments of budget and production resources; they all are sequestered and marketed in primetime; and they all bring with them significant corporate risks.[4] The high production values that pervade each of these event-status program forms—a factor that inevitably places style center stage—also makes financial loss an ever-present possibility.

Economic clouds across network horizons had, by 1989, cast a long shadow on the programming value of epic televisual events. The trades headlined the fact that ABC lost $20 million on *War and Remembrance* and $64 million on the 1988 Olympics. Cap Cities boasted that it would plug the "spend-spend-spend 'black-

Primetime's casts of thousands. *Marco Polo* and *Jesus of Nazareth* on NBC in 1982 and 1984. (Terrence O'Flaherty Collection, Arts Library—Special Collections, UCLA)

hole'" of the former ABC management.[5] Cap Cities assured the press that another mega-miniseries on the scale of *War and Remembrance* is out of the question because it is too risky. According to the network, not only did the miniseries put the network on the verge of bankruptcy, it also completely missed the network's ideal viewing audience: "The over-forty-year-olds loved it, but nobody under forty watched it."

Yet this eulogy proved premature. Even the venerable PBS network, never known for programming flash, aped the event-status prototype by airing Ken Burns's multipart documentary series, *The Civil War*, on successive nights during the start of its fall season in 1990. Wildly successful ratings proved that the obsession with event-status programs did not die in flames with the *War and Remembrance* "failure" in 1988. Critically marginalized but widely viewed cable entities like the Family Channel—a mainstreamed outgrowth of Pat Robertson's new right Christian Broadcasting Network—now make articulate public defenses of event-status programming as a key to their corporate and marketing morale. Even cable networks like CBN, then, whose defining, cash-cow genre has always been the talk show, have been born again. In breaking through the stylistic confines of the bread-and-butter talk-show formula, there is, apparently, now bottom-line wisdom in special-event programming. Finally, for the fall of 1993–1994, NBC released not one, but two new miniseries under the umbrella title Great Escapes. The first, *Trade Winds*, was set on exotic St. Martin, the second, entitled *The Secret of Lake Success*, utilized the same kind of extended family that made *Dallas* and *Dynasty* famous, but traded in Ewing oil for Fortune 500

Emulating and fabricating big-picture icons in primetime's *From Here to Eternity* and *Shogun*. (Terrence O'Flaherty Collection, Arts Library—Special Collections, UCLA)

pharmaceuticals. To the guiding creative executive at NBC Productions—apparently ignorant of the myth that reality programming ruled—the time was right, not only to bring back the glitz and the lust, but to "push past the formulas" of the miniseries.[6] Even the reincarnated primetime soaps of 1993–1994 were being pitched as viable because they were stylistically innovative.

Event-status programming, then, has regained its clout as a hook for bringing aimless viewers back into the tent. Given the financial risks of high–production value programming, event status shows function, for cable and broadcasters alike, somewhat like retailing loss leaders. Even if showcased spectacles—miniseries, MOWs, primetime soaps, and international sports—lose money because production costs outweigh spot sales, they typically provide marquee points of entry that entice viewers to sample more commonplace products in the programmer's store. Sometimes the value implicit in such shows is more widespread. Miniseries can, for example, create network identities as high-profile banner carriers intended to rally the troops. With one eye always looking over the corporate shoulder, financially anxious affiliates and fence-sitting independent stations that buy in syndication constantly need to be reaffirmed, to have the distinctive and original nature of the package sent down to them through the network pipeline underscored. Loss-leader shows provide regular transfusions of attention-getting marketing fodder for network seasons, and for sequential strips aired each evening on affiliate stations. Where would the generic Sunday television guide in any one of thousands of local newspapers be, for example, without the loss-leader showcases? Such events—fabricated as they are—provide prepackaged cover stories for each new fall season and for each of the seasonal sweeps.

In other cases, however, televisual loss leaders are designed for more immediate and explicit financial returns. Commodity tie-ins, for example, are not limited simply to much ballyhooed theatrical blockbusters, but pervade the televisual industry well, which is no real surprise since television was birthed by advertisers and sponsors in the first place. The growing power of non-network production entities has meant the return to a type of sponsorship reminiscent of the late 1940s and early 1950s, a situation where highly visible sponsors dramatized both their shows and themselves on television. Rather than simply selling ad space on network-controlled program time, companies like Tribune Entertainment have now created Target Marketing Subdivisions that bypass traditional coop funding entirely by opting for value-added retail tie-ins. Under this system, syndicated event programming is, from the start, self-consciously designed as a device that both markets a retail product and that is marketed *by* products in point-of-sale promotional campaigns. In 1993 this kind of two-way promotional scheme was used by Sprite soft drinks in its Soul Train Music Awards sweepstakes campaign, and by mega brewer and (apparent) civil-rights heavyweight Anheuser-Busch on *The Apollo Comedy Hour*'s sponsorship and promotion of Black History Month.[7] Televisual loss leaders are not, then, limited to the pantheon of Aaron Spelling and miniseries heavyweights like ABC Circle Films. Event-status programming has also become a widely discussed ploy in fairly mundane consumer goods merchandising. Televisual events programs are not therefore just *symbolic* loss leaders for network programmers. With mass-market patrons like Budweiser and Sprite, they have become *material* merchandising displays as well, as is evident in the 7-Eleven convenience store that anchors any local stripmall.

Academic theory should seriously reconsider the importance and economic centrality of special programming, for loss-leader televisuality undercuts at least three privileged theoretical orthodoxies. First, the pervasiveness of special television poses problems for those theorists like Stuart Kaminsky and John Cawelti that explicitly justify and celebrate television because it is a reflection or projection of vernacular and popular culture.[8] As we have seen, television frequently *denies* its origins in popular culture, by showing off its constructedness as both a spectacle and as a headline-making event. Second, this showcase form challenges those critics who base their theories on television's essential mundaneness or day-to-day regularity.[9] Event-status programs regularly masquerade as original, distinctive, and important events. Third, and finally, loss-leader televisuality undercuts the large body of critical scholarship that developed analytical methods from Williams's flow theory—a view that presupposes that the *television text is "boundless" and without borders*.[10] Loss-leader events programs make every effort possible, in fact, *to underscore and illuminate their textual borders*, no matter how big those borders are. The bounds of distinction are in fact a crucial part of the genre. As a programmer's marquee, loss-leader televisuality attempts to define and show off its silhouette, to signal that it sails *above* the mind-numbing flow. Unlike literary and cinema studies, television theory has never had a great-books or auteurist canon to contend with or to react against. As a result, television theory's sometimes uncritical celebration of the anonymous, the mundane, and

The Day After on ABC proved to be the prototype for 1980s event-status programming—a primetime spectacle about global annihilation. Epic history convulsed through personal lives in countless MOWs like NBC's *Secret Weapons*. (Terrence O'Flaherty Collection, Arts Library—Special Collections, UCLA)

the boundless as defining qualities of the medium have made theory blind to one of television's central components: "important" televisual events spectacles air like clockwork in any season's new line of programming.

Volume Discounts: History-Narrative
Exhibitionist History

Event-status programming—including the miniseries, special episodes of primetime dramatic series, and retrospective specials—tends to be preoccupied with two recurrent obsessions: the aura of history and the epic possibilities of large-scale narrative. If designer televisuality celebrates excessive intentionality and sensitive relevance, and franchise televisuality hypes embellished quality-control through digital packaging, then loss-leader televisuality utilizes and exploits two other forms of presentational fuel: excessive narrative and historical exhibitionism. History attracts and validates viewers; excessive narrative anchors and secures viewership across longer periods of time. Even as the miniseries and other loss-leader forms rip narrative discourse from conventional plot- and character-oriented moorings, they also rip history from any connection to the real world by endlessly ritualizing its formal permutations. It is important to note that television has, in fact, always exploited history: to justify its breakthroughs, to pro-

One of many shows that created exhibitionist historical flourishes in primetime. (NBC)

claim its import, to attract viewership. But historicity—the disembodied signs of history rather than history itself—helps loss-leader programs as well, by intellectually elevating the televisual marquee above the banal flow.

The emergent penchant for excessive historification evident in recent event-status and marquee programming can be seen as part of three institutional tensions: as a reaction to the dominance of reality/talk programming; as a symptom and construct of the mythology of liveness; and as a *mass-produced textual therapy* for what has been described as the crisis of collective memory. The *performance of history* has become increasingly popular in marquee shows like *Quantum Leap, Northern Exposure, The Wonder Years* and in the now-obligatory special event status and historical exhibitionism of series finales like *Knots Landing* and *Cheers*. The formal guises and the political implications of these self-conscious presentational rituals are, however, also as glaringly evident in embarrassing low-culture historical ecstasies like *Bradymania* and *Legend of The Beverly Hillbillies*.[11] In specials such as these, the viewer is inevitably invited to have a seat and "do some history" with the show's producers and celebrity hosts. "C'mon down, viewer," Televisual History 101 is waiting.

Event-status historicity—the televisual ritual of historical embellishment—can also be understood by noting several historical, generic, and theoretical comparisons. By way of historical comparison, George Lipsitz argues persuasively that 1950s television reinforced mass hegemony by creating "collective amnesia," "a loss of memory," and a "crisis of historical consciousness."[12] The family sitcom gradually elided both class-consciousness and ethnicity through shared dramatic operations. This gradual and generic erasure of race, ethnicity, and class specificity fit well the need for American cultural homogenization during the cold war era. Yet exhibitionist programs from the 1990s appear to do just the opposite. That is, they ask the viewer to engage in historification and ethnic-cultural identification in explicit ways, and they do so by invoking realist positions, guises, and artifacts—tactics completely absent in the sitcoms that Lipsitz uses.[13] It is worth considering the ideological implications of this reversal in textual strategy, for the change is symptomatic of important differences in mass culture in the 1950s and the 1990s.

Historical exhibitionism now frequently parades, rather than effaces, ethnicity.

Touted in mass-circulation ads as "must-see" programs—a sure sign that programmers are working hard to fabricate event status—prestige and loss-leader programming in 1992–1993 included a wave of episodes that self-consciously dealt with ethnic and specifically Jewish issues: *Homefront*, about a Holocaust survivor; *Picket Fences* about Jewish burial laws; *Northern Exposure*, about the importance of Jewish religious observance in a Gentile culture. *Seinfeld, Mad About You, Murphy Brown*, and *Love and War* all perpetuated the model of the Jewish comedian in urban-hip sitcoms. Some Jewish representatives complained about the superficiality of this trend. "There's a lot more to being Jewish than bagels," remarked the head of the Simon Wiesenthal Center.[14] Yet these prestige shows did not always flaunt the same shallow and one-dimensional ethnic clichés that audiences had grown accustomed to. Despite the ironies and hipness that pervaded these shows, there was a certain *earnestness* with which they were teaching Americans about Jewishness. On *Northern Exposure*, Dr. Joel Fleishman unequivocally tells his Native American assistant (and the mass audience), "I'm not white. I may look white, but I'm not. I'm Jewish. Jewish. A fellow person of color. A victim of oppression."[15] Even as the show toys with Joel's upper-middle class Republican values, it denies his whiteness in order to create an ethnic victim on par with Native Americans. Far from the suburbanizing, deracializing 1950s sitcoms that Lipsitz critiques—*Life of Riley, Mama*, and *The Goldbergs* gradually elided issues of ethnicity and class—1990s television flaunts the subtleties of race and ethnicity. Marquee television frequently invites the viewer, via historical exhibitionism, to imagine and celebrate ethnic origins and difference. American mass-culture now imagines itself not as a consensual melting pot, but as a hodge-podge collection of different ethnic and class oddities. It is no coincidence that this general ideological lesson and mythos is also identical to the basic premise of both *Seinfeld* and *Northern Exposure*.

Generically, historical exhibitionism also appears to stand in stark opposition to the reality and talk-show forms that now dominate daytime TV. The stylistic modes of the two genres, for example, are dramatically different. Cinematic style and narrative are favored in epic televisuality; live studio style and rhetorical discourse are favored in talk and reality programming. Yet exhibitionist television frequently creates a position of liveness and/or reality *within* its ritual form from which viewers and participants can look back and engage in historical analysis. In an Emmy-winning period-piece episode of *Northern Exposure* entitled "Cicely," the ensemble of characters relived the town's turn-of-the-century origins and past through a present staged via the documentary interview form. *China Beach, M.A.S.H.* and other shows have used the same live interview device to access the historical past. This televisual construal of a documentary present within the drama provides viewers with a great deal of textual and *historiographic power*, traits not normally associated with the medium in academic accounts that aim to define television's essential qualities—presentness, amnesia, and lack of context.

Finally, from a theoretical point of view, the historical exhibitionism of event-status and marquee programming implicates two important notions: the models of hegemonic consensus and therapeutic discourse. Talk shows, for example, clearly evidence what Mimi White has termed therapeutic value.[16] But they also showcase confessional modes that evoke Foucauldian penal and religious rituals: flagellation and absolution.[17] From a stylistic point of view, daytime talk shows represent one of the lowest forms of televisual artifice. Exhibitionist television histories, on the other hand, seem excessive and stylistically overcoded and flaunt cinematic forms of televisuality. Yet exhibitionist rituals can also be described as a kind of therapy—one in which the viewer is asked to reconstitute a personal and textual history in a way that is reminiscent of psychoanalysis. By reenacting textual origins that may or may not have occurred earlier in these shows—and by reconstituting historical memories—the viewer is allowed both the patient's catharsis and the power of the analyst. In therapeutic terms, this is less talk therapy—with the viewer as surrogate confessor in a volatile studio encounter group—than it is a highly directed visual and projective exercise, one that *overdetermines* the very kind of cultural origin and ethnic specificity that collective amnesia is argued to have erased.

Historical exhibitionism, one of two recurrent properties of loss-leader programming, should not, therefore, be written off as an example of postmodernism's preoccupation with nostalgia. Surely postmodernism's retro fetish only inadequately explains television's recent preoccupation with historification. The power and pleasure found in televisual rituals of historical exhibitionism seem less a result of the individual's alienation in mass culture—arguably a key to the popularity of talk shows with daytime viewers—than a result of some of dominant culture's prized contemporary rituals: the institutionalization of self-help, the legitimation of memory-based twelve-step programs, and the celebration of a progressive-looking but *empty* forms of multiculturalism. Television constantly flaunts historicity in its epic narrative forms. In some cases historical exhibitionism is financed as the sole basis for entire cable networks, like the new national cable corporation History TV. The network makes no bones about its commitments: "Introducing History TV. All of History. All in one Place. . . . History TV energizes the past. Through riveting documentaries, miniseries and movies, it *dives into history, grabs it shakes it and brings it back alive.* This is not your father's history lesson."[18] Cablecasters, then, have come to ritualize the very secret of event-status televisual programming, and they do so on a daily basis. Suffering apparently the same fate as realistic imagery and naturalistic narrative, history can be made material—physically grabbed, shaken and televisually assaulted, all in order to energize television. Historicity, like narrative discourse, has become a ritual of formal permutation and embellishment.

Case Study: Excessive Discourse in the Miniseries
Epic Narrative

If historicity sets the stage for loss-leader viewership, then epic narrative, a second obsession of event-status programming, seals the viewer's relationship in time within the virtual world of the programmer. An in-depth look at a single, but epic, loss-leader event helps clarify this process. Before television viewers, for example, have had a chance to see and experience the epic televised story of World War II in the miniseries *War and Remembrance*, or to decipher the complex web of characters in the docudrama, they are confronted by competing markers of discourse and contradictory frames that ask for different kinds of reception. Consider the following three divergent claims from the opening episode of the miniseries:

And now, the motion picture *event of a lifetime . . .* [19]

An ABC novel for television.[20]

War and Remembrance sweeps back the pages of history.[21]

In a way that can only be characterized as schizophrenic, the television miniseries suggests to viewers that it is all things to all spectators. It is a motion picture, a novel, a televised book, with episodes that are "pages of history." As if this contradictory aesthetic framing and claim-making were not difficult enough, the viewer confronts a formidable textual task in other areas as well. An endless opening to the first episode stretches far beyond normal programming constraints. The introduction, in fact, is over twenty minutes long and includes a total of 339 shots. The viewer is given neither story nor plot in this segment, but rather self-conscious and excessive displays of icon and image. In addition, the sheer quantity of narrative information and the fractured way that that information is delivered in the episode create a frantic mode of telling best characterized as manic. The net effect of these televisual strategies is that the miniseries does not *tell* a story. Rather, it flaunts and *disgorges* discourse.

This schizophrenic claim-making, manic pace of telling, and discursive performance raise several key issues. First, why does television work so hard, and with so many voices, to deny that it is television, to pretend that it is a novel, a book, a motion picture? Second, what kind of relationship exists between televisual excess and narrative? Can current models of narrative analysis help explain the presentational challenge of the epic miniseries? Finally, in cultural terms, what kind of *work* does the viewing and consumption of epic televisuality involve? In a narrative and generic sense, the historical miniseries harks back to the novel—in its characterizations and in its frequent adaptation from best-selling paperbacks. In practice, however, the experience of the epic spectacle works the viewer over with excesses very different from those of the novel.

War and Remembrance problematizes the doctrinaire view of the relationship between formal excess and narrative. In high theory, excess has come to stand for a potentially subversive trait or situation within a work of art or narrative. Since all texts are multidimensional, certain secondary aspects of the text, notably style, can threaten to overflow the boundaries of the work's central controlling narrative or intention. According to this view, the plenitude of style and the pleasures of excess can engage the viewer in ways that run counter to the official logic of the text. In short, stylistic excess can undermine the text's authority, that is, *if* the viewer chooses to read against the dominant grain of the text. This general view—that excess has subversive potential in relation to narrative—was derived mainly from the analysis of mass culture feature films. Its application to television is more problematic.[22] David Bordwell, for example, embraces a broad view that considers style as an element of narration, as part of a formal system that both "cue[s] and constrain[s] the viewer's construction of a story." Following Barthes, Bordwell ascribes to formally excessive phenomenon a "third meaning, one lying beyond denotation and connotation."[23] For Kristin Thompson, these third meanings are

Television as both a novel and as "the motion picture event of a lifetime." (ABC)

"casual lines, colors, expressions, and textures [that become] 'fellow travelers in the story' and that stand out perceptually but which do not fit either narrative or stylistic patterns."[24] Critics with political rather than narratological interests see the process and presence of formal excess as a key to counteracting the constraints of dominant cinema and as a potential site for counter-readings.[25]

Certain television theorists were quick to point out that excess may play a different ideological role in primetime television forms than it does in film. Jane Feuer shows how primetime soaps like *Dallas* and *Dynasty*, at least when contrasted to 1970s television, were defined in part by excessive style.[26] Although admittedly not as hysterical as, say, a Douglas Sirk film, such programs were nevertheless opulent compared to other primetime and daytime shows.[27] Feuer gives as an example of formal excess in this genre the use of exaggerated and scene-ending closeups. In the final analysis, Feuer argues that it is wrong to assume that the openness of television texts is a key to their radical potential. The lack of closure in primetime soaps, rather than exposing contradictory sites that can be exploited by counter-readings, is actually an essential *generic* part of television. As such, formal excesses can be subsumed within the dominant narrative logic that house them.[28]

In accounting for the miniseries, it is important to consider, first, whether there are stylistic devices in television that stand outside of denotation and connotation as Thompson and Bordwell imply, that fit neither narrative motivation nor stylistic pattern. Given the all-consuming appetite for stylistic pluralism in television and television's self-conscious foregrounding of discourse rather than story, the neoformalist view of excess as *autonomous and extratextual* seems suspect when applied to television. *War and Remembrance* uses a wide range of discursive devices and a hyperactive style that blurs the traditional boundaries separating diegesis from context, teller from story, and voice from mood. In short, this kind of television inverts the narrational process characteristic of classical cinema. With televisuality, a number of important critical terms are reversed. Plot now apparently supports the narrational act, rather than the narrational act producing plot. From the perspective of enunciation theory, *histoire* now apparently supports and presents *discours*, not vice versa, in the event-status epic miniseries.

Televisual excess has its own logic, as the analysis that follows will indicate. Rather than being extraneous to or autonomous from the dominant text, excess *is* the very substance of the text. A closer analysis of the opening segment of *War and Remembrance* shows how visual signs work to create these reversals between story and discourse. The extended and elaborate opening sequence shows that stylistic excess is not merely a by-product, overflow, or worse, a distraction to narrative as many critical theorists imply. Rather, excessive visual style is a fundamental way that contemporary television narratives are paraded and performed before viewers. After closely analyzing this performance in the miniseries, the ideological and cultural implications that follow from the practice and performance of excessive televisuality will be examined more closely.

The Perpetual Opening

An examination of the first episode of *War and Remembrance* discovers an introduction to the miniseries that is over twenty minutes long. Given the expensive cost of even a minute of primetime network programming and the viewer's familiarity with standard program durations of thirty or sixty minutes, this extensive commitment to a mere introduction is extraordinary. Since typical viewers neither count shots nor time minutes when watching, the more likely question that arises when viewing is: When will the program *actually* begin?

In fact, the firm designation of an opening picture frame and program starting point is almost impossible to determine. This is because each scene that follows after the initial twenty minutes *also* introduces new characters and typically starts with some form of orientating graphic or subtitle ("Italy, December 14, 1941"). Several factors, however, suggest that the point I have chosen in the miniseries functions as a de facto end of the endless opening. First, when FDR in this scene orders a retaliation in the White House against the Japanese, the basic time-place context for the miniseries that follows—the present tense of the story—is established. The scene also occurs after both major voice-over announcers have completed their lengthy orational tasks, after the major show titles, cast, and production credits have rolled, and after the main character (Robert Mitchum as Captain Victor "Pug" Henry) and his dilemma are introduced. At this point in the episode, two major structuring mechanisms have been set in motion in the text: Pug's newly assigned and battle-ready warship has just made way from the wreckage of Pearl Harbor and his triangular romantic entanglements have been suggested. More importantly, it is at this shell-shocked point that the U.S. military depicted in the miniseries finally begins to strategize its military response. The strategic retaliation set in motion here, after the introductory twenty minutes, also motivates plot developments in subsequent episodes of the miniseries.

Since various major subplots are introduced in sequence throughout the opening installment of the miniseries—complete with time-date graphics and establishing shots—it is conceivable that the *entire* introductory episode can be seen as an opening. Because the form of the epic miniseries is continuing rather than serial, this entire episode can be considered a mere two-hour narrational bracket. For the first extended evening of viewing, then, viewers are watching the nuances of a *textual boundary*—a beginning. The viewer's never-ending entry into what looks like a limitless text underscores again that the viewers are watching not a story, but a play of discourse—a textual performance that endlessly sets up and promises, but never really delivers, a story.[29]

Authorship, Authority

The narrational act depends upon a social agreement, one that sets in motion and utilizes a hierarchical relationship defined by power. Viewing, consuming, and deciphering are roles that assume deference to a master teller or maker. One recent debate among narrative theorists has focused on the extent to which narrative reception requires the perception and attribution of authorship.[30] David Bordwell has sought, among other things, to counter the influence of enunciation theory by arguing that narrative works can be received and interpreted by viewers without requiring that they conjure up a sending source or an implied narrator.[31] Bordwell argued instead that narratives work by cognitive paradigm, and so are not dependent upon whether or not a reader can imagine that an author is telling the story. These narrative paradigms are seen as cognitive maps for reading and interpretation, not as viewer fantasies about tellers.[32]

It is difficult to apply the authorless theory of the neoformalists to *War and Remembrance*, even though recent television *theory* also seems to oblige such a view of the medium. ABC works hard in this program, for example, to identify the viewer's experience as coming from the pen of Herman Wouk. It flaunts the show's origins as Herman Wouk's *War and Remembrance* in several titles and begins with an austere quote, not from FDR or Churchill, but from Wouk, who appears to ponder the gravity of the world's situation.[33] Why did the producers frame the entire miniseries with a quote from its *fictionalizer* Wouk, rather than from an actual historical figure like Churchill or Roosevelt? Surely the historical alternative would be a more forceful way to legitimize and intensify the import and veracity of the network's depiction. ABC, however, apparently wants the audience to realize that a *story* is being told.

This authorship consciousness undercuts the notion that film, in contrast to television, is more highly sensitive to issues of authorship and aesthetic origination. This view of authorship—of film as authorial and television as anonymous—can be traced, for example, in the work of John Fiske and John Hartley. Regarding the openness and mass quality of television they state, *"Thus we can say that there is no single 'authorial' identity."*[34] Influential theorist Raymond Williams, in reconsidering his earlier theory of the flow, states the need to "confront this notion of flow, because it really does belong to the medium in the sense that *you don't know the transmitter is there unless there is some sort of signal"*(italics mine).[35] To Fiske, Hartley, and Williams, then, television is structurally linked to anonymity because of its massive cultural scale and its abundant dispersion in culture. Clearly, the television-anonymity argument does not hold for event-status programs like *War and Remembrance*. Here there are, to use Williams's terms, many "signals" identifying transmitters and communicators. Authorship is an essential presentational part of *War and Remembrance* for the viewer. I will return to the reasons for this signaled privilege later. Suffice it to say that since *War and Remembrance* is thoroughly and overtly visual, formal, and stylish an analysis of it tends to follow more closely Bordwell's account of the art cinema than it does the classical narrative cinema. There is much more to this presentational and authorial package, however, than an artist. The claim of author(ity) here is clearly more than an aesthetic operation.

Redundancy, Autonomy, Discourse

The formal content of this segment, then, is worth considering in greater detail. The endless segment consists of multiple and repetitious elements, including: two different omniscient male voice-over announcers; two different, and lengthy, cast/actor credit sequences; several graphic references to the executive producer/director Dan Curtis; several visual and aural references to novelist Herman Wouk; numerous ads (spots that are first formally introduced and announced by the network as sponsors and then, only after this introduction, actually aired to viewers); and finally, an extensive variety of *War and Remembrance* titles for the miniseries—graphic displays that float, that reflect light and image, and that are visually choreographed alongside swelling theme music. In short, what one experiences in this segment is a seemingly endless barrage of visual images and sounds in a pattern that is both *complex and also highly redundant.* The question of redundancy is an important one. Why do even secondary actors need two different, isolated visual credits? Why do sponsors first have to be visually and aurally introduced, and allowed to perform their spots directly for viewers only after this "formal" introduction? In addition to the numerous voice-over announcers in the opening ads, the show *layers* different omniscient voice-over announcers *on top of other* omniscient voice-over announcers. A closer analysis suggests that what is being performed is not the story but the *production of the story.* Highly textual markers in the miniseries stand front and center; the story and plot are essentially absent.

Consider the main title sequence, both as an image of extreme visual density and kinetic movement, and as an indicator or the relationship between formal excess and narrative. The sequence is much more than meager graphic information. The title shot bears all of the traits of televisual excess—it is heavily layered, it is reflectively surfaced, and it includes numerous images that are simultaneously displayed within the parameters of its font. Although titles traditionally function as

Loading up the viewer with discursive maps—endless visual flash cards of sponsors, cast, and important locations around the world.

verbal and expositional information, here the viewer witnesses the title itself as a kind of spectacular gymnastic event. The mere television frame, in this sequence, is never quite able to encompass and depict the entire title in any one instance. Given the oversized scale of the title, the graphic is always slightly off-screen at any given time. The fact that the sweeping title flies into frame only at the end of the sequence suggests that the event itself has epic proportions, a scale that was heavily engineered. The sequence promises that the program the viewer is about to see, then, is not simply profound, but that it is expensive and complicated as well. With an entire montage of ever-changing imagery layered and matted *inside* the edges of these epic title letters, the viewer gets a heavy dose and sense of the scale of the megastory that is to follow. Here, different scenes are presented simultaneously inside a single image field. This device keys the viewer in to the kind of narrative density of time and space that will follow.

This main title sequence, and the abundant variations of it that follow throughout the miniseries, exploit a number of televisuality's favored presentational guises. Each manifestation of the title sequence, while related to the original, also attempts to distinguish itself from the other titles either through: (1) its pictorial components (the type of image or footage chosen and matted into the title letters differs in each title); (2) its gestural components (the dramatized action that the title is flown over or that is incorporated inside the letters themselves); or (3) its spatial aspects (the digital video effects designers work out numerous ways to float

Flying title sequence lands over photographic war memorabilia. Numerous title variations on a theme show that the visual spectacle cannot even be contained in the frame-dwarfing type that glides past the viewer's myopic window.

and move the image on all three spatial axes, x, y, and z). Although constructed with visual and filmic fragments ripped from the narrative, these recurring title sequences also boast autonomy from the narrative. Their formal ecstasies ask to be seen for their own sake. Through manic and constantly varying transitions in space, in gesture, and through matted and layered imagery, these recurring titles stand out and apart from the narrative flow and announce their own import. If formal excess is indeed antithetical to narrative, then the manic, overwrought ar-

Counter-tendencies. Discursive fragments—military orders, personal letters, voice-overs—work to narrativize the endless visual chaos of the first episode. Pervasive date-place flags make time and history very much the audience's business in an opening that has no time.

tifice of these narrative titles is surely and fundamenally at odds with the episode's story.

On the other hand, however, the presentational context within which these excessive titles are displayed works hard to narrativize them. Robert Scholes, in an influential article on narrative theory, has described the presence of "narrativitous counter-tendencies" in many works of narrative.[36] Even in many fragmented works, formal and stylistic devices can influence the viewer to experience the flow as narrative rather than as a formal exercise. According to this view, even unorthodox and modernist works—works not even meant to tell a story—can still bear the traces of Scholes's narrativitous counter-tendencies. The viewer's mind, in such cases, will take even minimally discursive clues, as well as fragments never intended as narrative clues, and work them over in order to fabricate a narrative.

Despite the various and extended formal indulgences in the sequence examined here there are also numerous counter-tendencies that work to narrativize *War and Remembrance*. For example, the segment is loaded with overt references to physical and dramatized acts of storytelling. Such discourse markers infiltrate the sequence and regularly make the viewer conscious that a story is at stake here. The entire episode is framed, for instance, by a love letter written by Captain Pug Henry's mistress. He reads the letter in close-up, and the audience is expected to experience his flashback through the reading. In another instance, sound-bites refer to the Jastrow family's forged documents and papers. One of the first naval scenes, furthermore, is based on a reading of written commands from the chief of naval operations, official orders that reassign Captain Henry to another ship. This authoritative document, these orders, initiate the voyage and mission to follow. Such examples are not just references to aspects of the narrative's plot—data regarding the story's space, time, and point of view—rather, they are overt visual and aural *reminders that a story is being told*, and that it is being *manufactured* from numerous narrative documents inside the text. These narrative-making fragments (letters, legal documents, written commands), all mark and announce that the endless montage is actually a story. Even if audiences lose themselves in the show's perceptual pleasures—graphic-title pyrotechnics, libidinous close-ups, or fiery images of violence—there are constant and repeated reminders that a story is being told and experienced. The abundance of these discursive markers is directly proportional to the excessive formal elements that must be constrained and controlled. Dense discursive markers seem to be the narrative's best hope for tying down the loose televisual cannon careening on-deck.

One last image in the sequence is worth considering since it raises an important question about the extent to which formal elements can be autonomous. A key point in the narrative opening occurs when Pug's son, Warren Henry, a junior naval officer on leave, arrives home to greet his wife and family. Coming almost at the end of the production credits, the heretofore rapidly cut opening sequence slows to a standstill by statically framing the image of Warren's young wife. In a frontal medium shot, with the body of the actress directly facing the camera, the scene unfolds without changing frame. Fried chicken crackles in a frying pan on the stove in the shot's foreground. The young woman is wearing one of her husband's khaki shirts, its tails tied under her chest, exposing her midriff in the center of the frame. She tends to the preparation of food in a daydreamlike fashion, looking neither directly at the camera nor at her husband, who enters the room in the background. After approaching her from behind, young husband Warren grabs her, and the two embrace and moan pleasurably. Following an exchange of nonverbal niceties the two finally exit screen left in order to see Dad (Pug) outside.

Dramatically different from the images of Auschwitz, violence, and high energy graphics that precede this scene, the viewer is here given an image rich in symbolism; one that overflows in *assigned* meaning. The woman's body is visually displayed as a target for numerous projected meanings. Suggesting perhaps the pleasureable plenitude that existed before the war, this single image of the female body connotes peacefulness, sustenance, sexual pleasure, and bodily maintenance. The image is the bearer of many gazes: the woman depicted here does not look, but is looked at, from many directions. The sequence appropriates the woman's body from at least four major directions. First, the officer-son, on entering the scene unnoticed by his wife is something of a voyeur; he then stalks and grabs his wife from behind. Second, the sequence allows the audience to view the woman's abdomen without being viewed themselves. Third, key production staffers, the director of photography and a producer, utilize her body at the end of the credit sequence to graphically inscribe, isolate and erect their own titles

Intercutting domestic images with the Holocaust, the episode hangs titles on pleasurable bodies that give the chaos of both the miniseries and the war a vague sense of backstory and causality.

and identities. Finally, the narrative uses the image of her young body, the chicken, the sense of pleasure, and the air of sustenance to set up the symbolic world of the story. This image is one of the first in the episode, for example, to stand out from the long and frenzied montage that leads up to it. Any narrative cause and effect that exists to this point *seems* to come to a spatial-temporal standstill. That is, the seeming urgency in the rapidly cut and unending opening montage creates a sense that the episode will deliver a maximum amount of dense narrative

information in a short amount of time. But on this particular image of bodily plenitude, the presentational urgency and informational burden evaporate. The camera, the production staff, and the male character all use the image of the woman's
body as a pleasurable resting place, as an aside as a textual respite within the manic
opening. As much as any image in the long sequence this one seems to bear the
traces of formal excess—it is set aside, given autonomy, and loaded up with multiple meanings.

If as Scholes suggests, however, the spectator is characterized by an almost
neurotic desire to give and ascribe narrative causality even to sequences that lack
causality, then the shot can be seen to have an important narrative function as
well.[37] Although it bears all the marks of the regressive male gaze that Laura
Mulvey describes, it also fills in important narrative information through retrospect. While in Metzian terms the endless opening looks like a fragmented succession of autonomous shots and bracketed sequences, this particular shot stops
the manic flow of these autonomous images, and retroactively gives some of the
earlier montage fragments a sense of narrative motivation and moral justification.[38] For example, the viewer only learns through the visual agency of the
woman's body who two of the main characters depicted in the preceding montage are (her husband and father-in-law). In this way the shot, even with its pleasurable autonomy, helps answer plot questions in a way more typical of filmic
narration in general. The viewer also gets some sense (or better, sensation) of the
apparent reasons that Pearl Harbor is to be defended: the home, the women, the
American way of life. Like the extensive sequence of shots in the opening that
focus again and again on the American flag, the woman's body—by contrast to
the air of contention and violence that precede and follow it in endless montage—
becomes a *narrative anchor and sign post*. Characters begin to make sense only
in relation to her bodily presence, even as the story is given a symbolic cause
and motivation for being.

Like the machinations of the flying and performing title sequences, this shot
at first seems like nothing more than an autonomous aside, an excuse to procras-
tinate on a pleasurable image, to linger under the pretense and task of providing
important production credits. But like the excessive title sequences, its context in
the episode also encourages a narrative reading by the spectator. Since both examples (the kinetic title sequences, and this female body–aside) work to encourage narrativity, the difference between these two examples is really only one of
formal tactic and degree. The show's title sequences, in particular, bear a kind of
narrative structure unaccounted for by Metz in his "grand syntagma."[39] Clues for
narrativity in the titles appear in a *layered and simultaneous fashion within a single
image* and are unlike anything in Metz's syntactical model. Even Metz's episodic,
bracket, and parallel syntagmas, while not linear in the conventional sense, are
defined for the viewer by their *ordering within an edited sequence*, not by simultaneous layering. Clues for narrativity in this female body–aside (the woman as
fried-chicken-war-therapy), provider of sustenance and comfort on the other hand,
come from the shot's cognitive context. Despite being isolated and loaded up
graphically with the credits of production staffers, the shot also offers a context
for narrative reading. Seen against what Metz would call an ordinary sequence (a
loosely linked succession of visual displays with temporal discontinuity), the abun-

dant image of the woman also allows the viewer to pause and retroactively make sense of the probable narrative environment in *War and Remembrance*.

Both sequences demonstrate how televisual signs work on multiple levels. On one level the title sequences flaunt authorial power by orchestrating a multitude of images within a single frame. One can no longer even talk of such images as shots, since they are layered and matted for decorative effect and embellishment. But like embellishments in the frieze of a Doric temple, these decorative elements also provide extensive narrative detail, such as relationships between characters and between the story and history. The fact that this loaded visual world is spinning and flying past the viewer *anoints* the perceived narrator with a tremendous amount of *authorial credibility*. In this way, then, even antinarrative graphic elements heighten the sense that this is a narrating instance, because the stylistic skill demanded by such flourishes focuses attention on the importance of an artistic or narrating authority.

Formal ruptures, like the hyperactive titles, announce to viewers that they are in the presence of an epic narrating source. Formal ruptures like the female body–aside, on the other hand, are used to pace the manic narrative, to anchor the characters, and to give retroactive causality to the narrative actions. In both cases the sense of narrative authority that results encourages the viewer to ascribe to the fragmented flow a sense of intelligibility and logic. The deference by the spectator to this implied author(ity) is essential if the viewer is to play the narrative game of the miniseries effectively. No narrative will result, after all, if the viewer is unwilling to cognitively rearrange the fragments into an intelligible or credible sequence of events and causes. Although narrative in any traditional or classical sense is absent in both of these examples, excessive visuality actually enhances the sense of narrativity since its sensory appeals make narrative spectatorship more overt and demanding. Televisual excess and narrative-making counter-tendencies are obviously interrelated. While the formal structure of the miniseries is clearly noncausal and nonlinear, the form has its own logic nevertheless. Although linear causality does not motivate the link between excess and narrative here, the interrelationship becomes clearer if one looks at the ways that space, time, and categorical information are visualized.

Foregrounding Structural Categories

Reminiscent of Pare Lorentz's 1930s documentaries, ABC repeatedly uses in *War and Remembrance* a form of verbal and visual litany. Interspersed between announcer hype and character sound-bites are lists and more lists of topics, places, and things that are ponderously recited by the show's announcers. "London, Paris, Berlin, Auschwitz" are intoned in one; battleships "the *California*, the *Oklahoma*, the *Maryland*, the *West Virginia*, the *Nevada*, the *Pennsylvania*, the *Tennessee*, the *Arizona*" in another. Visual images illustrate the terms in some of the litanies (the World War II cities); while in others (the battleships of war) the voice of God merely resonates across empty water and naval scenes. In the latter case—the omniscient voice over open waters—viewers are encouraged to imagine and construct the space and absent objects of the story. Other litanies, like the picture lists that stand in for war cities, create a different task for viewers. They provide a kind of memorization game for grasping and understanding the program. The recurrence of these lists, litanies, and categories was a marked feature of the premiere episode of the miniseries.

In data-processing terms, this mantra of categorical litanies in image and sound fulfills an inputting function for viewers. Setting aside twenty minutes to download important pictures suggests that the network narrators aim to first produce viewers and their expectations, before actually delivering the program. The presentational goal in the episode seems determined to prepare the ground of reception and to construct a useful horizon of narrative expectation. Even more than is the case with other meganarrative forms on television, such as primetime serials and continuing dramas, there is little guarantee that viewers here have had an extended viewing history with this particular story or the experience and familiarity necessary for facile viewer interpretation. Because of its special event status, the miniseries does not have the interpretive luxuries provided by those genres with a more continuous presence in programming. The longer lifespans that come with other and "lower" televisual genres, enable regular series, for example, to more easily exploit and stroke the audience for its awareness of the program's history.

War and Remembrance does two things to counteract this one-time event-induced structural disability. First, to compensate for the audience's relative lack of familiarity with the spectacle's backstory, the miniseries proudly announces to the viewer that it comes complete with its own built-in and substantive *aesthetic history.* This prepackaged history spills onto the viewer from the very start of the premiere, when *War and Remembrance* muses on its origins in the earlier 1985 miniseries *Winds of War.* Secondly, the miniseries works obsessively hard in the first episode—and especially in its extreme and endless opening—to bombard the viewer with a large-scale repertoire of interpretation-inducing viewing lists, schemas, and categories.

Some litanies are purely visual, such as the sequence that include recognizable shots of historical figures; other lists are stuffed into single shots, such as the frame of archival footage that depicts FDR, Stalin, and Churchill at Yalta. Since these characters are also dramatized in the fiction, their images are more than just stock shots. They establish a working network of human relationships that can be further utilized by the producers. Consider the multiplication and elaboration of this network of relationships in the form of actors' credits. In one of the *two* separate opening credit sequences, the actors' names are litanized over dramatized images of those same actors performing without sync sound. In the second actors' credit sequence, orchestral theme music plays as actors names are visually burned into their stills, stapled into either monochrome photographs of the main characters or graphics of the lesser ones. Thus *voice-defined* dramatic ascriptions in one become *visually-defined* past-tense artifacts (black-and-white photographs) in another. Televisuality here appears to stylize and litanize for both left-brain and right-brain audiences, and to place each character in both the historical past and the live present. If the audience does not get it one way, the segment serves up character relationships in another way. As a result, an informational *redundancy* occurs, even though the sequences themselves go through a variety of very different stylistic and perceptual *permutations.*

Table 6.1 The Viewer as Contestant: Game Show Categories in *War and Remembrance* (Televisual maps, discursive networks, and structural relationships marked in the first twenty minutes)

Context/Level for Viewing	Conceptual Items in Map

Cities of the war
  Displayed as stock footage
  with graphic subtitle as I.D.
	London ____ Paris ____
	     / \\
	Berlin ____ Moscow ____ Rome
	  \\ /
	____ Auschwitz ____

Historical figures of the war
  Displayed as archival photo-
  graphs in title sequence
	(Yalta Conference)
	____ FDR ____
	  / \\
	Churchill ____ Stalin

Characters in (hi)story
  Displayed as non-synchro-
  nized dramatic bits from
  scenes; then displayed as
  monochromatic photo-stills
	FDR ____ Hitler
	       |
	Capt. "Pug" Henry
	  / \\
	Mrs. Henry ____ Mistress
	  / | \\
	W. Henry____ Son's Wife ____ Tudsbury
	 / | \\
	etc. etc. etc.
	(21 total presented with footage)

Actors in miniseries
  Displayed as graphic titles
  lower-thirds, or as ponderous
  voice-overs
	Ralph Bellamy
	    |
	Robert Mitchum
	 / \\
	Polly Bergen ____ Victoria Tennant
	 / | \\
	Michael Woods ____ Robert Morley
	 / | \\
	etc. ____ Peter Graves ____ etc.
	(29 presented as graphic credits)

Sunken battleships
  Intoned over long empty
  shots of sea in harbor; shots
  invoking absence, time
	The *California*
	    / \\
	The *Maryland* ____ The *Oklahoma*
	 / | \\
	Nevada ____ *Pennsylvania* ____ *Maryland*
	  \\ | /
	The *Tennessee* ____ The *Arizona*

Major battles
  Displayed and
  decontextualized by images
  of explosions, and violence
	Philippines
	  / \\
	Guadalcanal ____ Singapore
	  \\ /
	Hiroshima

Emotional traits associated with
  war
  The opening claims to bring
  these unique traits "alive";
  illustrated by gestural
  performances (e.g., Hitler
  gyrating, troops cheering, a
  smoking corpse)
	The turmoil
	 / \\
	The tragedy ____ The passion
	  | |
	The fury ____ The glory
	  \\ /
	The horror

Sponsors
  First introduced by show
  over faux marble back-
  ground; then aired in full
  motion as displays of
  patriotism
	Ford
	  / \\
	IBM____ Nike
	 / | \\
	American Dairy ____ GE ____ Exclamation
	Association \\ | /
	Toys'R'Us

Narrative History as Game Show

While the informational redundancy that results from repetitive litanies might suggest that televisual complexity is aimed mainly at the "idiots" in the audience—at the inattentive who need repetition in order to grasp the information—the actual formal complexity of the lists and categories suggests just the opposite. Categorization takes place through a stylistically rich array of presentational guises. The privilege given this tactic suggests that viewing pleasure in the miniseries takes place as much through engagement with varying presentational and formal embellishments as it does through the transmission of redundant narrative information. It is clear then that the opening persistently works to articulate various structural relationships considered keys to the story. Stylistically this process involves both redundancy and variation, and simultaneously pitches dense but essential information to viewers via different perceptual channels: verbal, aural, visual. Consider the congruence formed by iconic layers and structural networks, all laid out in visual-verbal litanies that fill the first twenty minutes of the episode. (See Table 6.1.)

Faced with this barrage of categories, the viewer confronts none of the traits thought to be necessary for the production of narrative *in the traditional filmic sense.* Instead, the viewer faces an unrelenting atemporal sequence of televisual and aural maps. Through litanies, image-text combinations, and credit rolls, the viewer of this narrative is not given story, but is given structural *relationships* of the most abstract form. Relationships—between characters, battleships, cities, advertisers, and essential emotional traits—are visually and methodically displayed. Rather than developing a character or a hook in the traditional way that a dramatic fiction initiates diegesis, *War and Remembrance* instead tries to cue the audience in to important conceptual *fields* of information—fields that promise to aid and abet the viewer's subsequent reception of the narrative. One televisual field after another is spelled out in detail. Furthermore, the producers illustrate each of these fields and interpretive maps through pictures and displays of information deemed necessary for viewers to decipher or play the epic narrative. In the epic miniseries, history has become a game show that constantly parades categorical choices for the viewer.

This narrational overload of contextual information stands in stark contrast to the orthodox truism that classical narrative should be elegantly reductive. As any one of numerous scriptwriting books preaches, conventional production wisdom has it that motion picture stories or television drama should never have extraneous elements and that viewers should only be given the information they need to know to get to the next scene, and eventually to a clean resolution. Televisuality rejects this norm in its frantic attempt to *spell out all of the relationships up front.* Though some might view this epic opening as a mere exercise in collage, close analysis shows that the footage is anything but a mindless preoccupation with, and synthesis of, empty form. Rather, the segment reveals an extreme consciousness of narrative *structure and discursive needs,* and it flaunts that structure explicitly before the viewer.

The sheer volume of narrative involved in the miniseries may be the factor that makes this preoccupation with categories and litanies so central. While the excessive quality of the miniseries is striking, the trait is not alien to television in general. Soap operas also work under the burden of meganarratives. Other genres expect viewers to make sense of single episodes within a larger dramatic series. In both cases, viewers must have a broad paradigmatic field of reference within which to make sense of specific and more traditional narrative sequences.[40] In this sense, the miniseries may not be as unique as it first appears. The fractured performance of discourse and style may work because it gives viewers a paradigmatic frame of reference to interpret and place elements in the spectacular flow of images and sounds. This paradigmatic sensitivity may follow more basic and fundamental cultural logics: the family, the personal and social, the conflation of the private and the public. From this perspective, the recognition of cultural parameters and markers within the excessive televisual flow motivates and justifies the flow as a meaningfully organized experience. The excessiveness of a single episode in *War and Remembrance*, then, becomes symptomatic and typical of contemporary television in general. While the miniseries dramatizes its overdetermined construction of the viewer through an ecstasy of discursive downloading, many other television forms also struggle to organize the overwhelming flow of image and sound into meaningful paradigms and categories.

Icons of Space and Time

Spatial considerations may play a greater role than temporal aspects in the way that narrative is fabricated for the *War and Remembrance* spectator. In the opening segment there is no clear beginning, middle, and end to speak of. However, the historical event (World War II) is given temporal markers (time and date graphics) that start many of the segments throughout this episode and the miniseries in general. Given the obsession with graphic, temporal codification in this miniseries, it is ironic that the episodes themselves do not seem to progress much in historical time. Without a clear sense of narrational or story time (to use Gérard Gennette's concept for the time and duration it takes to narrate the story), it becomes impossible for the viewer to easily construct a sense of historical time (the time represented and alluded to by the fiction). In this miniseries, flashbacks and flashforwards are empty concepts since there is never really any clear sense of a present time from which to concretely judge the temporal movement forward or back. The tremendous flux in time and place evident here prevents the application of any conventional hierarchy of time, of any frame anchored upon an unequivocal sense of the present. Here present, past, and future intermingle on a playing field that is more or less level.

All one really gathers from the constant shifting back and forth between characters and scenes is a great and abiding sense of *simultaneity*. In fact, what this fiction perpetuates as much as anything else is the idea that all of these people are feeling and experiencing the epic qualities of life *at the same time*. This trait is one of the most engaging and distinctive aspects of the miniseries form. That is, the genre is able to take a small slice of historical time—for example, December 14, 1941—and explode it into epic and unfathomable proportions. Since the miniseries purports to represent the lives of scores of characters, time seems to stand still as characters repeat the actions that would otherwise have taken place

Long, nonverbal network promos construct a loving, emotional network family identified by, and accessible through, the days of the week. *War and Remembrance* uses its marquee to teach the viewer the intricacies of programming time.

at the same time. This then is a fundamental difference between forms like *War and Remembrance* and many traditional dramatic forms on television. With the exception of the soap opera, viewers are well versed in the extreme *condensation* and reduction of historical time that characterizes shorter half- and one-hour formats, since the process typifies many genres on television. In the epic miniseries by contrast, time explodes rather than implodes. This exaggeration of time cues the viewer into the experiential power of the miniseries. A sense of textual empowerment results from the stretching and expansion of time beyond its common, personal scale.

While time is important, it is not necessarily story time that counts. Redundant references to times and dates clearly mark the start of numerous sections in the premiere episode of *War and Remembrance.* The idea and ideal of time is, for example, invoked in nonprogram forms throughout the episode. ABC uses the long opening segment to set up and promote its new season's offerings. In a rapidly cut and highly visual promo within the segment, the entire network lineup is condensed into frenetic montage cut to an upbeat and sentimental theme song. There are no spoken words in the sequence. All that is given to the viewers, alongside a massive display of network talent (stars), are the days of the week, "Monday, Tuesday, Wednesday." In short, the ABC cache and capital of star power is inextricably linked and identified by calendar time. Without recourse to verbal narration, this linkage is made primarily with images, for the network clearly idealizes time

Star personas are also invoked via time. Robert Mitchum then (in *Winds of War*) and Robert Mitchum now (in *War and Remembrance*). Peter Graves then (in *War and Remembrance*) and Peter Graves now (in the *New Mission Impossible*).

as a principle that governs both individual and programming strategy. ABC betrays its unwavering allegiance to time in this sixty-second season promo, an homage where the network's good feelings, along with its stars, are organized fundamentally around time. It is far from insignificant, then, when *War and Remembrance* boasts to viewers that it is an event of a *lifetime*. Not only is the show structured around time, it is also promotionally anchored by ABC because of its specific place and function in programming time.

Other traits underscore the sense that the episode parades time as one of its governing stylistic principles. The actor Peter Graves, who had a part in *War and Remembrance*, is also featured in a preview for what was then an upcoming episode of *The New Mission Impossible*: a close-up of Graves in one of the opening credit sequences of *War* is mirrored by a close-up of Graves in the revived *Mission Impossible* series. Visual shorthand makes the connection clear: this is the very same shot of Graves, but in two different shows. By juxtaposing and multiplying single icons in multiple shows, the promo showcases time on two levels. First, this promo sequence, embedded within the opening sequence, can be seen as a flashforward to a viewing time that follows the *War and Remembrance* event. Viewers are made conscious both of their present viewing time and asked to remember an important time in the near future. Second, *The New Mission Impossible* clones and re-creates an earlier series. The show—even apart from the miniseries—has its own detailed history, and the promo makes this history clear by reintroducing what it refers to as an old cast member. *The New Mission Impossible* touts its newness to the viewer by simultaneously drawing attention to the old and the past as well to the new and future.

This repeated homage to time by the network—to the old and new and to the importance of narrative development over time—was operative in *War and Remembrance* as well. The miniseries *War and Remembrance* followed its precursor, *Winds of War*, by several years. The earlier miniseries featured Robert Mitchum and many others. Good viewers of the later miniseries no doubt found the actions of the characters meaningful precisely because those character actions three years earlier helped motivate events in the sequel. If Mitchum's character, Pug, feels claustrophobic due to his love entanglements in *War and Remembrance*, viewers may consider it just retribution given his loose view of the marriage vows dramatized in the earlier miniseries, *Winds of War*. These, then, are the dialectic frames that reify time—*Winds of War* and *War and Remembrance*, *Mission Impossible* and the *New Mission Impossible,* Robert Mitchum then and now, and Peter Graves then and now. The segment spreads out to embrace nearby programming forms, promos, and previews, and keeps the issue of temporal revival and history on the viewing agenda.

This performance of time and historical exhibitionism weakens cause-and-effect relationships within the actual plot of *War and Remembrance*. The story itself evidences a weak linear chronology and little causality, thereby suggesting that the segment depends instead upon causality of a different kind. That is, an alternative logic repeatedly works to display connections between the viewed sequence and other "special" programming forms, an extratextual logic that overshadows connections within the episode itself. In this regime of self-reference, history, the network's nightly schedule, originals and remakes, precursors and sequels are all assigned a causal and natural relationship to the miniseries. A heightened consciousness of structural time is celebrated throughout this special-event showcase, a preoccupation that betrays television's institutional obligation and programming agenda.

Space, as much as time, is a crucial factor in the reception of this miniseries. The discursive networks and lists described earlier map out, cue, and orient the viewer to the epic *scale* of the conflict. Litanies of major cities, of battles, and of historical figures representing different countries all offer to the viewer spatial and geographical schemas. These spatial schemas work to define parameters for *where this event takes place*. The spatial map in *War and Remembrance* is by nature a gross reduction of real history since it cannot possibly include all of the important sites (battles, generals, cities), but the televisual maps do create a credible enough *place* for the personal narratives to unfold. Whereas the credibility of classical film and television depended upon the illusion of a three-dimensional *space* for the narrative action to occur, *War and Remembrance* creates instead a credible *geographical place*. The convoluted plots, narrative scale, and manic telling of the miniseries are clearly matched by an epic spatial scale, by proportions derived from global geography rather than from the more diminutive proportions that come with a typical sound-stage or set. The question of why the epic miniseries transforms narrative space into narrative geography is worth addressing. It may be that such transformations merely echo the importance and signifi-

History—cut loose from its moorings—is everywhere, even in stock-footage fragments optically ground into titles. Global geography is repetitiously invoked to give intimate personal lives an epic scale.

cance of the network's program offering. The fiction of the miniseries thus takes on sobering geographical proportions as a marketing tool and as an intertextual strategy providing ABC with a facade of global distinction beyond that of its programming competition. The net effect of the transformation, however, also has a crucial intratextual function.

Spatial maps scaled to historical and geopolitical import directly affect the characters represented in the miniseries. Geography legitimizes and elevates the personal lives of those who travail within this televisual arena, for the personalized playing field utilized by the miniseries is overdetermined by style and transformed into epic and tragic global proportions. Consider how this process contrasts with that of another genre. Soaps also intensify their everyday characters, but do so by dramatizing their personal excesses and deviations from everyday norms of morality, civility, and decorum. The epic miniseries, by contrast, can intensify its characters in much less personal ways, without reference to psychological or behavioral aberrations. Space and geography intensify epic miniseries characters by enlarging them to worldly proportions. Epic transformations of space, then, do not just orient the viewer in narrative space-time, they also textually intensify and globalize the lives of the characters.

Geography is not the only agent of spectacularization in the miniseries. As we have seen, history also plays a fundamental role in making the space of the narration epic. Invoking history with visual signs, war footage, archival shots, and graphic data helps establish an ideological as well as narrative place for the fiction. This place in history carries a much different kind or level of authority than does the narrative space that governs classical production. Establishing shots, rules of continuity, and the placement of privileged fictional characters in a credible place where they can see the story unfold work together to achieve an aesthetic experience that is bracketed-off from everyday experience. *War and Remembrance*, instead, uses televisual forms to invoke and exploit the very notion of history itself. Various televisual operations strip, manipulate, and refabricate historical images from archival fragments in the form of keyed opticals, subtitles, and military reenactments. The net effect of this televisual refabrication of history is the textual promise that the characters the audience sees function in a world that is *more ontologically real* than the world surrounding the more strictly fictional characters in other genres. Hence, docudramatic miniseries and fictionalized historical accounts depend more upon space verified by history than do many other televised fiction forms. Conventional narratives frequently announce their status as fictions before the viewer enters into a willing suspension of disbelief. The epic miniseries, by contrast, complicates this act of denial by continually claiming— both through televisual performance and verbal avowal—its status as history. This particular segment rifles and combines historical imagery with personal imagery in a continual flow intended to construct an epic space on the screen and an ideological context in the mind.

In one sense all of *War and Remembrance* can be viewed according to the twin terms of its title. They are, after all, terms that directly raise the issues of narrative space and time. The miniseries flaunts structural relations of space (cities, battles, places), but it also places personal lives in those spaces. A key structural division occurs between personal space and social space, between personal time and historical time. It is useful to consider the show along the following two axes.

War	Remembrance
Space	Time
Violence	Love
Public	Private
Political	Domestic
Power	Desire

War and Remembrance constantly negotiates between the terms of these structural oppositions. The segment itself ceaselessly shifts between the public and private, violence and love, politics and domesticity, history and memory. There are clearly two regimes of narrative at work here. On the one hand, there are many dramatic scenes that focus on desire and romance. These recurring scenes of passion regularly break through into the knowable space of history and politics. Only after a cathartic confrontation with violence or tendentious politics does the show retreat into the murky, ambiguous world of memory, love, and personal thought. The extended and excessive opening leaves no doubt as to the links between these two regimes. The early establishment of the two schemas—of the historically known and the psychologically desired—is significant given the fact that the large-scale opening functions, through its visual excess and categorical schematization, as a mechanism for bringing the viewer *up to speed*. Over twenty minutes of air time and elaborate stylization have gone into establishing these two broad interpretive regimes. Numerous visual clips depicting romance or passion are inevitably followed by scenes of violence or fiery explosion. These constant structural shifts suggest that the viewer gains some pleasure from persistent and manic collisions between desire and violence.

Televisual narratives like *War and Remembrance* allow viewers to extend a highly privatized realm—a psychological world of parental guilt, lust, remorse, adultery—out into the external realm of worldly power, and vice versa. At the same time, the regime of the knowable and powerful—the world of FDR, Churchill, and Yalta—can also be conflated with and disguised as interpersonal relationships, as passion and desire. This conflation is one of the fundamental mythic operations at work in the narrative of the miniseries. *War and Remembrance* constantly transforms the psychological and personal domain into the world of the social and the powerful. At the same time that the episode's style reifies time and fabricates historical space, it also reifies and overvalues the personal world of romance and memory by turning it into an issue of power. The psychological world of the self is falsely congealed with the world of power. In the epic miniseries, televisuality and excess help privatize the world of power. It is worth noting that the privatizing textual and stylistic operations of the epic miniseries during its golden age in the 1980s mirrored a much broader conflation: the cultural ideology of privatization celebrated during the period of Reaganomics.

Conclusion

Confined by the academy to texts whose *textual limits* were rather self-evident, narrative studies of classical literature and film could endlessly articulate the intricate fictional structures and layers embodied within such works. With the coming of television to the arena of academic respectability, however, narrative study (à la Gérard Gennette or Seymour Chatman) confronts a text forever on the run, one that ceaselessly ruptures the rather arbitrarily applied brackets of implied and distinct narrative agents and one that confuses even simple assignments of point of view. The text that results from fictions that are infiltrated by televisual form does not readily oblige analytical attempts at focus or delineation. Given this textual messiness, I have analyzed a marquee program form in a way that treats the show not as the presentation of story, but as a specific and overt display and performance of discursive activity. The epic miniseries is also a site where multiple narrative impulses, competing visual signs, and cultural determinations are interwoven in a continuous but ideologically problematic way. These embellishment and discourse effects invite the viewer to participate in a game of narrative and historicist hyperactivity.

The new critical theory on television that emerged in the early 1980s brought with it a marked shift in interest away from narrative story toward modes of discourse.[41] Important work by Margaret Morse, first on the "discourse of sport" in television, and then on the television news personality, extended Emile Benveniste's view that discourse is "every utterance assuming a speaker and a hearer, and in the speaker, the intention of influencing the other in some way."[42]

Whereas discourse in the classical narratological model (Gennette) concerned the manner in which a narrative was presented, this view of discourse (Benveniste via Morse) made the narratological model more social and persuasive. The verbs and address of the latter tradition are more active, the intent of any discursive process being to influence and to change. My aim has not simply been to show that discourse is privileged in television, for this has been demonstrated by others. Rather, event-status televisuality exploits a different kind of *consciousness* than that which accompanies the discursive performance in film or classical narrative. I have used the discursive framework from narratology to analyze the epic miniseries, because it helps delineate and isolate within the dense televisual flow important aspects of the presentational *telling* and engagement. Televisuality can be accounted for in a narratological sense as discourse, but to see the phenomenon as a kind of performance—as a visual spectacle and form of exhibitionism— also suggests a different and problematic relationship to the viewer-consumer.

The new visuality in television cannot simply be explained sociologically— that is, by seeing it as a result of consumerism or as a phenomenon of self-promotion—although to some extent it clearly is a function of both. Rather, as the epic miniseries suggests, television may also continue to change the way we see and conceptualize narrative. Unlike traditional novels or film, the epic miniseries tends to wear its structural patterns on its sleeve; it leads with and privileges graphic arrays of character relations, and it flags the viewer with overt markers of discourse, rather than concealing its marks of enunciation. It takes what Lévi-Strauss or Noam Chomsky would call the deep-structure of myth and story—not the surface elements one normally associates with dramatic narrative—and *parades it* persistently before the viewer. Story and plot pale before this televisual performance of discourse and style. In this way, story becomes a mere excuse for the display of structure, fodder to be fed to the televisual programmers. Since these viewers are positioned as consumers and analysts of *deep structure*, rather than manifest story, one could argue that the viewers of *War and Remembrance* are idealized and positioned as anthropologists of passion and history, rather than as narratees. In the kind of manic narrative that ensues, viewer pleasure and engagement depend upon the skilled discrimination and consumption of narrational fragments and stylistic flags. Engagement is not apparently dependent upon narrative resolution or denouement in any case, since such events are endlessly deferred and never really promised at all. The viewer is instead enamored by an ambient world of image and sound, rather than by one developing toward a final cause. The experience of *War and Remembrance* is one of endless setting up and structuration rather than of goal-driven development. Telos is absent.

This preoccupation with the televisual performance of discourse, rather than with story, implies that television viewers are good in part because they know and can value certain aesthetic and cultural distinctions. They know and value the difference between the personal and the social (between war and memory, passion and history), between serious dramatic art and a weekly series, between prestige authorship and mundane programming. The televisual flow demands active decipherment and critical facility on the part of the viewer. It prizes and reinforces the trait of discrimination—a factor that gives members of the mass audience the pleasures of distinction and the powers of taste.[43]

In two ways, then, the televisual performance of discourse is political. First, by challenging the audience to bring to bear their proven skills at interpreting history and epic narrative, event-status programs bestow on viewers an air of textual and conceptual distinction. By intellectually flattering viewers in this way—no matter how empty the gesture may be—televisuality also surely makes them better consumers, for an ideology of discrimination drives consumerism. But epic televisuality also empowers broadcasters in important ways. It assigns to them the stylistic prowess and power one might reserve for an impresario or bard. Television exploits and announces its ability to convolute story with excessive visual style and to complicate telling with self-conscious displays and utterances. This form of self-aggrandizing announcement—that television can do very *complicated* things—is really a nomination of authority. Stylistic and discursive announcements are important, for no narrative can take place without an initial claim of authority, a badge that allows the narrator or producer to speak or to tell. The narrative issue, then, is clearly more than one of authorship. It is really a question of cultural authorization. Clearly, the brokerage of authority is a cultural and political issue of fundamental consequence. The means by which aesthetic authority is granted is arguably one key to cultural hegemony. Through stylistic legitimation and the pretense of consensus, televisual excess allows the producer and consumer to misconstrue their aims as analogous, even though their presentational and narrative tasks are very different. That is, the viewer gives over to television the authority to work the excessively visual spectacle in deference to television's ever-present ideology of expertise—an expertise that television is more than willing to show off—in both technical and aesthetic ways. The cult of technical superiority discussed in chapter 3, then, has a narrative cohort: *the industry legitimizes itself as much by overproducing and complicating narrative as it does by overproducing and complicating high-production values.* At least as it functions in epic event-status narrative, stylistic excess appears in practice as anything but radically progressive or potentially disruptive. Rather than the *collective transgression* that Dan Harries refers to in the subcultural readings of cult films—a genre rife with stylistic excess—televisual exhibitionism in the primetime miniseries suggests and encourages *collective legitimation.*[44]

The case of *War and Remembrance* also shows that the amount of discourse increases in direct proportion to the relative quantity and density of formal excess. If one thinks of a text's preoccupation with style as an enemy of narrative, then it is easy to see why the voices and guises of discourse multiply. Discursive incidents and markers proliferate in the face of burgeoning formal excess. Discourse seeks both to constrain and exploit textual overflow. The relationship is dialectical and complementary and serves well the needs of programmers. The performance of discourse constantly makes the viewer aware of the sender's power. Over and against the initial sense that the heterogeneous and dense flow of epic televisuality might offer the opportunity for plural counter-readings stands the realization the stylistic density of the event spectacle actually overdetermines the text. As a result of endless structuration and manufactured artifice, the speaking and presenting presence becomes stronger and more credible within the dense play of televisual style that the miniseries marshals.

When viewers are told by television to stop what they are doing in order to come watch "the motion picture event of a lifetime," they are also told to bring to the screen their proven skills at narrative and history. Once the images and graphics actually start flying at the viewer, however, the broader implications are clear: "Hold on to your seats, you are witnessing a boundless narrative spectacle, an epic history, choreographed in ways never before seen." Glazed eyes and deafened ears not withstanding, night after night after night the networks and their loss-leader competitors hawk the same adrenalized claim to event-status distinction. The new cablecasters, then, are right. "This ain't your father's history." It is, rather, a manic rollercoaster of overwrought telling.

7 Trash TV
Thrift-Shop Video/More Is More

And now for something completely different—a man with three buttocks.
—*Monty Python's Flying Circus*

We've tried not to talk down to children. [Many] Saturday morning
cartoons . . . assume kids aren't intelligent and won't notice if [they] are
inconsistent, illogical and poorly made. We respect kids—their intelligence,
their humor. . . . We're trying to make a better product.
—The producers of *Rugrats*[1]

Trash has always gotten a bum rap from television critics. I would like to salvage
the designation "trash," for it offers one of the best ways to describe and understand televisual excess. If we dispense once and for all with the notion that trash
is a moral judgment and consider it an art historical and iconographic tool, then
one of television's favored guises makes more sense as a historical phenomenon.
Peter Jennings not withstanding, trash television did not begin with *Rock-and-
Rollergames, Thunder and Mud*, or *American Gladiators*. These were merely manifestations of television's increasing penchant for loading up the space in front of
the camera with as much clutter and cast-off material as could be mustered. The
white-trash country-and-western spectacle *Hee-Haw* in the 1960s, Monty Python's
imported *Flying Circus*, and *Sanford and Son* in the 1970s were but three early
examples of the same televisual mode that would be later celebrated by academics as cutting-edge in *Pee-Wee's Playhouse* in the 1980s. If signature television
exaggerates authorial intent and loss-leader television exaggerates history and narrative, then trash television baits the viewer by exaggerating space into overwhelming proportions. Its exhibitionist mantra? "Let's get physical." Its aesthetic maxim?
"One man's bohemia is another man's garbage dump."

To the sensitive boutique of the designer, the domineering marquee of the spectacle, and the mass-market packaging of digital franchising, we must admit, once
and for all, to the importance of an additional televisual paradigm: thrift-shop
video. Within this guise, the power of a show is directly proportional to the sheer
number of objects, items, surfaces, and bodies that a producer can stuff into the
studio space. Liberating trash from its moral definition also reveals that the thrift-
shop paradigm is free from the bias bound in traditional notions of high and low
culture. Seen from a semiotic point of view, then, *Pee-Wee's Playhouse* is little
different from low-culture schlock like *Hee-Haw, WWF Professional Wrestling,
Guts All-Stars*, and *Knights and Warriors*. They all seek to overwhelm the viewer
not with narrative or history, but with physical stuff and frenetic action. From a
systems theory point of view, this kind of communication is dominated by informational noise. There simply is no background. Everything is channeled up front.
Seen from the perspective of the garage-sale aesthetic mastered by many discriminating viewers, however, this kind of spatial overload is one of the chief pleasures that comes from the televisual tube. As any weekend warrior will tell you,

Trash precursors. *Hee-Haw* (its "Fit as a Fiddle Dancers" shown here) and *Sanford and Son* (its 1986 *Redd Foxx Show* clone here) cashed in on *horror-vacui*—excessively loaded space and an unabashed "clutter aesthetic." (Terrence O'Flaherty Collection, Arts Library—Special Collections, UCLA)

garage-sale foraging demands as much cultural distinction and buyer discrimination as any form of televisual auteurism.

In terms of mode of production, trash television privileges techniques deprecated by the higher televisual guises. Whereas cinematic forms of spectacle depend on sensitive photographic imaging and subtle transfers to electronic stock, and videographic televisual forms create virtual electronic worlds, trash television merely collects foraged *things*. If cinematographers, directors, and DVE artists rule the other worlds, then the lower production crafts rule the trash television guise: art directors, set decorators, costume designers, claymation artists, and model builders. Trash excesses have as much to do with carpentry, sawdust, clay, and paint, as they do with electronics or photochemistry. With shows like *Fraggle Rock, The Jim Henson Hour, Pee-Wee's Playhouse* in the mid-1980s, and MTV's *Remote Control,* sets took on an importance they had not been given before.[2] Their complicated physical surfaces, clashing colors, and disjointed angles all provided an experience not seen since *The Cabinet of Dr. Caligari* ruled the expressionist mind in the 1920s. Many other shows shared this fascination with space as an

Hustling hyperactivity off-primetime. *The Electric Company* on PBS and the *Solid Gold Dancers*. (Terrence O'Flaherty Collection, Arts Library—Special Collections, UCLA)

expressive vehicle, even if they did not ape trash's preoccupation with splattered paint and latex molds. The game-show genre, never one to shy away from its own importance, evolved a studio style also meant to overwhelm the audience. *Wheel of Fortune*, one of the most popular shows in the 1980s, proved that audiences would pay to lose themselves in an interior world of flashing lights, bright colors, and artificial surfaces. Overweight visitors from the "midwest" stumbled over their primary-colored and bannistered confines to heave the enameled wheel—even as the ever-mute and vogueing Vanna surfaced her abdomen with metallic sequins of teal and aqua.[3]

While Vanna and Pat Sajak emceed soundstage electric conglomerations modeled on Las Vegas and automall showrooms, the MTV network continued its de facto role as R&D for trash television's neobohemian excesses. Their never-ending succession of bumper and station material proved an endless mine for uglier but hipper versions of trash television. MTV's haste to spin out ever newer physical looks throughout the season meant that the paint and the clay digitally crunched into their spinning graphics never seemed to be around long enough to dry. As with all of the major televisual guises, trash television seldom comes to the viewer clean—its painterly, architectural, and sculptural ecstasies are frequently combined and infiltrated with other modes: videographics, motion photography, animation, and filmed clips.

As the fashion of postmodernism passed through academic circles, shows like *Max Headroom* and *Pee-Wee's Playhouse* faded from the theoretical spotlight. Yet this transience is illusory for the trashy visual excesses of both shows lives on in a variety of ways across the broadcast and cable spectrum. Even economic downsizing and the recession have not ended the programming viability of trash TV. *Dinosaurs*, a Henson clone on ABC, weekly brings to primetime a rubbery and sculpted virtual sitcom world, and a nuclear family that is both paleolithic and postmodern. *Beakman's World* resuscitates Pee-Wee's junk-ridden plastic world and combines it with educational and scientific lessons for network consumption on CBS. *Sea Monkeys* turns comedian Howie Mandel into a performance artist, ricocheting through a fantasy world of crowded technosets, assemblage, fisheye lenses, and characters disguised as lower nautical life-forms. For the younger post-X generation set, Nickelodeon's *Roundhouse*, staffed by a large ensemble of frenetic adolescent performers, hybridizes the discarded sets and props of trash television, together with street-culture hip-hop, and the amphetamine-driven pace of *Laugh-In*.

Far from dead, trash television continues to hybridize itself and infiltrate other forms. The handmade look of trash television became a part of Arsenio Hall's opening in 1993. The sequence boasted "seventy-five cuts in twenty-five seconds," for a mind-numbing subliminal pace of one-third of a second per shot. Each image was hand-painted, pixilated, and animated with wet paint. Like experts on art historical hybridity, the producers argued that the jagged edges that dominated the *Arsenio* sequence were both modernist and African.[4] Even Fox's *Roc*, touted as a renaissance in live ensemble drama, invoked the aggressive violence of trash physicality. It's opening collaged images created from jagged and contrasty linoleum cuts—one of the lowest and crudest forms of the printmaker's art, a technique frequently explored in elementary schools. *In Living Color* mastered trash physicality as well. As its opening sped through a sequence of frenetic shots, performers threw cans of paint at the lens of the camera, an abstract expressionist gesture that invoked both Jackson Pollack's action painting and street-smart graffiti tagging.

Many of these trash shows defy conventional demographic categories. Even those intended for adult audiences, like *American Gladiators*, exploit remedial appeals more typical of children's programming. Many shows, like *Pee-Wee's Playhouse* and *Dinosaurs*, crossed-over in attempts to work both adult and child audiences, a trait that left theorists wondering if children got the loaded and ironic adult references. Such a question could now be directed at almost any trash or children's show on television. An afternoon episode in 1993 of the Amblin/Stephen Spielberg/Warners' series *Tiny Toon Adventures*, for example, masqueraded from start to finish as a *Citizen Kane* clone on Fox.[5] Do kids get the reference to Welles? The producers of the animated series *Rugrats*—having announced their commitment to high artistic standards and to not speaking down to the audience—suggest that children can indeed interpret these references.

By 1992, the apparent emptiness evident in infantile trash programming became an issue with the government. The FCC began to pressure the networks to

Warped architecture, carpenters and set decorators on acid in early, pre-Rubens 1980s. *Trivia Trap* (ABC) and *Out of Control* (Nickelodeon). (Terrence O'Flaherty Collection, Arts Library—Special Collections, UCLA)

insure that a certain percentage of children's programming provided educational value. The official responses of some broadcasters cynically reported that the *Flintstones* and the *Jetsons*, because of their interpersonal dramatizations, were indeed educational programs. Even acting FCC chairman James H. Quello reacted with skepticism by counseling such apologists that if he had to choose which program to air, he'd pick a show that was specifically meant to be educational.[6] Worried about the continued emptiness of children's fare, then, the FCC delivered only an innocuous warning—one that still clearly tied educational status to broadcaster intent and interpretation. In many ways, this ambivalent regulatory incursion was really a smokescreen, for many of the new animated series (*Tiny Toon Adventures, Disney's Ducktails, Rugrats, The Simpsons*) and iconographically trashy shows (*Ren and Stimpy, Beavis and Butt-head, Liquid Television*) were packed with smart cultural references, ironies, parodies, and intertextual paradoxes. Intellectually loaded, yes. Educationally valuable, who knows? With trash television, education is in the eye of the beholder. One lesson, however, is obvious. The spatial and temporal excesses that define trash television, also inevitably flood the viewer with knowing references. The accumulation of junk and gestural marks swirling around the performers in these shows is matched only by the thickened flood of smart cultural codes given off by the very same objects.

The best indication of the continued financial clout of the trash television guise can be found at NBC. The network banked on the look for its corporate image in 1993–1994, when it announced to the public that it was commissioning an array of "important artists" to redo and reinterpret the NBC Peacock, a corporate symbol first created in 1956. Among those artists showcased to "draw attention to the new fall schedule": multimedia director David Daniels (from *Pee-Wee's Playhouse* and Peter Gabriel music videos fame), computer animator Mark Malmberg (MTV's *Liquid Television* and the film *The Lawnmower Man*), animator Joan Gratz (creator of the clay-painting technique), Pop artist Peter Max, caricaturist Al Hirshfield, *Beavis and Butt-head* animator J. J. Sedelmaier, and *Ren and Stimpy Show* animator John Kricfalusi. NBC claimed that the *"new logos express the [network's] new spirit of rebirth, revitalization and, above all, fun."*[7] The hallucinagenic and hyperactive look of trash was publicly positioned as a key to this network's personality overhaul. Even as Paul Rubens's public image was only slowly rehabilitating from his indecency indictment and acquittal, his creative people—and the excessive look he helped popularize in *Pee-Wee*—were being brokered for much broader corporate and network interests.

Case Study: Pee-Wee's Bourgeois Bombshelter
Modernism Meets the Mass Media

In the fall of 1993, Fox premiered *Mighty Morphin Power Rangers*, a show that was hyped by the network as a "live action–science fiction–comedy." Combining "elements of *Teenage Mutant Ninja Turtles, Transformers, The Monkees*, and toy commercials," the new series packed as many different presentational modes as possible within each weekly episode: live action, animation, martial arts, videographics, "interplanetary sorceress Rita Repulsive," and motion-effects robot models.[8] By simultaneously assaulting the viewer through so many stylistic channels and by directly tieing-in the experience to the robot-toys hawked during ads in the show, *Power Rangers* was merely the latest in a line of hyperactive trash programs that followed the success of *Pee-Wee's Playhouse*. Shows in this genre are defined by throwing as much radical-looking form at the viewer as possible, even as they unabashedly promote specific consumer products. This hybrid stylistic and consumerist flurry raises some important issues about the cultural implications of televisuality. On the one hand, the genre bears all the marks of radicality proposed by twentieth-century avant-garde polemicists. Yet the same shows do not wear well the revolutionary mantle woven by prescriptive theorists and aestheticians. In fact, the continued success of *Pee-Wee's Playhouse* on commercial network CBS from 1986–1991 suggests that this particular exercise in avant-gardism was exactly what the mass audience and, one assumes, the dominant culture wanted.

In addition to its mass television viewership, the Pee-Wee phenomenon also worked over ancillary markets. A line of Pee-Wee fashions was introduced and marketed by retail heavyweight J.C. Penney. Other companies sold miniatures of Pee-Wee directly to Americans in a manner proven successful earlier by dolls with less confusing gender traits (like Barbie and Ken). The series also had a direct and innovative influence upon television production methods and visual stylistics.

The production company for the series, Broadcast Arts, simultaneously produced nonprogram forms and commercials that surrounded and were interspersed throughout the show. Clearly derivative of the series in look and technique was an entire genre of rapid and frenetic thirty-second commercial spots for breakfast cereals and junk food. Many of these spots and transitional devices depended upon the same pictorial overload and semiotic abundance that characterized *Pee-Wee's Playhouse*. Pee-Wee had taken the narrative pace and visual density of Saturday morning network programming and cranked them up several notches.

Theorists did not wait long to embrace *Pee-Wee* as quintessentially postmodern. *Camera Obscura*, a film journal with a history of engagement with Lacanian psychoanalytic theory, devoted most of one issue to the psychosexual significance of the show. In the "Cabinet of Dr. Pee-Wee: Consumerism and Sexual Terror," Constance Penley argued that "the interest of this show lies in the way it represents masculinity and male homosexuality" around the problematic and off-limits terrain of infantile sexuality.[9] Gay camp, male hysteria, and sexual inquiry were argued to be keys to the meaning of *Pee-Wee's Playhouse*. Other essays in the journal repeated many of the very same examples to critically justify a Freudian reading of *Pee-Wee*.[10] While industry trade magazines focused on special effects and production technology in the show, other academic journals isolated explicitly homosexual themes within the show.[11] The delimited approach that these articles take to the show (that the *Playhouse* is an allegory of sexual inquiry) and the broader issues that they overlook or downplay (political and cultural issues, the play of semiosis and consumerism) suggest that more is at stake in the new genre than a crisis of heterosexuality. To argue that spoken lines and actions in the program fit, say, a Freudian paradigm for sexuality is to engage in a type of criticism that may be inappropriate for the televisual experience that actually unfolds for the viewer of *Pee-Wee*. This kind of reductive interpretation works well with classical television and cinema, precisely because those forms intentionally conceal their formal and stylistic elements in order to privilege plot and narrative. Plot and narrative elements are precisely the kind of verbal and allegorical components that the *Camera Obscura* critics target, isolate, and extract. *Pee-Wee's Playhouse*, by contrast, barrages and engages the viewer with multiple and simultaneous image, sound, and graphic signs. Viewers face, then, not plot or narrative, but excessive composite signs comprised of sensory elements that otherwise seem unrelated or contradictory. Many of the show's formal combinations are, in addition, explicitly cultural and political rather than psychoanalytic in nature. Rather than acknowledging the overt density of the text that the viewer of *Pee-Wee* actually confronts—a density that is both stylistic and ideological—reductivist and psycho-allegorical analyses verge on the kind of content analysis one associates with empiricist broadcasting studies.[12] In its hyperactive formal and narrational construction, *Pee-Wee's Playhouse* presented a new and different kind of perceptual relationship between the viewer and the television box, one that placed heavy non-narrative demands on the thrift-shop spectator.

The analysis that follows suggests how the perceptual demands of *Pee-Wee* complicate the ways that a television viewer constructs meaning. Such demands result from the program's active preoccupation with visual style, a tendency evident in many presentational aspects of the series. As much as any other show during the decade, *Pee-Wee's Playhouse* established a fundamental *pictorial and icono-graphic* dynamic as an essential part of its engagement with viewers. The analysis that follows will show, first, how the program can be simultaneously seen as both modernist and postmodernist. This confused status thus makes problematic a fashionable critical dichotomy and categorical distinction popular in post-structural theory. Second, the analysis will attempt to describe the peculiar ideological dynamic involved in this excessive form of pictorial engagement. Such an engagement depends not only upon the symbolic nature of the character of Pee-Wee, but also on his commodity-filled domestic environment. The analysis suggests, among other things, that the designation postmodernism incompletely explains the net effect of *Pee-Wee's Playhouse*. A closer and in-depth look at the formal components of this series helps clarify both the continuing pleasures and the ideological stakes that typify the genre as a whole.

Formal Components of the (Play)House That Brecht and Godard Built

Even a cursory application to *Pee-Wee* of the seven tendencies of radical modernism articulated by Peter Wollen shows that this children's daytime genre is an outright act of "counter-cinema."[13] It is immediately clear in analysis that the form of the show strikingly illustrates the progressive norms laid down by avant-garde patriarchs Brecht and Godard.[14] Before considering the show's significance in the 1980s and 1990s, it is essential to show how the program arms itself with the stylistic weapons of radical modernism. The following examples follow Wollen's seven canonized principles of radicality.

Narrative intransitivity. Classical narrative depends upon cause and effect, or what Wollen terms transitivity. With narrative causality, one action, scene, or event seems to justify, lead to, or to motivate subsequent events in the fiction. *Pee-Wee's Playhouse*, on the other hand, dramatically ignores and avoids causality in favor of textual fragmentation. Like other broadcast forms and genres, the program's narrative is necessarily ruptured by commercial breaks and advertisements. But unlike many other broadcast forms, the fragmentation is not limited to the tension between program and commercial. Rather, the process of intransitivity and randomness is intratextual as well. Loaded with a variety of elements, each show is characterized by an extremely unmotivated and loose linkage between individual stories, segments, and program parts. The frantic intra-episodic flow and mix of various skits and performances (the King of Cartoons, Jambi, Cowboy Curtis, Penny's claymation narratives) imply that there is no underlying narrative causality at work—nor is one apparently needed for the show to work.

Estrangement. The Pee-Wee character is not cast or postured as empathetic. Since empathy or pathos are requisite parts of classical drama, the reasons for Pee-Wee's presence as a plastic and infantile child-man are worth considering. Whether Pee-Wee's performance evokes childish hyperactivity or the performance art of Russian futurism, the net result is the same. Both hyperactivity and futurist performance are based on forms of estrangement and objectification. Neither are behaviors that allow for easy engagement and viewer identification. In each episode, Pee-Wee is made an object of the gaze through pixilation and high-speed

Stylistic fragmentation (neoprimitivism, Art Deco, 1950s linoleum, and classical statuary) mirrors the show's disruption-driven narrative form. Pee-Wee, a human projectile, flies past the camera and toys with cooking show conventions. (CBS)

filming. Pee-Wee is also repeatedly filmed in wide angle. Such a lens further distorts his face by magnifying facial features close to the lens (like his nose), while stretching the rest of his cranial ovoid back into the distance. Aggressive blocking of the actor in the opening segment shown each week causes Pee-Wee to fly and orbit the camera. The camera can only frantically pan in an attempt to follow this human projectile.

Secondary actors in the series, like Miss Yvonne and Cowboy Curtis, are also objectified and caricatured through overstylization, costuming, and gaudiness. They function in the playhouse less as sites of consciousness and feeling, than as caricatures and one-dimensional stock types. This performance demeanor, then, is the antithesis of either depth psychology or Stanislavsky method acting. There is no emotional center and little potential for the subjective identification one associates with empathetic and caring daytime hosts like Mr. Rogers. Personalities are reduced to mere surfaces and types. Relationships are subordinate to a textual logic that is more plastic and sculptural than interpersonal. Invariably, these types physically clash, rather than relate, when they make contact with each other. Both in relationships with the other characters and with the audience, Pee-Wee brings new meaning to the ideal of estrangement.

Foregrounding. While more traditional styles work to serve and emphasize content, modernism has come to be associated with foregrounding and an explicit emphasis on form. As much as any other tendency, the inversion of the hierarchy of content and form that dominated the arts from the classical period on has become a hallmark of modernism. While for Clement Greenberg this foregrounding norm involved a reduction to essential formal elements and self-referentiality, others, like Godard, utilize foregrounding as a ritual of reflexivity and self-critique. *Pee-Wee's Playhouse* works on both levels. First, there is no background. Pee-Wee's person is comprised of a world of props, sets, and electronic graphics. Content or subject matter is no longer a meaningful distinction because the formal stuff of television is also the subject of this show. Second, Pee-Wee's foregrounding also provides a ritual of self-reference and critique.

At one point, the actor looks directly overhead and into a ceiling-mounted camera. The viewer is given a bird's-eye view of Pee-Wee's faux-gourmet skills in a clear parody of PBS-type cooking shows. This direct address in no way shatters the filmic illusion. Rather, Pee-Wee's knowing look suggests that generic play is at the root of the playhouse experience. In addition, videographic devices and funky electronic effects are used throughout the program, like the keyed electronic halation effect glowing around Jambi in his box. Even the repeated transitions to commercial breaks foreground form. To get to and from commercial breaks and advertising spots viewers must pass through a television control room replete with preview monitors and images of the show itself. Artifice is front and center and self-reference is pervasive in the Playhouse.

Multiple diegesis. Whether or not most classical works have a *singular* narrative world, they typically have a single *dominant* world that underlies and organizes various subplots. In *Pee-Wee,* however, the entire experience is comprised of subplots and secondary tellings. In one of the three episodes analyzed here (the "missed invitation to the party" episode), there are thirty different parts and segments within the show itself, plus an additional twelve commercial spots. This means that the viewer faces forty-two discrete narrative events within a thirty-minute period. Each diegetic segment or part, then, is less than a minute in length. In addition to the flux and transition caused by these competing diegetic worlds, there are multiple narrators. Magic-Screen narrates lessons in the text, the animated Penny, Jambi, and the King of Cartoons do so as well. Narrating responsibilites are shuffled from agent to agent. Multiple and competing narrations are an integral and regular part of the series, rather than an alienating and disruptive countertactic.

Jambi's funky video graphics, splintered shards of resolution-less 1930s cartoons, and audience address that reifies sight. (CBS)

Aperture. Along with narrative causality and motivation, final resolution is a fundamental component in the orthodox wisdom on narrative. Works that do not fully and finally resolve themselves fail according to such standards. By contrast, irresolution and openness, termed "aperture" by Wollen, are touted as viable countertactics for radical media makers. Significantly, narrative closure and resolution are not a part of the world of *Pee-Wee's Playhouse*. One episode orbits around the issue of alienation by making the viewer guess whether or not Pee-

Wee will be invited to a party. Ironically, this question is never fully answered, since Pee-Wee does not finally make it to the party at the end of the episode. More traditional shows, even on recent television, work hard to answer any underlying fictional questions or premises once they are raised. This obligation to tie-up loose ends is an important part of screenwriting orthodoxy. The *Playhouse*, however, is satisfied to raise central questions, and to move on without fully resolving them. Even the cartoons screened within the show each week are merely random *fragments* of primitive 1930s cartoons—never complete or resolved productions.

In the spirit of the *Playhouse*, the cartoons are shown without narrative beginnings or endings. One can assume that such things, along with many other aspects of narrative thought to be obligatory, are no longer necessary in this form of trash television. Finally, like American television in general, the program switches to commercial breaks only at points of greatest aperture or irresolution. The number of commercial spots, then, also guarantees many secondary points of irresolution. As Sandy Flitterman has argued, this pervasive trait—the cliff-hanging segue to commercial spots—works according to its own devious logic.[15] The constant irresolution of the text, and the viewer's corresponding anxiety, are only answered in the falsely resolved world of the commercials. Because resolution is only available in the ads, the agent of resolution is always extratextual, and the power that comes with resolution is always assigned to the ads. The connection between the world of the fiction and the world of the product is systematically exploited in trash televisuality.

Displeasure. In what must be ranked as one of the great achievements in recent sophistry, the Brecht-Godard-Wollen tradition suggests that pain is an essential part of viable and progressive art. Good and emancipatory works do not pander to the easy sensual appetites of bourgeois consumer/viewers. It is important to note, however, that this brand of politically motivated asceticism was promoted by the theorists before the televisual age. Obviously, a mass market show like *Pee-Wee's Playhouse* would not be on the air if there were *no* pleasure for large numbers of viewers. Even Brecht recognized the need for some emotional engagement along with instruction. Nevertheless, this show is filled with distancing and painful elements. Pee-Wee throws at least two major tantrums in the alienation episode. At one point in the tirade, even the floor yells at Pee-Wee to stop the noise! Pee-Wee also goes ballistic in a later scene, when struggling to write a letter to the Advice Lady. The viewer is, notably, never allowed to see the contents of the letter being written, even though such access would normally be deemed appropriate from a director's perspective. Instead, the shot lingers on Pee-Wee in medium close-up and forces the viewer instead to deal with the emotional trauma (his high-volume vocal outbreak and protest) rather than with the subject of the scene (his formal request for advice). Finally, one recurrent event disrupts these episodes as it does every other installment in the series. On a weekly basis, Pee-Wee tells the children at home to scream at the top of their voices when key theme words appear in the show.[16] As if the displeasure that child viewers inflict upon parents early on Saturday mornings were not enough, all of the cast and props within the diegesis also join in the behavior by erupting on screen in a frenzy of screams and tantrum. Modernism's device of promoting displeasure to achieve distancing here becomes more mainstream and celebratory—modeled less on Brecht than on primal scream therapy.

Reality. Like Godard, Pee-Wee makes repeated use of direct address by facing the camera and speaking to the audience. In conventional terms this sort of address should disrupt both fiction and mindless consumption by showing the televised event to be a concrete act of communication or exchange. Here, however, the device does anything but disrupt or shatter the illusion of viewing. Continued viewership suggests that the mass audience is not apparently threatened or challenged by such confrontations with reality. Other examples of the show's penchant for reality therapy and explicitness abound. In one scene, Pee-Wee is criticized and defensively responds by retorting that he "has a million [TV] friends." One weapon, then, in Pee-Wee's textual arsenal is his ability to acknowledge and exploit the reality of his star persona.

Ruptures and manipulations of reality are also found in more subtle examples throughout the show. Many of Pee-Wee's reaction shots and comments exist clearly for adult consumption. More mature viewers get it, of course, when Pee-Wee calls Ms. Yvonne "real busty."[17] Throughout the episode, Pee-Wee also enacts sophisticated and sometimes subtle ironic facial reactions. These ironic gazes and reactions are directed variously at the characters and the camera. Knowing looks like these suggest that such guises exist for the benefit of learned viewers who know the realities of the media and of the depicted issues. These overtly signaled mannerisms are really a method of acting more related to David Letterman and live stand-up comedy than to fictional film and television acting. Like the aggression operative in much comedy, such references work at the expense of a third party. In Pee-Wee's case the *victim* of the reference is the mock show itself. Ruptures of reality like these, then, are less counteractive and disruptive than they are tactics intended to seal an alliance with the viewer-consumer.

Pictorial Aspects of Postmodernism

In examining two episodes of *Pee-Wee's Playhouse*, my analysis so far has focused on two questions. First, what are the favored pictorial and stylistic components of televisuality; and second, how do these modernist strategies promote or resist viewer oppositionality? Given the fact that *Pee-Wee's Playhouse* so comprehensively fulfills the aberrant ideals of avant-garde radicality, it is worth examining why such shows fail to threaten or disrupt dominant culture. How and why do these kinds of avant-garde strategies survive on network television?

Periodization in arts history can be a trap, as this example clearly indicates. One solution to understanding televisual exhibitionism may be to resist compartmentalizing such works under discrete stylistic categories in order to shift the analysis to the logic or signification process that governs stylistic choice. Much can be gained by considering postmodernism according to its *ideological* rather than *formal* logic. Postmodernism as characterized by Baudrillard, Jameson, Hal Foster, and others gains much of its power by dissolving traditional distinctions between subject and object, public and private, and between different historical periods.[18] One common thread among such theorists recognizes the catholic taste and pluralism of postmodernist culture. Postmodern institutions like television seem ever eager to assimilate even oppositional or antagonistic forces and recast them within a new dynamic of consumerism. Counter-cinematic strategies then, do not guarantee oppositionality, at least under the shadow of postmodernism.

Trash television, and the *Playhouse* in particular, can be further understood by applying the four categories that Jameson describes as central to postmodernism. According to this view: postmodernism turns reality into images; it reduces time to a series of perpetual presents; it utilizes and favors pastiche or blank parody; and finally, postmodernism employs a schizophrenic form of communication and cultural behavior. Consider, then, *Pee-Wee's Playhouse* as a stylistic exercise in postmodernism instead of modernism.

Reality into images. *All* of Pee-Wee's world is artificial. The excessive visuality and artifice of the *Playhouse* makes German expressionist films like *The Cabinet of Dr. Caligari* pale by comparison. The sweeping introduction to each episode is filmed with an optical snorkel camera that roams wildly across Pee-Wee's miniature world. The technique works to present his world as an abundant visual spectacle, so excessive that the viewer must fly over it. In addition, the lighting in each episode is high key and flat. As a result, two-dimensional surfaces are emphasized—rather than the three-dimensional depth and modeling one associates with more directional or expressive forms of lighting. Highly chromatic images are composed of bright pastels or primary colors. Even foods—merely objects of physical sustenance in the "real" world—are seen here as circuslike performers that dance before the spectator. The refrigerator, for Pee-Wee and the viewer, is *framed* as a proscenium. The food becomes a kinetic spectacle that the audience must look passively into and onto. The lighting, color, sets, camerawork, and performance all present the fictional world not as a world, but as pictures. *Pee-Wee's Playhouse* introduced and exploited visual artifice to an extent only hinted at by earlier shows on network television. The *Playhouse is* a spectacle.

Perpetual presents. Time is only slightly less important in the *Playhouse* than visual artifice. The character Pee-Wee is served up to the audience by the show as someone with no readily available history. Pee-Wee has no parents, no siblings, no pedigree, no personal background. Furthermore, Pee-Wee is a walking contradiction of bodily time. He is a boy-man, a person difficult to categorize in terms of normal developmental stages: he has the personality of a child in the body of a man. Time, though, is not simply twisted biographically or developmentally. Sets and props also complicate and problematize the desire for a coherent sense of time. Pee-Wee's room is loaded with various items from the past. Toys from the 1950s, 1960s and 1970s are all pulled out of their historical context and recombined in the *Playhouse* present. In addition, Pee-Wee is shown on or before each station identification spot, frozen on a *preview monitor*. The character is always, as it were, on hold, never over or away from the place and time. Without past or future, Pee-Wee either uses the present for his excessive performance or is depicted in limbo, waiting for the present. Significantly, the room clock in the main set spins wildly out of control, suggesting that the playhouse is out of sync from culture's standardized system of time. Pee-Wee as a boy-man is a temporal odd-

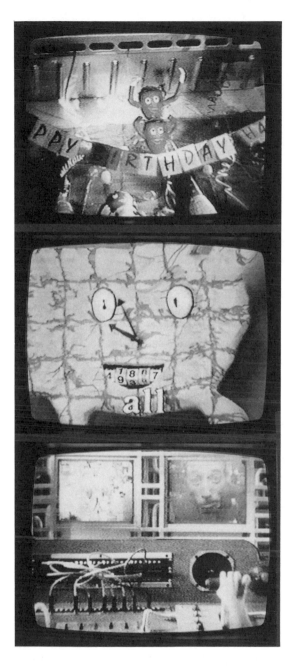

The visual spectacle of carnival devours food behind the refrigerator door. The studio clock spins wildly in a perpetual present, and Pee-Wee is frozen perpetually in CBS switcher's preview mode. (CBS)

ity, a person without natural time, and the playhouse is no less atemporal, for any and all references to time are simultaneously denied, appropriated, and paraded in the perpetual present of Rubens's performance art. Pee-Wee's dramatized atemporality stands in stark contrast to the linkage broadcasting inevitably makes to consumer time—to the program's thirty-minute weekly airing each Saturday morning and to commercial breaks. The relationship is complementary. Televisual atemporality of the *Playhouse* fuels the conscious temporality of consumer time and vice versa.

Pastiche. As Jameson points out, even high modernist works utilized parody. Pastiche, by contrast, is argued to be a blank or disinterested form of parody that typifies postmodern culture. Pastiche operates in contemporary works of mass culture that base their form on retrospective styling. That is, many contemporary works style themselves after earlier periods without any regard for the original intent, logic, or significance of those styles. Retrostyling is also a fundamental semiotic part of *Pee-Wee's Playhouse*. Scores of distinct artistic styles are ripped out of their meaningful contexts and placed in the playhouse environment. A leveling process occurs that decontextualizes all iconographic components of the mix: a surrealist collage is offered to Ms. Yvonne as a portrait; a Greek Ionic column supports a classical portico next to the barber's pole on the playhouse; an aboriginal totem is juxtaposed with a 1950s tear-drop lamp; a red Naugahyde-padded art door—a castoff from the bowling alley subculture—exists in the same room as Gainsborough's classical paintings. All of this overt and self-satisfied aestheticism pervades a room filled with three decades of postwar consumer toys. Pee-Wee's environment, then, uses art history (all of it) and low culture in combination. The result is, to use Jameson's term, an extreme example of "historical amnesia." Historical styles are merely toys that humor Pee-Wee and his friends. In the playhouse, art history gives way to antiquarianism, and style is valued because it can be collected and accumulated. The aura of the playhouse is based less on an art-world aesthetic, than it is on the antique store, the garage sale, and the swap meet.

Schizophrenia. Jameson means by this term a kind of disorder wherein signifiers are no longer coherent or continuous, a view that he bases on Lacan's theory of language. *Pee-Wee's Playhouse* is certainly worth considering in this regard, since it is preoccupied with the play and manipulation of signifiers. Signs and objects are transformed and given new meanings as an automatic and regular part of the show. Signs, inevitably, no longer function as they were intended to. A sculpted Greek discus thrower, for instance, becomes a mere door handle in the Playhouse. By this act, the signifier for classicism is transformed into an item one might purchase at a hardware store. In like manner, one episode taught children how to make "ice-cream soup." Such a "lesson" creates a recombinant sign that disavows the difference between hot and cold food products and that disrupts the designations that regiment cuisine.

Words as well as objects are part of the schizophrenic game of signs. In the dinosaur family vacuum cleaner episode, Pee-Wee teaches everyone to add an "o" automatically to the end of their names. This is clearly reminiscent of word-salad games actually played by schizophrenics, where words are merely spoken and linked based on their sounds and physical characteristics. Other rituals in the series are centered around verbal re-creation and play with language. Children and objects convulse frenetically and gongs go off when nearly meaningless secret words are spoken. Eruptions of behavior based on coded stimuli constantly raise the specter of Pavlov in the playhouse. These weekly word-instigated convulsions might be seen in other contexts as pathological. Here, however, they indicate that the power of the show is based not just on the violation of social decorum but on

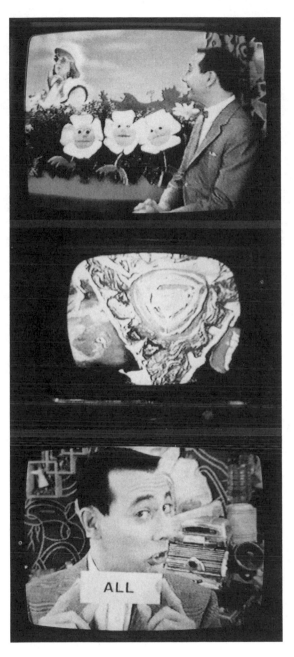

Ms. Renee's departure for a retro-1960s "groovy shindig," collides with undulating abstract expressionism and meaningless Pavlovian word lessons that send the kids at home into sleep-wrecking convulsions. (CBS)

violations of language—its rules, conventions, and meanings.

Another indication of the power of the show's play with language comes in the form of the playhouse "gang," a group that combines emblematic names from mass culture. For example, the real Bud Melman, from *Late Night with David Letterman* fame, is grouped with child characters named Cher and Elvis. This combination of names is an eclectic hybrid of pop culture itself, one that makes

Greek statuary as garage door opener. Pee-Wee is forever locked on the inside looking out, and retreats even further into his mediated womb at the moment of greatest alienation. (CBS)

blank reference to the 1950s, 1960s, and 1980s. On whom, one might ask, is this name game played? Do four year olds get the reference to David Letterman or Cher? In any case, the name gang, the word-salad skit, and the Pavlovian verbal convulsions all show forcefully how any one thing or sign can stand for any other thing in this show. Signifiers are ceaselessly appropriated, shifted, and toyed with in the schizophrenic world of the playhouse.

Thematics: Pee-Wee's Specular Womb

Stylistic analysis alone, then, is a trap. This is especially the case if analysis does not try to articulate the reasoning and logic behind the choice of specific formal codes and stylistic displays. If one function of stylistic analysis is to categorize and periodize, then *Pee-Wee's Playhouse* presents an interpretive problem. Its pictorial roots and formal devices are essentially modernist, while its attitudes and organizing principles are postmodernist. The show performs its style, then, in contradictory ways. Its *look* is oppositional, but its *attitude*, if one follows recent theory, is mainstream.

Stepping back from the show's explicit formal make-up allows us to consider it in a psychological and anthropological sense—that is, as a problem-solving operation. In other words, the process of making formal choices within the show—not merely the traits themselves or the story—reveals an alternative world, a possible way of being, that viewers can engage and identify with. The process of stylizing is a very active and overt phenomenon, a ritual that presupposes and suggests an agent or persona responsible for the acts of stylization. Thus, the workings over of text and object can be taken by viewers as a paradigm for dealing with lived situations, as a model for resolving certain social and psychological problems. This perspective, of viewing the process of stylization in a way that structuralist anthropologists view myth making, makes two things possible: first, Pee-Wee can be seen by his style and (life)style choices as emblematic of a male figure dealing with gender/racial/cultural traumas. Style and artifice, then are not just distractions for Pee-Wee, nor should they be considered ornamental obstacles to be cut through and discarded by the critic and analyst obsessed with plot and narrative content. Such things are instead crucial parts of lifestyle. Secondly, and perhaps more importantly, the narrative evidences a mythic and ritual conflation of the social into the psychological. To show how this ideology of conflation operates, I will describe eight characteristics that work as explicit oppositions and repetitions in the show. Viewed through these textual relationships, analysis shows that the artifice of *Pee-Wee's Playhouse* is also socially symbolic.

Interior versus Exterior

Problems are repeatedly articulated in this show by opposing key terms and positions. The oppositions inside versus outside, interior versus exterior, and personal versus social are inscribed in various ways throughout the show. In many ways, the home is a refuge and a shelter for Pee-Wee. Because all of the other characters are represented only by their irregular incursions into the home-playhouse, they are always identified and set up for the audience as outsiders. Pee-Wee apparently has no family, only acquaintances who may or may not choose to enter his single-unit dwelling. The series positions not only friends, but also threats and enemies, as outsiders who impinge upon Pee-Wee's personal space. Messages also regularly come in from the outside in various forms. Therapeutic advice comes in. Invitations come in. Even aliens from outerspace threaten to violate the secure and womblike space of Pee-Wee's playroom in the "Zyzzyballubah-Alien" episode. Evil also lurks on the outside in less fantastic forms. Door-to-door salesmen are frequently and abusively warded off by Pee-Wee in many episodes. The implication is clear. Evil operates outside the walls of his home, while the inside offers security, reflection, respite, and defense. The walls of the home are symbolic psychological barriers. Even the things that lurk behind and inside the walls are dangerous and dark. In each episode a middle-class dinosaur family unfolds its violent traumas within the abscesses of the homeowner's walls. The unknown is out there, while the known and comfortable is inside. Pee-Wee is fortified within his private property by a wealth of middle-class accumulations. Home-owning here depends more on psychological fortification than economic ability.

Extreme Specularity

Given the marked spatial and symbolic gulf that exists between inside and outside, the show compensates for this heightened distance by overemphasizing sight. This gambit is logical, since the relative power of sight mitigates against spatial distance, especially when aided by optical devices. The privileging of sight is underscored throughout the series by formal tactics that emphasize sight or by scenes that are about the act of seeing. Specularity, then, plays an important part in the narrative as well as the mise-en-scène of Pee-Wee's playhouse. Many of the subordinate narratives and skits center on the act of seeing. Throughout the show the viewer sees numerous screens within screens and televised people watching screens. Spectating, then, is the behavior of choice. The King of Cartoon's performance, for instance, consists merely of announcing and presenting the screening of a film. The Magic Screen provides its own television show, actually a lesson, within each of the televised episodes. The Penny cartoons are viewed by the actors as well as by the viewers. All of these segments firmly elevate sight and viewership as the most fundamental behaviors in the playhouse. The weekly dinosaur installments also center on scenes that refer to spectatorship. In the "alienation" episode of Pee-Wee and the series as a whole, the dinosaur group is pictured as a popcorn-munching, couch-sitting nuclear family. Pee-Wee stoops to a rodent's eye view in order to watch the dinosaur family's spectacle unfold in the stagelike space inside the playhouse wall. Chaos then rains upon the television-watching dino family, as an out of control wind-up robot creates havoc within the domestic sphere. In another episode, the apron-clad mother dinosaur barely avoids violent disaster by wrestling an out-of-control domestic machine—her vacuum cleaner—to a halt. The net result of these oft-erupting spectacles is that the viewer, along with Pee-Wee, assumes an ideal position inside the house, looking out onto any and all "realities." By intensifying and dramatizing sight, *Pee-Wee's Playhouse* teaches and reinforces good spectatorship.

Extremely Mediated Communications

A second dynamic, the pervasive use of extremely mediated communications, also compensates for the heightened spatial distance between interior and exterior. Very little communication occurs on an interpersonal, or face to face, level in the series since characters frequently only interact with each other via electronic devices and communication aids. Very little communication passes directly from

The show is a parade of specular texts: guests emcee special screenings and rodent-sized dinosaur families gaze at the television screen in the walls of the playhouse. Mediated communication makes Pee-Wee the perpetually distant listener-subject. (CBS)

person to person. Instead, characters in the playhouse use message balloons, picture-phones, letters, and mail. Such devices mediate between sender and receiver, addressor and adressee—in a recurrent ritual that substitutes for direct communication.

This repeated privileging of distance and the need for secondary codification in communication suggests that information management is an ideal and privileged skill for the generation that watches this show. Mediation rather than personal

Human and interpersonal contact is only made here through a video camera. Claymation Penny plays out regressive mothering roles, and the dinosaur family feigns domestic bliss in its suburban paleolithic living room. (CBS)

contact is dramatized as the fundamental way by which people deal with the outside world and other persons. Good viewers, and perhaps good children, know how to translate and send their messages and impulses via the media. Skills at codifying speech through secondary sign systems are highly valued in Pee-Wee's televisual universe. This penchant for constant codification and information management provides one implicit message: *children should learn how to package and promote themselves.*

Reversals of Race, Class, and Gender

Since this series clearly takes place within the domestic sphere, it is worth considering—regardless of how far Pee-Wee bends his gender—why the main character is a male. It may be difficult, for example, to accept Pee-Wee in conventional adult male terms, since he talks in a genderless whine. The array of people who come into the space are either people of color (the King of Cartoons, Cowboy Curtis) or women (Reba, Ms. Yvonne) and also challenge traditional roles. Vocational roles repeatedly suggest gender and ethnic ambivalence and ambiguity. Cowboy Curtis is black, but has the mannerisms of the white, western cowboy archetype. The cab-driver is actually a tough, jazz-playing female trumpeter. The closest thing one has to the traditional white middle-class nuclear family much popularized by television, is the dinosaur family, who are forever lurking behind the wall. The dinosaur mother, father, and two kids are always shown within the TV environment of a subterranean living room. Death threatens in many of their episodes, but not from outside. Typically some middle-class commodity goes haywire (a toy robot, a vacuum cleaner) and threatens the very lives of the family unit. Dark in humor, these cynical claymation sketches do anything but support middle-class suburban notions of the family or America.

Pointed depictions of gender and racial roles are also regularly found in the Penny claymation cartoons. In the duck episode Penny describes "girl" ducks as having to stay and watch their eggs while "boy" ducks make a lot of noise. The images that accompany this description show that the boys (ducks) get to listen to boom boxes, and sit by the pool. The girls (ducks) get to mother their young. On one level, that of explicit plot, these representations of family and gender are regressive, since they restate the gender ideals of the status quo. But there is also a great deal of ambiguity lurking under the surface. Do children understand the cultural rigidity and lack of conviction depicted in these jokes, or do they go along with the norms that the jokes voice? Whether or not one can deduce the effect of this ambiguity on the spectator, the trait certainly helps construct an emotional and ideologically ambivalent world in *Pee-Wee's Playhouse.*

In one sense, Pee-Wee defines himself by the constant parade of threats from the outside. The threats are, in effect, incursions from forces outside his control. Women, people of color, and domestic items give Pee-Wee a frantic sense of always trying to cope with new things. If the audience has anything to identify with in traditional dramatic terms here, it is Pee-Wee's perpetual *anxiety*—an anxiety created by the constant incursion of others, of outsiders, and of people defined by difference. Pee-Wee is the emasculated male of the house, posing unconvincingly as its patriarch. At one point even the vacuum cleaner nearly sucks his head off in an attempt to pull him into the dinosaurs' living room. *Pee-Wee's Playhouse,* in the final analysis, takes no partisan stand on race, class, and gender; it simply indulges in an atmosphere and aesthetic of male anxiety, induced by a flood of unstable "others."

Commodities, Consumption

If *Pee-Wee's Playhouse* takes on pointed cultural issues with ambivalence and anxiety, the same cannot be said for its attitude toward objects and property. Without

Pee-Wee nearly gets his head sucked off by a runaway domestic appliance. The playhouse is a commodity ark, celebrating and anticipating the same kind of consumption that is highlighted in ads embedded in the broadcast. (CBS)

question, the textual foundation for this program is built on commodities, toys, and the act of consumption. Pee-Wee exists in a playroom that has *everything*. There are toys from many decades: Indian tee-pees, pirate flags, mechanical horses, robots. Along with these thrift shop finds, there are domestic notables including touristy souvenir china, lava lamps, and highway roadsigns. Pee-Wee communicates with worker ants in his toy ant farm. They communicate back with a hip written message that says, "Dig it." Commodities are everywhere. Significantly, the alienation episode opens with Pee-Wee frantically looking through his toybox for something important—"a real cool toy." Also emblematic of the issues involved in this preadolescent culture of consumption is the Penny segment. Situated here within the "Playhouse Gang" segment that sets it up, Penny describes and defines the entire Chinese race on the basis of its relationship to china—the tableware, that is, the middle-class commodity, not the country. This ethnocentric displacement of signifieds shifts the meaning of household objects to the new and exciting world of xenophobia.

What is the net effect of all of this stuff, of this endless accumulation of manufactured goods? Consider Susan Stewart's explanation of the cultural meaning of "collections":

> We might therefore say . . . that the archetypal collection is Noah's Ark, a world which is representative yet which erases its context of origin. The world of the ark is a world not of nostalgia but of anticipation.[19] Here we find the qualities involved in the apprehension of the miniature: the distanced and "over-seeing" viewer, the transcendence of the upper-classes, the reduction of labor to the toylike, and the reification of interiority.[20]

The playhouse is not a museum, where objects are entombed, and codified, nor is it an archive where time-place origins are clarified and made explicit. No, the playhouse is more like Stewart's symbolic ark and collection of miniatures—a showcased assortment of representative objects stripped of their context and brought along for some as yet unknown future use. Pee-Wee clearly betrays no nostalgia about the toys of the 1950s and 1960s. As a person without time, he seems little interested in either the past or historical consciousness. Rather, Pee-Wee invariably wants to know what this collection of toys can do for him—*now*.

Effective consumerism, in a culture based upon mass consumption, depends upon one pervasive attitude and sensibility—the *desire* to consume more goods. Desire, in turn, is predicated upon the recognition (or illusion) of need or lack and is constantly triggered by the planned obsolescence of manufactured goods. With the mountain of consumer detritus in the playhouse constantly signaling its obsolescence, the viewer is trained and encouraged to anticipate the acquisition and accumulation of more goods, more stuff. In a market economy, the anticipation that Stewart refers to is evident not in any disembarkation from a representative ark or in the propagation of species. Rather, anticipation shows itself in the endless process of accumulation, obsolescence, and consumption. Lack and desire fuel this process. Shows like *Pee-Wee's Playhouse*, along with television advertising in general, work to reinforce this lack and to construct this desire in viewers. By way of summary, then, the series does several things: it celebrates Pee-Wee's masterful accumulation of goods; it reinforces lack with endless visual signs evoking obsolescence; and it structures desire in viewers by skillfully performing and valuing consumption. *Pee-Wee's Playhouse* is not just about infantile excess. It is also a kind of boot-camp training in consumerism.

Pee-Wee as Recombinant Bricoleur

If sheer accumulation structures consumer desire, then bricolage shows viewers how to manage and utilize manufactured goods in a personally fulfilling way. Since commodities form the concrete textual base of the show, Pee-Wee's privileged narrational place is to animate, orchestrate, and recombine those objects. By doing so, Pee-Wee becomes what Lévi-Strauss or Barthes would term a bricoleur. In its original designation, *bricolage* referred to the creative process by which members of tribal cultures synthesize new objects and functions from existing, even discarded, objects. Bricolage was contrasted to engineering, the process by which technological cultures invent and fabricate new objects and machines according to rational and functional rules. Pee-Wee, although himself a product (or castoff) of technological culture, is also very much an intuitive expert at the thrift-shop aesthetic and a bricoleur of postindustrial junk. Each episode shows him accumulating discarded toys and possessions, and mastering their re-presentation or hybridization.

Bricolage, a kind of semiotic mixing and matching, makes possible endless meanings for existing objects. The history of the various domestic objects, toys, and knick-knacks in the playhouse fights the very logic and depersonalization of consumer culture typically reinforced through the process of accumulation. Consumer culture, to reiterate, is based on the idea that anything one purchases is already obsolete once the commodity changes hands. Pee-Wee's bricolage counters this logic by dehistorifying discarded objects, personalizing them, and giving them new use value. Bricolage, then, *should be an oppositional practice,* for it resists the logic of commodity culture. Unfortunately, like car customizers or art-clothes fashion designers that mix and match past styles, Pee-Wee's bricolage actually functions to celebrate mass culture. It is a game to be played well and inventively, rather than a game based on disrupting and antagonizing mass culture. Pee-Wee seems less of a shaman, therefore, than a resale and performance artist. He catalyzes and transforms groups of inanimate objects into new functions. By doing so, he also markets and promotes himself and his mastery of commodity culture. In addition to privileging mediated communication, the program gives premier status to the person who can maximize meaning by managing the material of consumption.[21] Commodity bricolage and creative consumption are skills that the viewers of Pee-Wee are practiced in and rewarded for. Such things are not skills that can be acquired through cultural separatism or asceticism, for such lifestyles mitigate against the direct experience of shopping and its many and diverse pleasures.

Site of Extreme Intertextuality

The concept of pastiche discussed earlier is part of a more general process termed *intertextuality*. This is because pastiche appropriates styles and looks from other periods and times for its own textual use. In addition, specularity, or texts that are about spectating, also prove to be intertextual, since they make reference to, and enact, formats and audience relationships found in the other arts—film, theater, sculpture, and painting. But these two tendencies suggest only part of the

Playhouse's utilization of intertextuality. The character Pee-Wee himself exploits numerous and sometimes esoteric intertexts, including the performance and demeanor of the child-man archetype and convention of Buster Keaton, Charlie Chaplin, and Jerry Lewis. Consider also in this episode the mixture of high–art world styles in set construction, along with Mrs. Renee's pop culture promotion of 1960s "groovy shindigs." There is also a collaging and collapsing of mixed performance styles and generic references—the conventional stand-up comedic approach along with *Late Night* irony, the soundtrack parody from *Jaws* over the vacuum cleaner attack; the parody of *Oz* during the playhouse abduction in the "Zyzzyballubah" episode. These references are wide-ranging and eclectic. *Pee-Wee* appropriates texts from classical Hollywood, the art museum, and the club scene alike.

The density and complexity of this intertext raises questions as to whom *Pee-Wee's Playhouse* is directed. The level of engagement demanded by the complex aesthetic origins of the show assumes a kind of familiarity and cultural capital that few children would possess. Sociologist Pierre Bourdieu has argued that aesthetic sophistication and awareness are required by this kind of complicated intertext. Obscure aesthetic references demand a somewhat privileged class-based education in the arts. It follows that works that utilize such references are by nature elitist, since few viewers have the skills to decipher such works. But since young children watch and enjoy the show, this explanation seems implausible. A reading based on the notion of aesthetic demands is only one possible explanation, for children of many ages watched the show. There are in fact a number of possible ways to read *Pee-Wee*, since it combines multiple discursive and semiotic levels, and it does so for an audience that is anything but homogeneous. This then demonstrates the trap of sociological interpretations that try to reduce such shows to singular ideological readings. After all, the audience inevitably engages in multiple, and even contradictory, readings. The hip, art world chic and intertextuality of this program are not necessarily the only pleasures of viewership, as any child can attest. Such tactics are, however, incessant parts of the show's performance of style.

Social-to-Psychological: *Pee-Wee* as a Therapeutic Discourse

Underlying the spectacle of *Pee-Wee's Playhouse* is a ritual that conflates the social and the psychological. The recurring textual oppositions between inside and outside reinforce this polarity. So do the mediating factors that bridge the poles between the social/public and the psychological/private world. Consider, for example, how extreme specularity intensifies the sense of distance by constantly requiring that the viewer look; or how the incessant mediation of all forms of communication actually emphasizes the separation of individuals from others. The irony is that while both traits—incessant looking and mediated communication—are presented as devices by which to bring the social/public other closer to the private, the artificial way that they do so betrays a great and significant distance. To summarize the premise of this show in its most simple terms: Pee-Wee is forever, socially alienated. His acquaintances literally do not invite him to the party,

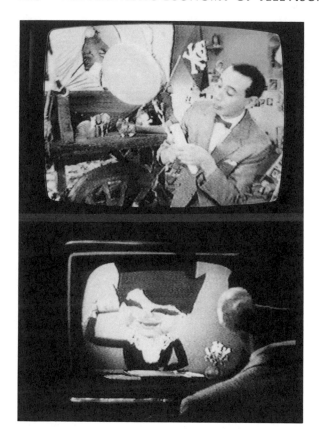

Engineering a communique from the detritus of Western culture. Pee-Wee's concrete social problem is couched only in therapeutic and privatistic terms as he "tunes-in" for professional help. (CBS)

and the show expends great energy pondering the significance of this traumatic alienation.

Pee-Wee's repeated, cathartic tantrums appear to be but one way of countering separation and isolation at home. Pee-Wee also attempts various diversionary therapies.[22] At one point he prepares food and binges. At another point he tries to solve his problem by escape, by watching images. All of his efforts at self-therapy are to no avail. Ultimately, he seeks professional advice from the Advice Lady, whom he dials up on the picture-phone. Pee-Wee, in effect, is told by the therapist to be honest and to "get your feelings out." This he immediately does in an outburst over the phone to the host. His resulting agitation offers only further alienation. Pee-Wee's dilemma is a social one. He is cut off from other persons, alone in his room, fortified and surrounded by his commodities and spectacular fantasies. The solution offered by the episodic narrative is that he should seek professional advice, professional help.

Significantly, Pee-Wee can only enter the therapeutic relationship through a situation involving extreme mediation and isolation. He cannot personally interact with the professional. Instead, Pee-Wee is strapped into the claustrophobic picture-box and forced to communicate to the therapist through vintage garage-sale audio and video apparati. The therapeutic hook-up—in what should be a deeply intimate relationship with the therapist—ends up having the technical quality of a distant satellite hook-up. "Tuning in" means visually wading through animated abstract expressionist paintings, listening through a tin can, and yelling through a picture phone. Alone in his media box, preoccupied with himself and his needs, there can be no better image of Christopher Lasch's *Culture of Narcissism* than this.[23] Although the show has established that Pee-Wee's basic problem is a social one—he is cut off from the social/public group—the episode transforms and reduces Pee-Wee's dilemma to a private, psychological, and self-centered issue. In short, the episode is a ritual that psychologizes what is otherwise a concrete social and communal problem.

Consider the origins of Pee-Wee's hysteria. First, it involves the almost psychotic practice of personalizing and anthropomorphizing objects and commodities from the outside world. Second, Pee-Wee is traumatized by the ambivalent cultural depictions of race and gender that invade his home from the outside. Given these threats and mental operations, it is easy to see how the conflation of the social and psychological works. Pee-Wee's problem is not answered in any cooperative, practical, or communal way. Rather, he goes outside himself, not to make contact with other persons, but only to find a professional, an expert who can inform him of proper notions regarding selfhood. A professional therapist, à la Dr. Joyce Brothers, gives Pee-Wee emotional solutions for his social problems. This episode only resolves itself, in a narrative sense, by deference to outside therapeutic expertise. In the final analysis, even this therapeutic gesture turns out to be woefully incomplete. Pee-Wee frantically escapes from the playhouse—as he does at the end of every episode—on his customized scooter. Although this exit is preceded by discussion suggesting that he may finally go to the party, his escape at the end really goes nowhere. It turns out to be the same stock end-title sequence that is used week after week. Pee-Wee is, in short, in an endless loop of retreat. Pee-Wee never really makes any human contact after all. Pee-Wee never leaves the extreme mediation and artifice of his world.

It is possible to view the show's artifice and Pee-Wee's excessive manipulations of style and commodity as his frustrated attempts to deal with and make sense out of the world. The irresolution at the end of the episode suggests that the kind of personal therapy enacted here is without end. Nothing is changed, nothing is gained. Pee-Wee will simply act out his hysteria and psychosis next week in an endless play of embellishment, style-consciousness, and excessive visuality.

This televisual transformation of the social into the psychological, of style into lifestyle, may be the key to *Pee-Wee*. If media practitioners interested in change are ever to get past the kind of televisuality popularized by series like *Pee-Wee's Playhouse*, it is unlikely that they will do so in a purely formal sense. *Pee-Wee* is an extremely complex operation of style and embellishment, one that makes a mockery out of the notion that radical innovation is possible in formal terms. The networks, in short, have mastered a favored tactic of the avant-garde. Given this fact, alternative producers might consider resisting the dominant media system by reversing the ideological dynamic and *logic* that drives this brand of televisuality onward. Instead of locking people into sheltered and specular wombs defined by material consumption and therapeutic discourse— that is, into media worlds where the social is made psychological—alternative media makers could attack the very site of this conflation and contradiction.

By symbolically and narratively rupturing the psychological and forcing or reappropriating it into the domain of the social—that is, by depicting psychological problems and giving them social solutions—alternative television productions might recast and regain control of the terms in the cultural equation. If one looks at television narratives as cultural problem-solving operations, then the institutions that get to dramatize and enact the terms of the problem, also delimit potential solutions. In this sense, the new televisuality is also a political issue. By questioning and challenging the aesthetic paradigms that we operate under, even production practice can be seen as political. As long as excessive televisuality embellishes and pervades television, the sort of contradictory, ideological dynamic at work in *Pee-Wee* also provides a site that is open to engagement by producers. Excessive visuality and formal radicality are now legitimate properties of the dominant media, even in its trash variants, not the avant-garde. Independents have at least two choices. They can either continue theorizing alternative looks and different ways of seeing, or they can embrace the new and excessive televisuality while at the same time sabotaging the therapeutic logic that drives television toward its self-conscious performance of style—and its trashy pleasures.

8 Tabloid TV
Styled Live/Ontological Stripmall

You cannot, in search of news and profit, break into people's houses this way. It is simply intolerable.
—U.S. District Court Judge Jack Weinstein, ruling against CBS *Street Stories*[1]

I think there's an opportunity in the marketplace for a user-friendly magazine show, but we've got a lot of work to do.
—Derk Zimmerman, president, Group W Productions[2]

Lock your doors. When breaking and entering and user-friendliness are considered parts of the same documentary genre, then no one is safe from stylistic exhibitionism. Even live and documentary television forms now bear the scars of televisuality. During the February sweeps in 1993, the *Geraldo* show lead with a story about men who murder their wives. The program began by televising the assassination of a woman by her estranged husband. The spousal execution had been recorded "live-on-videotape."[3] Gathered on-stage were other victims of point-blank assassination attempts. The announcer's hype and the number of victims collected on stage suggested that this particular matrimonial ritual was not uncommon. The first on-camera question was directed to a woman who survived "five gun blasts" to the face and abdomen. Ever the sensitive host, Geraldo asked her not about her own experience as a victim, but about how she felt when viewing the televised camcorder tape.

The videotape, then, was clearly the real showcase—a bloody videographic artifact bought and appropriated from a local station. Although the participants on stage were primarily there as witnesses and confessors—to muse on the status and nature of the videotaped icon—the show also punished them. By forcing the first interviewee-victim to view a televisual execution—one that the producers had helpfully conflated with her own—the show locked the woman into a Pavlovian reprogramming vise reminiscent of Kubrick's anarchist victim in *Clockwork Orange*. Although such shows typically feign therapeutic value, they also exploit and manipulate liveness in important ways. Videotaped liveness has become a charged apparatus by which personal behaviors converge in public spectacle. Portable tape in contemporary television does not really function like the personal, epistemological pilgrimages in Antonioni's *Blowup* or Coppola's *Conversation* either. The narrative brackets that set aside taped imagery as a special mode of modernist experience are gone. Tape, masquerading as liveness, is now the recurrent mode by which the public intervenes in the private.

Yes, even liveness and reality are televisual constructs. Having stepped through the higher televisual modes—the signature boutique, the loss-leader marquee, and the videographic franchise—we arrive finally at the one form of television that promises most to resist the long-arm of artifice: reality programming. Along with the deprecation "trash," hostile critics in the late 1980s also coined and flaunted

Amy Fisher was a tabloid phenom long before all three networks cranked out "reality" epics on her bloody affair with Joey Buttafuoco. During the standoff in Waco, NBC frantically built a duplicate Branch Davidian compound for their Movie of the Week. Tabloid interest in Koresh faded almost as fast as the smell of smoke left the air in Texas. (NBC, NBC)

"tabloid" television as a favored form of castigation. Like *trash*, *tabloid* is a term that can be salvaged for iconographic ends. The term aptly describes two tendencies imported from tabloid print journalism: the heavy emphasis on pictorial stories and illustrative matter and an obsession with short and sensational topics. Like the obligatory one-sentence paragraphs that define print tabloids, tabloid television takes singular but aberrant headlines from the news and turns their one-shot punch lines into endlessly mutating video wallpaper. Despite the fact that a segment on *A Current Affair*, *Unsolved Mysteries*, or *Hard Copy* may run five, fifteen, or twenty-four minutes in length, very little additional information beyond the original headline is usually released during that time—that is nothing is added to the promotional angle that was heavily hyped and repeated throughout the afternoon programs and newscasts that preceded the tabloid.

Since there is, in fact, little or no actual journalistic investigation involved—a defining absence that saves production costs and makes a show dependent upon preexisting stories otherwise buried in the wire and satellite subscription services—tabloids exploit the only viable presentational process left to them: the endless elaboration, dramatization, reiteration, and re-creation of some aberrant event or sensational hook. Lacking even journalism's fetish for context and backstory, tabloid and reality producers have become masters of the closed representational loop. That is, a story's success is dependent upon how many different ways the headline can be said and resaid, shown and reshown—via live videotape, slow mo-

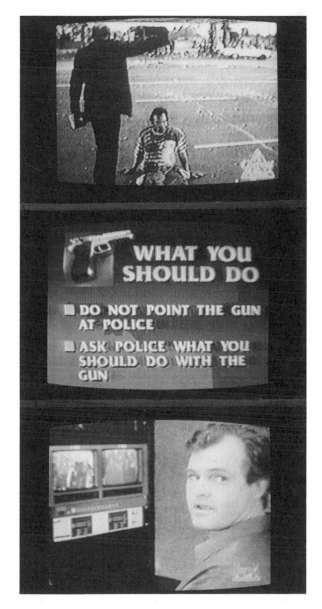

Current Affair scooped the networks when L.A. burned. Other shows taught viewers how to avoid being shot by police at home. *Beverly Hills 90210* did an episode critiquing tabloids on the very network that had helped birth them, Fox. (*A Current Affair*, Fox)

tion, black and white, color, archival collages, and dramatized visual reenactments. Electronic postproduction rules this mode of production, for the only way to stretch topical tabloid fragments into a segment is by endless videographic permutation, not by factual substantiation. There is both an aesthetic and economic logic to this stretching, for syndicated producers face the Herculean task of filling hundreds of nightly half-hour slots, year in and year out, with reality material.

By isolating productions in simulated newsrooms, by stripping the journalistic origins from any story, and by centering production in on-line postproduction suites, producers benefit from both economies of scale and mass-production. Even

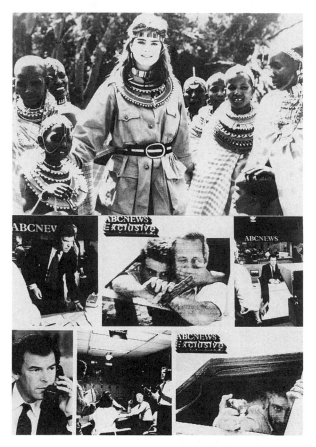

The participant observation question. Brooke Shields goes native on *Exciting People, Exotic Places*, while Peter Jennings keeps a watchful eye on Islamic terrorists—6,000 miles away. (Terrence O'Flaherty Collection, Arts Library—Special Collections, UCLA)

though syndicated tabloid shows are marketed nationally, they face few of the bothersome burdens that come with real national news networks: the assignment of regional and international reporters and bureaus, production crews, and the massive deployment of production and transmission equipment. No wonder real broadcast journalists hate this stuff. Labor has been erased along with journalistic method. Many tabloid and reality programs are really just reality clearinghouses. Together, each night in affiliate markets across America, they comprise what might better be called "ontological stripmalls." From the early afternoon talk shows, to local newscasts, to pre-primetime tabloid and cop shows, the embellishment of reality and liveness rules television. Timewise, these programmer's strips now overshadow any other form on television, including soap operas, a genre that academic theorists once considered television's quintessential narrative form. Always available down at the pre-primetime "corner," tabloid and reality strips now promise the viewer quick-stop shopping to meet any and all tastes: for the exotic, the kinky, the violent, and the voyeuristic.

Although critics tend to write off syndicated near-primetime, magazine shows like *Current Affair, Unsolved Mysteries*, and *Hard Copy* as tabloids, the presentational forms of those shows apply as well to many ostensibly more serious network offerings, like *Inside Edition, Street Stories, 48 Hours*, and *20/20*. While

Robin Leach's *Lifestyles of the Rich and Famous* and *Entertainment Tonight* provided financially successful prototypes for tabloid reality programming in the early and mid-1980s, CBS's *West 57th Street* provided the same kind of viable prototype for the neworks' subsequent attempts at niche reality journalism. Aimed at a young and quality demographic, and produced by the same network that had seen senior citizens Mike Wallace and Harry Reasoner dominate yearly Nielsens in *60 Minutes, West 57th* was seen as CBS's flirtation with America's twenty-something yuppie culture. In its dynamic and hip opening montage and transitional segments, young and attractive network reporters dashed past cameras, tossed their manes of coiffed hair, and sensitively emoted in close-ups. The same shirtsleeved and hip reporters appeared on camera, in stage-lit tableaus, with red and green gelled light, expressionist silhouettes, and production gear in the background. The segments that followed came as close to the condition of music video as conceivable on the network. The hip reporter-hosts functioned more like video DJs than wraparound correspondents there to set up a story. *The Selling of the Pentagon* or *NBC's White Paper Specials*, this series was not. Its segments were short and always flashy. The somber specter of Edward Murrow had, apparently, long since vanished from the halls of CBS's network news division.

By Fall, 1993–1994 NBC, CBS, and ABC had recognized and cashed in on the value of this genre. Nine different hour-long primetime news specials were scheduled and hyped per week. Cheaper than hour-long dramas, such series were also owned outright by networks—not by production companies or studios who could share in profits or savings—for the life of the show. Idealized as a more respectable variant of reality programming than the tabloids, network news specials have in many cases actually become poor imitators of the lower tabloid form. Roger Ailes, executive producer of the shock talk show *Rush Limbaugh* and consultant on *Hard Copy, Entertainment Tonight*, and *The Maury Povich Show*, notes with relish that the network news divisions now frequently plagiarize the tabloids.[4] *ABC News* ran with *A Current Affair*'s L.A. riot story in primetime. *Hard Copy's* original "hot" story on Long Island Lolita Amy Fisher was subsequently cloned by most of the network newsmagazines. Long after escaped convict and former Playboy Bunny Bambi Bembenek had made the rounds of the tabloids, network heavyweight Diane Sawyer took her turn with Bambi on primetime.

Not only do serious network offerings take from tabloids without shame, they have also learned well the programming power of tabloid stylization and sensationalism. *First Person with Maria Shriver* is one of the best examples of the network news special pushed over the televisual edge. Heavily promoted and scheduled to premiere immediately after the Super Bowl spectacle in 1993, *First Person* promised an incisive look at "gays in America," one that would cut past the stereotypes. What the predominantly male (and, one might assume, drunken) audience carried over from the Super Bowl saw, however, was an ecstasy of contradiction. An elaborately produced opening showed Shriver's high-heeled and strutting silhouette cutting through the gleaming space of a gentrified but empty Sohoesque industrial loft that had been painted entirely white. An omniscient establishing shot slowly craned down to show this fashion statement, even as it revealed the crew of another camera pulling their track-mounted dolly backwards in the face of Shriver's incursion. As Shriver advanced and laid down her heavy

West 57th Street's stylish angle was a breakthrough for yuppie demographics at conservative CBS in the mid-1980s. Loft-lit, dolly-announced Maria Shriver strutted her designer stuff even as she gay-baited in her "documentary" on *First Person.* (CBS, NBC)

rap, the show left no doubt as to what it meant by "first person": indulgent bodily narcissism and couture rather than journalistic point of view. Shriver had apparently gotten her journalism degree at the same school that trained supermodel Cindy Crawford of MTV's *House of Style.* Once into the living rooms of gay couples, however, Shriver gushed with empathy in a ritual of concerned nods that would give most people migraines. Although she was lit with more expressionism and attention to detail than any of the couples being featured, one thing was underscored repeatedly: the ever-progressive Maria "cared." Unfortunately, this empathic guise was repeatedly shattered by music-video segments comprised of frenzied and partially clad gay rights activists demonstrating and gyrating in ways destined to send middle-America running for cover. Maria was, clearly, either confused or duplicitous. This reality news special in primetime, then, was no more than meat-market voyeurism, disguised as a sensitive and serious look into gay living rooms. Shriver was clearly the self-satisfied subject. Gays were merely exotic background embellishments, meant to showcase the ever-sensitive primetime artist.

These examples—the syndicated tabloids, primetime news specials, and shock talk shows—clearly suggest the degree to which reality programming has become possessed by exhibitionism. Even the cop reality genre, in shows like *America's Most Wanted, American Detective,* and *True Stories of the Highway Patrol,* consistently come across more like Rod Serling than Frederick Wiseman or Rickie Leacock, more like the *Twilight Zone* than cinema verité. For all of these subgenres,

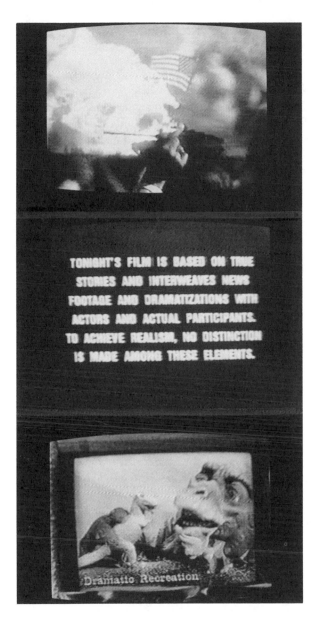

Constant markers make the viewer aware of the theoretical tension between realism and liveness and dramatic re-creations of everything from the Gulf War to reenactments of it on *Dinosaurs*. (ABC)

reality has become raw material to be mined and reworked. Reality shows now tend to transform the world into a succession of "altered states." Not only has televisuality changed the nature of reality and news on television, it has also impacted two allied notions and traditions. The following section will examine the impact of televisuality on the documentary form, a guise frequently imported to television from film. A subsequent analysis in chapter 10 on liveness—a reality guise seen by many as unique to television –will attempt to demonstrate how changes in the conceptual frame and industrial base for liveness has altered the viewer's relationship to the both the televisual image and the real.

Case Study: The (Tele)visionary Documentary
(Re)Inventing the Wheel

Consider a statement by two newsmaking marketeers and architects of the postmodern spectacle, Mark Frost and David Lynch of *Twin Peaks* fame: "People have fairly rigid notions in their heads about the term 'documentary.' . . . The idea is to create a very lyrical, visual style that lets the pictures tell the story. . . . And there is a voice-over narrator, but no on-screen narrator. No one comes barging in to dominate the screen. It's the narrator as a social anthropologist from outer space who is visiting."[5] With these earnest words the coproducers speak not about their approach to low-culture psychodrama on primetime, but about their notion of documentary. This description defines for the two the ideal "docu-poem."[6] It is a self-styled concept, one that described and promoted the 1990–1991 documentary series by Frost and Lynch, *American Chronicles*. The statement suggests that the documentary is both visual (about "pictures") and extraterrestrial (like "anthropology from outerspace").[7] The definition is both revealing and enigmatic. Although the statement seems oblivious to the many historical precedents for poetic documentary, Frost's words incompletely account for a new and broader phenomenon—a genre that might more accurately be called the "televisual documentary." The following analysis aims to account for the formal traits, spectatorship, and ideology of this new style-bound documentary form in order to demonstrate the relationship between televisuality and impressions of the real.

Surely two of the most influential film theories during the last two decades were those centered around concepts of the gaze and the apparatus. Various studies focus on the place and import of technology, voyeurism, and ideology in the construction of meaning.[8] Documentary theory, of course, has its own privileged master paradigms and ideals—actuality, the fly-on-the-wall myth, verite, social action, documentaire engage. Brian Winston, for instance, has argued convincingly that the field of documentary theory has overlooked the fundamentally scientific method and paradigm in the formation of the documentary mode.[9] Few documentary critics, however, address in the same way the concerns of Jean-Louis Comolli, Jean-Louis Baudry, and Christian Metz on the apparatus and Laura Mulvey on the gaze. It is important to consider the extent to which the master paradigms of the gaze and the apparatus are useful in analyzing contemporary documentary.[10] Mass media has changed much in the 1980s and 1990s, and recent developments in documentary and in television greatly complicate the task of generalizing about the ideological effects of the basic documentary and televisual apparatus.

Given the central role that television plays in the current production and reception of film documentaries, and the increasing pressure to produce and distribute documentaries on video, it is important to examine the ways that television challenges traditional notions of the gaze, reality effects, and the apparatus. Not only does television challenge the idea of the gendered gaze, but it makes problematic theories of reception based on perspectival centering and voyeurism. A survey of the paradigms that feed into documentary television, and analyses of several documentaries broadcast on television in 1989 and 1990, will suggest alternative ways that the televisual documentary spectator is positioned.

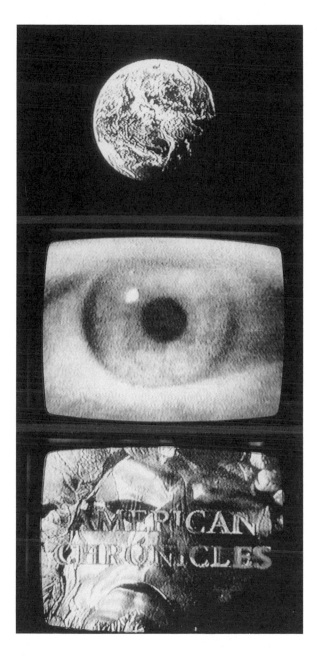

Alien vision and astral travel in the omniscient eye of Lynch and Frost's "visionary" documentary, *American Chronicles*. (Fox)

Television has recently rediscovered documentary in several important ways. Documentaries are increasingly shot on film and postproduced on video (in series like National Geographic's *Explorer*); film documentaries are coproduced with public television and Corporation for Public Broadcasting (CPB) funds (the documentary consortium at WGBH that feeds *Frontline*); documentaries are revised and packaged in different forms for broadcast, as part of thematic series (the Learning Channel's *The Independents*, and WNET-13's *Independent Focus*); and many

American documentary productions are funded by presales to foreign television like Germany's ZDF and Britain's Channel 4. Film documentaries are frequently only seen *on film* at festivals. The promotion and marketing of documentaries takes place almost entirely on video at places like the Independent Feature Project (IFP) Market in New York. Admittedly, an appetite for the cinematic image has affected the development and design of video imaging, the Rank-Cintel, and the CCD Betacam. Even though this influence of film upon video has been significant, the influence has not taken place in isolation. Far from it. Television has also fundamentally altered the way we see documentary films and documentary. In short, any concern for the ontology of the documentary image must take into account how current documentaries are actually seen.

Theorizing about the nature and meaning of the documentary is a risky task. One danger is that because it generalizes, theory tends to speak of the genre as an autonomous and identifiable form rather than as a convergence of heterogeneous practices and interrelated codes. Ironically, the power of Baudry and Comolli's and Mulvey's theories was that they offered universalizing explanations about why and how the cinema works. Their analytic tasks were complicated but their conclusions, their ideological effects, seemed uniform and universal. The technological apparatus was not merely technological; it was a pictorially codified device by which dominant capitalist culture bound the spectator in a dreamlike state of consciousness. The gaze in classical cinema, on the other hand, was the privileged, formal way that spectator sexual desire was channeled and appropriated for fictional effect. The power of Mulvey was that she tied ideological theory to concrete stylistic practice (for example, the image). The problem was that cinema was not as homogenous as she implied, nor as clearly male, pathological, and heterosexual. One can only generalize about the gaze or the apparatus, however, if one talks more specifically about a social context. In the case of contemporary documentary this specific social context must include, or at least address, the issue of television. As distasteful as this may be to film purists, film documentaries are transformed by video and television in substantive and complex ways when viewers actually watch them.

Television has not been absent in the literature about documentary. In fact, it has assumed a privileged role at several key points in the form's history, as an example of socially critical journalism with Murrow and others; as a catalyst for the development of cinema verite with ABC, the Drew Associates, and the Maysles. But a survey of recent important texts on documentary like Alan Rosenthal's is indicative of the limited role that theorists grant for television in the development and reception of documentary.[11] Chapters in the recent anthology by Rosenthal, for example, are mostly preoccuppied with issues of veracity, journalistic coverage, censorship, and bias. Not one chapter deals with the presentational ways that documentary texts are actually seen on television. By contrast to the decade that film studies spent describing how film formally positions the spectator, documentary studies tend to avoid issues of text and spectator altogether in favor of extratextual issues. With the exception of important work by Nichols and Renov, documentary studies tend to view television through the lens of politics, ethics, history, and journalism.[12]

Given this context, much remains to be done in closely analyzing the peculiar *videographic* ways that documentary codes are bundled and synthesized when they are broadcast and cablecast. In a semiotic sense, the televisual adaptation of documentaries is more than just a factor that creates a new social or viewing context. The process also modifies the documentary text itself. It is not enough, therefore, to describe how documentaries create the impression of reality or reality-effects. One must also ask whether other effects are also constructed and how those effects alter the viewer's ontological relationship to the image. As I have argued, a major shift in aesthetic paradigms has occured in television at large in the 1980s. The industry has seen a marked shift from programs based on rhetorical discourse to ones structured around the concepts of pictorial and stylistic embellishment. This general shift in the conceptual frame implicates documentaries as well as other genres.[13] In contemporary television, televisual and semiotic overabundance now describe documentaries as much as they do music videos. Consider the following four examples of what I term the "televisual documentary": *American Chronicles* by Frost and Lynch; the nationally syndicated documentary magazine *Eye on L.A.*; the broadcast of Christian Blackwood's *Hotel* on the PBS series *POV*; and a special episode of *American Playhouse* comprised of four short films, including *Tribes*, directed by Matt Mahurin. Such works demonstrate that the televisual documentary is a genre marked by a semiotic overcoding that traverses the traditional split between "mainstream" and "independent."

Four Televisual Documentaries

The documentary series *American Chronicles,* produced by David Lynch and Mark Frost, premiered on the Fox Network in fall 1990. Since the televisual documentary is characterized by stylistic indulgence, it is no coincidence that the art school–trained director of *Eraserhead, Dune,* and *Blue Velvet,* a director known for his eccentric style and vision, would prove valuable and attractive to programmers. Not only were Lynch and Frost riding a commercial wave of popularity started earlier in the spring by *Twin Peaks,* but they also offered something artistically valuable to the up and coming fourth network—a distinctive look and a signature sensibility that could help give Fox the programming product differentiation it needed to battle the big three networks.[14] Since Fox was not financially equal to ABC, NBC, and CBS (that is, in terms of numbers of affiliate stations and total advertising revenues), much of its early programming attempted to exploit distinctive shows and cutting-edge attitudes.[15] In this light, the shift to excessively styled documentaries like *American Chronicles* was not just a result of the aesthetic evolution within documentary form, but was rather a tactic tied to a broader commercial agenda. With Lynch now hyped by the network as a documentarist, the terms of the reality genre had obviously changed.

One episode of *American Chronicles* entitled "Biker Nation" ponders the international Harley motorcycle culture that annually descends en masse upon the tiny rural town of Sturgis, South Dakota. Typical of the series, this episode was not a direct depiction of the actualities of the subculture nor a representation offered by an effaced and unobtrusive observer. Rather, the viewer confronts an excessively stylized and ironic pose toward a curious subject. Although presented as reportage, dry, offbeat, and understated ironies emerge throughout the work.

Native-American "warrior" hails oncoming Harley bikers as his peoples' spiritual heirs, before dissolving into the Frost-Lynchian landscape. (Fox)

Somber words by a Native American—who scans the contours of the great plains horizon looking for the coming force—consciously evoke *bikers* as the true spiritual descendents of his people. Richard Dreyfuss, then, ponderously recites narration over visual apparitions of the migrating Caucasian biker hordes. The choice of Dreyfuss, of *Close Encounters of the Third Kind* fame, brings additional generic and cosmological baggage, for that movie was also about the convergence of extraterrestrials in the northwestern great plains.[16] The documentary's images

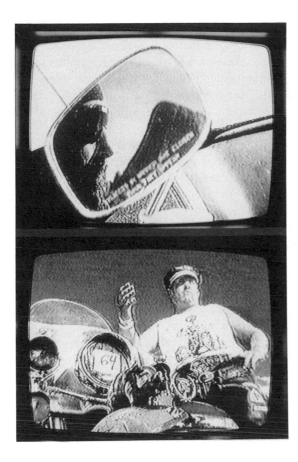

Every image is overcomposed. Obsessive optical details, visual tangents, chrome, and low-angles of metal and leather-clad road sages. (Fox)

and pictures, in addition, are overstated, heavily composed, and self-satsified manipulations of found events and footage. Formal and photographic references to films like *Koyaanisqatsi* are clear—the series' opening includes optically rich slow-motion footage of arc-welders in an assembly line—but the attitude of this episode as a whole is not. While the earlier *Koyaanisqatsi* promoted a universalizing metaphysical principle, the low-culture phenomenon here seems merely an opportunity and excuse for the director to delve into his own personal, and apparently dark, musings on life. The net result of Lynch and Frost's approach is not to know the world any better or clearer or even to care. Instead the episode pretends to sense the dark side of the world through what would otherwise be innocuous and banal. This new style, then has nothing to do with ontology or engagement. Its net result, arguably, is not presence, but estrangement. Its method for achieving this end lies in the pictorial and in an ironic soundtrack that plays against the imagery, not unlike Buñuel's dark *Land Without Bread*.

The style of *American Chronicles* includes obsessive and distracted close-up shots of tangential subjects (an authorial trademark of Lynch), of insects, of hairy male anatomical features, and chrome detailing. In addition, a slow-motion effect is used on many subjects, including even the ostensibly unimportant ones. The privileged and pervasive use of slow-motion technique is possible because the recent and widespread development of dynamic tracking in video postproduction. The motive for this obsession with slow motion is a more interesting question. These formal strategies suggest that in the televisual documentary the world no longer must fit documentary's traditional hierarchy of content and value. The world is merely fodder to be gazed at and entranced by. Native American culture, the narration of Richard Dreyfuss, bikers, Harley hogs, and Americana are all rolled into an image-oriented mix issuing from the hip Fox network.

Such a preoccupation and obsession with style might normally signify heightened seriousness and import. Here, however, it betrays indifference to the world's normal hierarchies of content. Style, or rather stylishness, *is* the subject here. An attitude of feigned seriousness results. Documentary series like *American Chronicles* give figures like Lynch and Frost the license to flaunt and parade image, to further fabricate their personas—in short, to merchandise an attitude. Televisual documentaries, to borrow Baudrillard's word, do not attempt to "simulate" reality, but to simulate and parody seriousness, to simulate and mimic the documentary gaze. Documentary has become merely one more generic excuse for the sale of a public pose.

If one argues, as I have, that Fox utilizes Lynch to individuate its network programming, it might seem that the question of documentary is not really the issue at all and that the show's signature excesses are merely a programming exception. It is clear, however, that even less pretentious and less prestigious offerings of documentary on television have evidenced a very similar shift in structuring paradigms. Consider the nationally syndicated and curiously titled half-hour weekly documentary series *Eye on L.A.* (curious because the series almost never dealt with subjects from Los Angeles). If Lynch's documentary work is about a visionary pose and a feigned seriousness, then numerous magazine documentary shows around the country like *Eye on L.A.* and *P.M. Magazine* are about the unadulterated visual pleasures of style and image. Driven less by the sense and presence of an artistic persona, and more by a drunkeness of picture, color, and motion, *Eye on L.A.* seemed to overwhelm itself with visual effects.

The on-camera talent and producer may seem earnest and deferential toward various social topics and interview subjects, but in the end the show crawls over its subjects like a ravenous semiotic machine. It annoints itself an all-seeing apparatus with an insatiable appetite for image effects. Documentaries like this one actually seem to offer digital video equipment companies prime sites for research and development. From the very start of each episode, it is clear that a coherent set of effects or privileged looks have *not* been isolated for the show, but that *every* effect on the switcher panel or the digital video effects device is used, regardless of content. Pity the poor female model's body that gets wrapped, spun, split, and disembodied in this orgy of style. The net result, however, is not Mulvey's fetishistic voyeurism; it is instead a kind of digital effects sadomasochism.

Effects come to bear no motivated or causal relationship with the show's subject. Rather, effects are used as ends in themselves. As the show continues over the programming year, new and exciting effects are added to the show's documentary repertoire as they become available. The endless formal permutations spin out like higher graphic life-forms, emerging in evolutionary sync with the ever

Eye on L.A. flies through images and worlds. Its series' logo digitally explodes as viewers cross into alpine terrain. (*Eye on L.A.*)

newer machines that presume to make those effects possible. This stylistic rev-erie of visual effects is pervasive in other versions of the documentary magazine genre as well. *Entertainment Tonight, P.M. Magazine*, and *Hard Copy* all bear the digital scars of this effects fetish. As we discovered earlier, the trend increasingly appears to encroach upon even serious documentary magazines like *20/20* and *Frontline*. The televisual paradigm, with its semiotic appetite for effects, looms over the documentary even as it threatens to infiltrate every genre on television.

One might protest that my examples thus far are not documentaries in the tru-est sense, but are rather bastardized tabloid clones of documentary promoted for market share on broadcast television. Granted *Eye on L.A.*'s view of Japan is not like Chris Marker's in *Sans Soleil*, but both works use the culture as an excuse to muse and embellish. The criticism that documentary magazines are not documen-taries is an unfair write off however, for the televisual paradigm has infiltrated more conventional, straight, and serious documentary shows and series as well. It has done so, however, in a more restrained sense. Consider executive producer Marc Weiss's series, *POV.* Surely this series must be acknowledged for helping to keep alive the modes and genres of contemporary film documentary.[17] Verité, di-rect cinema, ethnography, diary films, and other forms have made regular appear-ances on the series in recent years. Yet the original films that *POV* licenses for broadcast change in subtle ways for broadcast. Television consistently betrays a paranoia about unrestricted access by the viewer to documentary content. *The Independent*, a journal of the national Association of Independent Video and Film-makers, has frequently documented the editorial abuses of public television, an institution that persistently modifies and alters independent documentary films intended for broadcast. Partisan films, for instance, are typically aired with postproduced wraparounds or, worse yet, visual and aural disclaimers from public television regarding content. These tactics actively distance and mediate the documentary before and during airing.

The broadcast of Christian Blackwood's independent film *Hotel* on *POV* revealed the ideological stakes involved when film documentaries are broadcast. Because Blackwood's style is typically tied to his personal reputation as a film-maker, not to the parameters of television, the numerous instances of televisual mediation and infiltration throughout the opening four minutes of broadcast fundamentally alter his work. Corporate funders are graphically piggybacked onto Blackwood's production. *POV*'s series logo flys, spins, and attaches itself to the film's opening. Voice-overs define the series and explain basic film theory: "point-of-view is a filmmakers term for . . ." Additional narration summarizes and condenses the film's story into several seconds. This tactic of summarizing means that there is the film *Hotel* itself, along with visual and aural condensations or models of the film during broadcast. Pictures are ripped from the film's content and re-edited into graphic-image film summas or are used as postcards in the main title sequence of the series. By the time Christian Blackwood's meager film credit slides by, the audience has been bombarded by multiple and conflicting claims to meaning and ownership—PBS, *POV*, the MacArthur foundation, the NEA.

The wraparound, now an obligatory part of the genre when independent films are broadcast, here marshals the film's maker out from behind the safety of the camera, and into the television studio to face the audience. The director answers questions about "where the film came from" and "what the film is about." The studied responses of Blackwood disarmingly offer "people" and "love" as very personal answers. Unlike the viewer of the film in a theater, the viewer of the film on television is repeatedly and formally set up to understand and interpret the yet-to-be-aired film. In this way, televisual adaptation not only actively presents the film, it also actively tutors the viewer of the featured film. The televisual restructuring of the documentary in effect provides premature closure for the viewer, since the televised documentary comes with its own analysis, summary, and interpretation. In addition, by using a wraparound with the director in a *confessional* mode, the documentary film is recast as a personal, artistic expression. Through graphic opens and pretitle credits, the show is stylistically developed and discursively *overidentified*. Montages of excerpts from scenes in the film are wrenched out of context to create new scenes and montages that are stitched to the opening. These multiple strategies all work to change the meaning and effects of the film for and on public television.

Televisual operations, then, are not necessarily benign or passive. Graphic and visual though they may be, such operations alter the ideological effects of the documentary gaze as well. Consider the results of this televisual transformation. PBS offers viewers premature narrative closure, and engagement with effects rather than with ontologically real film space. It offers an identifiable artistic source and persona rather than producer anonymity. It prejudges and constrains meaning rather than perpetuating the openness that one experiences with verite documentary and observational *discovery*. Even film documentary purists like Blackwood must succumb inevitably to the operations of televisuality if they want to have their works aired. The televisual paradigm constantly works over adapted film sources, thereby distancing, mediating, and controlling the possible meanings of those sources.

A fourth and final example of the televisual documentary comes from a special episode of the *American Playhouse* series on PBS in which four short personal films are anthologized into a single hour-long episode. While one segment is a dramatic short, the other three traverse the generic no-man's land that exists between documentary and experimental film. Ed Lachman directs an episode on Route 66, Mustapha Kahn directs an episode on rap culture in New York, and Matt Mahurin creates a powerful image/sound tapestry about American culture symbolically called *Tribes*. This program as a whole departed from the serious stage- and theater-bound generic context one normally associates with *American Playhouse*. It also demonstrated the extent to which a certain kind of documentary—the highly visual and stylized documentary diary—has come to be privileged and featured for its own artistic and visionary qualities, rather than for its depiction or representation of a social, political, or cultural other that one normally associates with most PBS documentaries.[18] Therefore, traditional dichotomies and generic distinctions that have been used to define the documentary—the actual versus the fictional, the social versus the personal, the real versus the artistic—provide inadequate explanations for these four anthologized films. Nevertheless, in order to gauge the effects of the televisual documentary, it is worth considering at least one of the films (Mahurin's) first within the context and tradition of the documentary.

Although Mahurin's work is a brooding and beautiful meditation on life in America, it is comprised of seemingly found images gathered from around the country. Studio shots are included, but are interspersed with numerous handheld, black-and-white, documentary-type images. The latter images are taken from neighborhoods, on sidewalks, at the beach, and at the work place. If one is to write this off as art rather than documentary, it is perhaps because of Mahurin's stylistically impressive skills at controlling the subtle compositional elements that he collects.[19] He weaves found and disparate images of the world out there into a complex and sensitive image-sound pattern in postproduction. In the context of Mahurin's stylistic and visual performance, words, verbal, exposition, interviews, and narration would be inappropriate or counterproductive at best. Whereas Walter Ruttman and Joris Ivens in the 1920s focused on physical action and materiality to create an impressionistic sense and slice of society at large, Mahurin turns such images into a heavily symbolic and psychological experience.[20] The found image and event are made metaphysical through obsessive stylization and estrangement. As in Lynch's work, the eye and consciousness of the filmmaker are obsessed with the visual. But unlike Lynch's work, excessive visual stylization here evokes and invites subjectivity from the viewer, not bemused irony. Both visions are alike, however, in their dark, brooding, and primordial visual origins.

These four examples suggest the degree to which documentary has changed on television, even though they evidence that change in very different ways. In

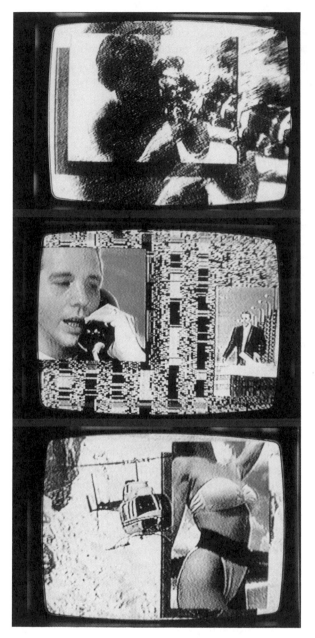

The televisual documentary as R&D for equipment manufacturers. Every visual effect is used. Bodies are digitally scarred and extruded into material surfaces. Male machines—cameras, helicopters, trucks—cruise the female landscape.

the case of Lynch's *American Chronicles*, network television imports an eccentric film director to give itself market share and reality programming distinction. In another case, the syndicated tabloid magazine *Eye on L.A.*, the documentary magazine genre in television masquerades as an extreme exhibition of special effects looks and styles, one that transforms documentary footage into grist for videographic postproduction. Even completed purist documentary films by independent producers, like *Hotel* on *POV*, are encroached upon, condensed, wrapped-around, and televisually predigested for viewers before and during airing. Finally, even those bastions of high culture in public broadcasting committed to serious drama, like *American Playhouse*, have found a new interest in the programming possibilities of highly personal and visionary documentaries. In all of these contexts, the self-conscious stylizing process fundamentally dominates collected or recorded actualities.

The Pictorial Effect (versus Fiction and Reality Effects)

Given the diverse origins of these documentaries on television, it is striking that they share certain presentational and structural similarities. Chief among their shared formal strategies is the pictorialization of spatial illusion. The effaced Bazinian frame, with its illusion of deep space, is rejected in favor of images that cast shadows and fly like objects through space (as in *Eye on L.A.* and the *POV* opening). Even when not using graphics, documentaries like Lynch's *American Chronicles* and Mahurin's *Tribes* loudly proclaim their footage *as compositions* and designs, not as real or mirror images. There is a self-conscious artificiality to this imagery that constantly works to deny what Baudry or Metz would call a fiction effect.[21] The same artifice and self-consciousness also deny what William Guynn calls the "reality effect."[22] Instead of these ideals, the televisual documentary might more accurately be described as being preoccupied with *the picture effect*: images become artifactual objects and pictures, not replications of the real. Production strategies aim to produce an impression of the picture rather than an impression of the real.[23]

A psychological dynamic forms the basis of the fiction effect. The viewer's pleasure in illusionism depends on a willingness to deny his or her own presence and to overidentify with characters in the filmic fiction. Pleasure comes in part from the scopic power the viewer senses in the visual construction of narrative on the screen. This scopic power and the accompanying shift in consciousness are seen as very much analogous to the process of dreaming. The fiction effect, then, depends both on a denial of the film's structure and artifice, and also on a denial of the viewer's own presence. Theorists relate this denial and splitting of the self to Lacan's infantile mirror stage of narcissism. While such interpretations are typically used to ideologically critique the fiction effect, one must remember that the viewer enters this process intentionally, and enters it because it is pleasurable.

With the televisual documentary, however, the viewer is never allowed to deny his or her presence. Instead, a thickened graphic flow draws attention to the screen, to the viewer's set, to the space, then, in which the spectator views. The televisual spectator is also denied the scopic control and sense of power one gets in classical cinema over the gaze of numerous character points of view in the filmed fiction. Instead, the viewers see images as physical pictures and kinetic objects traveling before their eyes. Some cast shadows, as if they are objects. Other film frames are grabbed and digitized, then propelled through the picture space. The televisual documentary, unlike film, does not depend upon darkness for viewing nor upon the evocation of a dream state. If any visual power is assumed by the viewer, it is not predicated on the narcissistic assumption that comes when one

The pictorial effect eats global imagery in a world where even video noise is an effect. The televisionary documentary pays homage to its master paradigm: the safari.

imagines being *inside* an ideal fictional space. Rather, the picture effect packaged in the televisual documentary is out front, on the surface. Televisual spectatorship does not necessarily produce a subconscious or unconscious state. There is instead an explicit performance of style that directly addresses the viewer's presence in some way. Any pleasure in the televisual process resides as much in the promise of stylistic permutation—endless repetition and the illusion of plasticity—as anything else. The televisual documentary makes real-world images both plastic and malleable, as part of an alienated observational trance.

The Collection Effect: TV as Safari, Electronic Post as Taxidermy

The gaze theory belief that pleasure in cinema results from a male heterosexual penchant for voyeurism and fetish may fit well with the operations of *classical* cinema and television. Such forms, after all, utilize classical perspective and space, both historically codified systems that center and overprivilege the spectator. Voyeurism needs effective illusionism to work, and the notion that this viewing state approximates and induces a sexual fetish of the woman's body is the particular spin that Mulvey gives in her critique of illusionism. As I have argued, however, the new televisual documentaries do not work to produce illusionism. Instead, they actively work to deny realism and illusionistic representation within imagery. Unlike the cinematic voyeur, the viewer of the televisual documentary is not placed in an ideal position from which to view the televisual flow. Televisual documentaries, furthermore, make no attempt to deny that the viewer is actually watching the flow. Because the spectator is neither ideally positioned and centered by Renaissance perspective nor titillated by Mulveyesque cropping or framing, an account other than one based on the gaze and scopophilia must be developed if one is to provide a fuller account of the peculiar pleasures afforded viewers of the televisual documentary.

A collection-consumption effect, rather than voyeurism or actuality, best characterizes one dominant thrust in televisual documentaries, since the term describes well the hyperactive pace at which images are gathered before the viewer. If there is power in this genre, it is power to see and be in control of an immense *quantity* of imagery— not in the psychosexual illusion of dominance that one has over the gaze of characters or of being able to see and not be seen in return. The collection effect, then, describes instead the power and pleasure of connoiseurship, of aesthetic discrimination and accumulation. The most basic indicator of the collection effect in the televisual documentary is the persistence by which a succession of images are stylistically diversified and episodically shelved. Given the semiotic density of this form of electronic taxidermy, direct cinema and visual ethnographies are austere and minimalist by comparison.

The gaze question then returns: can this process of conscious consumption and connoisseurship be seen as a gendered process? At first, it seems unlikely that the stylistic brokerage and stockpiling of imagery produces gendered behaviors or effects. On the other hand, local cultural bias might implicate gender in uneven ways. If gaze theory is based on the pathologies of voyeurism and fetish, then the collection effect is based on the pathology of consumption, of shopping, if indeed such a behavior is gendered or pathological.[24] The adrenalin-pumping mode of televisual documentary accumulation mimics one cynical bumper sticker: "When the going gets tough, the tough go shopping"—at the ontological stripmall.

Conclusion

The label "postmodernism" might explain some of the formal characteristics of the genre, but it does not explain *why* documentary is so prone to televisual embellishment: why the ideological effect of the televisual documentary is consumption not voyeurism; and why the new televisual documentary is about videographic embalming and not fetishistic illusionism or realism. I would argue that because

Sight is intensified in every image, even in many that are less than one second long. The televisual documentary gaze is a numbed, alienated trance.

postmodernism has removed restrictive categorical and stylistic norms—like those between drama, comedy, and documentary—generic distinctions have become even less important, and in some cases even meaningless, in mass culture. Since postmodern practice also privileges endless recombination and assemblage, *documentary can arguably be seen as postmodernism's most privileged mode of production.* Documentary has always accumulated and collaged. Both documentary and postmodernist recombination collect and hybridize images and sounds from the world. No longer enslaved, then, to the hierarchy of high and low content that typified past documentary—with its traditional emphasis on serious subjects and socially meaningful issues—postmodern television merely appropriates and adapts one of documentary's most basic modes, gathering and display. All the world, for the televisual documentary, is merely semiotic grist for the stylistic mill. Embellishment now rules even actuality.

Documentary theorists and advocates might benefit by shifting attention away from innovative film auteurs and independents who inevitably must struggle for recognition by television and mass culture. Instead, a focus on the ways that television alters the documentary gaze, and the privileged role documentary has in postmodernism, might do much to stimulate a new kind of documentary theory and practice. As distasteful as shlock journalism and the magazine format may be on television—because they are neither objective nor genteel—they are changing the way we see documentary in a broader sense. Documentary producers, on the other hand, at least need to consider the stylistic appetite that the new television has. This appetite, after all, affects the kinds of documentary that can and will be broadcast. There has been a major shift in aesthetic paradigms governing the field, and anyone concerned with ideology would do well to address the subsequent shifts in signification and viewer expectation. Such changes have monetary, practical, and political ramifications as well.

Proud of giving birth to the new docupoem form, Lynch's televisionary partner Frost castigated the rigid notion of documentary that most television viewers have: "They think it's going to be sixty grim minutes about a welfare mother." By this cynical remark, the producer meant to set his visionary documentaries apart from the mass mind and the numbing flow. But in the process, Frost also continued Hollywood's most demeaning valuation of the traditional social documentary form. This statement of exclusion clearly meant to create an aesthetic rather than a social space for the producers' new reality work at Fox. Social issues, a central and continuing source of important documentary work in the past, give way here to stylistic fashion and art world bravado. The tactic surely enhanced the artistic aura assigned to the pair by the network that year, but it did so at the expense of the very genre they now embraced and appropriated. That is, *American Chronicles* was touted as a good documentary, precisely because it was *not*, apparently, a documentary. For while the producers wrote off the blank uniformity of the mass television viewer, they failed to see that the very qualities they promoted as unique and visionary in their own work, were now actually commonplace in the everyday world of *Eye on L.A., Lifestyles of the Rich and Famous, Entertainment Tonight,* and *Hard Copy.*

By the summer of 1994 one outsider series, *TV Nation,* had begun to show that the televisual documentary could be worked for progressive ends. Posed as a low-culture, post-Bolshevik, man-in-the-streets, indie producer Michael Moore proved on network NBC that the safari mode of the tabloid and televisual documentary could indeed have both the taxidermist's stylizing bark and the activist's political bite. I will reconsider again, in the next chapter, two topics recurrent in the posturing of both the tabloids and the televisual documentaries of Lynch and Moore—"liveness" and the televisual audience. The formal permutations by which liveness is styled clearly implicate and presuppose not just a different vision of reality, but a different kind of viewer as well—a viewer attuned to the optical spoils of television's new hunter-gatherers.

Part III

Cultural Aspects of Televisuality

9 Televisual Audience
Interactive Pizza

Television is a vast, phosphorescent Mississippi of the senses, on the banks of which one can soon lose one's judgment and eventually lose one's mind. The medium itself is depressing. The shuddering fluorescent jelly of which it's made seems to corrode the eyes of the spectator and soften his brain.
—Jonathan Miller, *The New Yorker*[1]

ABC's *Wild Palms* speaks on many levels, frequently in code. There's even a dense tome called *The Wild Palms Reader*. But for a quick primer, clip and save this glossary of the series' symbolism and hidden meanings.
—Rich Marin, *TV Guide*[2]

The folks over at *The New Yorker* obviously live on a different planet than the one inhabited by *TV Guide*—a world, apparently, that is populated by viewing morons. Was their critic on acid? Granted, Rick Marin's viewing guide to Oliver Stone's miniseries, *Reading Wild Palms*, is not Louis Althusser's *Reading Marx*, yet both articulate the viewer's complicated relationship to mass culture. In fact, *TV Guide*'s lexicon, by describing primetime's complicated "codes, techno-fantasies, and political allegories," actually seems theoretically more up to date—that is, closer in spirit to Donna Haraway—than to certain of the Althusserian applications that dominated media high theory in the late 1970s and early 1980s.

Television critical theory since the mid-1980s can be characterized by a certain amount of desperation; by a frantic attempt to revive the audience that it had first helped textually deaden. This desperate intellectual attempt to rediscover and *activate* the audience comes both as a conscious reaction to indictments brought against enunciation theories and from a sense of text-bound methodological claustrophobia.[3] That is, having ostensibly corrected the ideological and formal glosses of mass media social science—by criticizing effects theorists who construed viewers as passive victims, and needs and gratifications theorists who overestimated the influence of viewing subjects—many critical theorists from the European traditions of psychoanalysis, semiology, and poststructuralism now seemed imprisoned by the artificial constraints of their own texts.[4] Audience breakthroughs in critical theory followed chiefly from three influences: from the deck-clearing populist and political wake of British cultural studies; from David Morley's sociological *Nationwide* study; and from the newfound popularity of viewer and fan ethnographies.[5] At least in an anthropological sense, media theory had in some ways come full circle. The new work of Ian Ang, Constance Penley, Henry Jenkins and Lisa Lewis was actually a throwback and emulation of a kind of ethnographic fieldwork that predated by several decades one of the founding patriarchs of critical theory.[6] Structuralist anthropologist Lévi-Strauss showed very little interest in either cultural specificity or field informants, even as he ranged far and wide in search of pregnant cross-cultural examples that would prove his generalizations. His poststructuralist theoretical descendents, however, eventually recognized and embraced the explanatory power possessed by those very informants and viewers. The ethnographic conversion of critical theory is somewhat ironic, then, for it meant that the field was discovering the very same methods that had pervaded American mass communications studies for decades—audience surveys, focus groups, the idea of controlled observation, and self-reports had all been castigated by earlier critical theorists as naive.

This theoretical burden to activate the viewer is also ironic, for the television viewer in practice has never been passive—nor even typically theorized as passive by the industry. Broadcasters from the start did not see the viewer as a couch potato, but as an active buyer and discriminating consumer. Coupons, sweepstakes campaigns, fan clubs, and viewer call-ins all reflected the urgency assigned audience response by broadcasters and advertisers. As long as advertising revenues and Nielsen numbers ruled broadcasting, audience activity and consumption would be crucial and overt parts of the television equation. The textual activity targeted by enunciation and psychoanalytic theorists was, then, always but one minor part of a much broader viewing ritual, one that was much more material and consumer oriented. Given the historical context of audience activity, therefore, it is worth considering how televisual programming makes activity and interactivity a part of its appeal to the viewer. What, for example, is the relationship between television's performance of style, its market, and the viewer?

Stylistic Individuation and "The" Viewer

Patients allowed to watch MTV during gynecological exams became less agitated. This is important because patient anxiety prolongs examinations, doctors said.
—Doctors at Arkansas Children's Hospital

MTV may provoke violent behavior among the criminally insane. Patients seemed transfixed by MTV and behaved violently after prolonged viewing.
—Psychologists at Connecticut's Whiting Forensic Institute[7]

When medical science weighs-in for a bout of media analysis—as in this debate over the neurological merits of MTV—the results can be schizophrenic. As long as science and its applicators continue to force conclusions about the viewer through a polar frame driven by the need for positive-negative attributions, then they will continue to perpetuate three of communications studies most persistent myths: first, that there is a single type of viewer that television affects; second, that television is itself a single or unified force that can wield these influences; and third, that the audience is a victim or target worked upon by this unified force. Although MTV and the critics who exploited these studies wished it otherwise, it is in fact impossible to generalize about the audience from the above studies— except to say that music videos make violent people violent and gynecological patients relaxed. The only real link between these two studies is that both examined institutionalized populations, one hospitalized, the other incarcerated. That fact insures that any results gathered during the study would be inapplicable to noninstitutionalized viewers and thus largely meaningless. Such results define, then, not what most viewers are, but what they are not. Studies such as these say nothing, therefore, about either the form of specific programs or about the television audience in general. Research studies like these are not contradictory, they merely describe very different, delimited, and specific settings and populations.

Any discussion of the televisual audience, then, must look more closely at the specific appeals that such shows make to specific viewers. As we have seen, televisual exhibitionism can be defined as a self-conscious process of stylistic individuation. The individuation and semiotic heterogeneity evident in televisual excess means that such shows are from the start defined by, and pitched at, niche audiences who are flattered by claims of difference and distinction. A celebration of audience specificity rules the world of televisuality. What I am calling for, then, is not simply a return to the text and a reconsideration of how it *implies* a viewer, but for a reappraisal of how the televisual text and the televisual industry open themselves up to and engage viewers. Since televisual appeals are rarely effaced and deferent—as they were in the classical film forms that enunciation theorists analyzed—televisual programs frequently demand a more conscious form of viewer negotiation. With point-to-point rhetorical communication and direct address rather than narrative diegesis, contemporary primetime narrative forms gladly engage the viewer in conscious forms of gamesmanship. After looking closely at one specific example of this kind of televisual gamesmanship in *Northern Exposure*, I will return at the end of this chapter to a different but related concern: the way the industry, not a program text, positions and accommodates the televisual viewer. When even local news executives acknowledge that "gone are the days of obligatory viewership" and that most broadcasters don't give audiences the "credit for being used to that little box in their room," the lesson is clear.[8] Savvy and self-consciousness—trademarks of the televisual viewer's pose—are lucrative sites for televisual engagement, even at the local affiliate level.

Cliff Notes' Western Civilization: Yuppie Night-School

The first time I set eyes on Cicely . . . was like the unveiling of Botticelli's Venus.

Is this some kind of parallel universe?
—Turn-of-the-century western characters in *Northern Exposure*

In one episode of the hour-long *Northern Exposure* aired in September 1992 about the origins of the town of Cicely, the references to intellectual, artistic, and philosophical concepts and traditions were so dense that they formed the stage against which each character played out the episodic drama.[9] This episode generically masqueraded as a turn-of-the-century western and as a complex ritual of memory that allowed its contemporary characters to reenact the cultural origins of their town. Dramatic dialogue was littered with philosophical fragments, including expositions of the Hegelian dialectic, the nature versus nurture conundrum, the nature of free will versus free verse, Rousseau's noble savage, Dostoyevsky's *Crime and Punishment*, theological paradoxes, nihilism, and aesthetic existentialism. By characterizing the town in *Northern Exposure* as the "nineteenth century Paris of the North," the episode gave free rein as well to many literary flourishes. Joel Fleishman appeared as Franz Kafka, a western sojourner; Rilke, Poe, W. B. Yeats,

Lesbian performance artists, tableaux vivants, feminist reform, and nineteenth-century beat poetry readings tame the savage male beast in *Northern Exposure.* (CBS)

and the "world's literati" were all invoked, along with expositions of literary hubris ("wanton insolence that brings down the wrath of God") and the stand-up reading of symbolist poetry anxiously entitled "Between Antigone."

This textual and dramatic mixage of literary, theological, and philosophical discourse was only matched by the episode's parade of art historical simulations enacted for the viewer. Numerous scenes were staged with the visual look of nineteenth-century German romantic paintings. Heavily composed tableaus also evoked the gilt-lit Olympian world of romanticist painter Maxfield Parrish, while other scenes embodied the Renaissance aura of Botticelli. When Cicely and her lover, Roz—characters set up by both the narrative and town as lesbian messi-ahs—come to the northwoods to "bring civilization," they really bring with them the history of Western art. These cultural missionaries and aesthetic patrons trans-form the bawdy and dirty saloon, originally a smoke-filled male den, into a per-formance space presided over by the women. But since this is Alaska, and not Soho, the space is modeled after the "salons of Parisian culture"; the world of dandies Guillaume Apollinaire and Alfred Jarry. When stand-up poetry readings and performance are formalized as a nightly social ritual, the faint and classi-cally feminine Cicely choreographs and stages her dance in the statuesque and deliberate manner of Isadora Duncan. When Kafka (Fleischman) experiences writer's block, he leaves Europe for the baroquely lit rooms of backwoods Ci-cely, where old friend Roz salves the male's creative blockage and mothers him toward completion of *Metamorphosis*. All of these stylistic embellishments are woven with the icons of the western genre and are further framed within a con-temporary present that the producers rendered as sensitive on-camera interview footage. Yes indeed—the art director and set decorator had done their homework here. This was a veritable showcase of production design facility, and also an ac-complished menu of Western art and intellectual history.

Identifying these intertexts does not, however, adequately explain the demands that shows like this make on viewership. No matter how you cut it, the excess operative here is more than just the kind of formal overabundance that theorists have traditionally associated with excess in, for example, the Hollywood movie musicals of the 1930s and 1940s. There is also a kind of intellectual excess in contemporary television series like *Northern Exposure* and *The Larry Sanders Show* and in their predecessors *Moonlighting*, and *thirtysomething*. Intellectual surplus also spills over in comedies like *Seinfeld* and *Murphy Brown*, but laugh tracks and other markers manage the spillover in more explicit ways. Smart iconic references, visual embellishments, tableaus, and historical masquerades—not just dialogue—make up this intellectual surplus. Gone, apparently, are the days when television scripts inevitably become pared down to essential components intended for wide access to most viewers. Because *Northern Exposure* does not even un-derline or draw attention to the flood of these references, many of them become little more than fleeting asides even to viewers who might actually know or care about their significance. By rewarding this degree of recognition and discrimina-tion, hip and prestige primetime shows like *Northern Exposure* also help justify the yuppie-requisite college educations that have helped produce the show's quality demographics. Such audiences feel good, that is, because they "get" the string of smart and stylistic references that read like the general education requirement or the great books list at a liberal arts college. The viewers' student loans, it seems, were well spent.[10] The ideas and imagery may be lofty here, but the fragmented amalgam of references really makes this a kind of Cliff Notes intellectualism—conceptual categories to recognize and discard—framed within the primetime epi-sodic hour.

If televisuality helps create an intellectual surplus for viewers in *Northern Ex-posure*, it does not leave the viewer with the kind of blank disinterestedness that Jameson characterizes as quintessentially postmodern. There is something terribly earnest about this episode. The historical reenactment in this primetime episode teaches the viewer about lesbian feminism in long expositions that question the value of heterosexual relationships with men. Roz consoles a distraught revivalist worker, anxious about her failed relationship with men, by offering a psychoanalytic explanation based on separation from the mother. "Men are confused." Their "ambivalence goes back to their relationship with their mother; the center of the universe; the object of both fear and love. See, the question is, do you really want a man? For the man, it's natural. But for the woman it's unnatural—for she has to transfer her affections away from the mother in order to love a man." Looking down the barrel of a gun in another scene, this same nineteenth-century character psychoanalyzes male aggression with a therapeutic discourse more on par with daytime talk shows than with the suffragette ideology that would have typified this period. "It's acceptable to have feelings of anger," she counsels, "but unacceptable to act [violently] on those feelings."

The whole episode, in fact, can be seen as a very knowing example of historical revisionism, and an ambivalent critique of the ideology of separate spheres that characterized gender relationships during the Victorian period. The private sphere, the world of the woman, emphasized the arts, culture, enlightenment, cooperation, and solidarity; while the public sphere, the world of the man, favored the excesses of work, hedonism, the ego, competition, and dominance. From the mythological perspective of Vladimir Propp or Northrop Frye, the episode can also be seen as a Christological redemption narrative. The world before Roz and Cicely was one of debauchery and animalism. All the men were mired in the darkness of immorality or anarchy. Ed was merely a mud-spattered wolf-child, raised like Romulus and Remus by an animal in the woods. Abe was a sexually impotent and powerless lacky. Kafka was an aberrant artist with migraines. Mace was a violent despot. Kit was a Nietzsche-quoting antichrist who morally shadowed the female revival workers blow by conceptual blow. Only when the messianic lesbians land in town and fulfill their aesthetic mission are the men converted, elevated, and civilized. The last vestige of male aggression is finally subdued only through the sacrificial death of the innocent redeemer figure, Cicely. A remorseful voice-over intones the lasting moral influence of her sacrifice: "The town was born that day, and everyone knew that it would be called Cicely. Kit—who always was a metaphysician—gave up banditry and became a preacher man. Roz mourned for a while, disappeared into herself—and finally vanished. Rumor had it that she went to Spain to fight the fascists." Having touched everyone's life, that is, Roz went back and vanished in the only kind of heaven that the politically correct producers of this intellectual excess could profer: the Spanish Civil War.

In shows like this, televisuality is also frequently a form of historical exhibitionism. Four historical narratives are embedded on top of each other: the history of the town's origins; an archetypal and theological history of worldly and personal redemption; the history of emergent gender distinctions and sexualities; and, finally, a revisionist look at the grand narrative of America's westward historical expansion in the nineteenth century. The very idea of manifest destiny was drawn and quartered, then repackaged for consumption in this particular variant of primetime televisuality. The anarchy of the uncivilized, testosterone-driven West, became the sounding board for a much more contemporary consideration of sexual politics. Rather than decrying the domestic-artistic ghetto assigned to women in the Victorian ethos, this show actually salvages it for progressive ends. *Northern Exposure*'s women-redeemer figures overhaul the very historical basis for American westward expansion—that is, the inherent right and freedom to own property—into a more refined kind of Americanism: the freedom to be artists, the "freedom to express our art . . . and our love." Who said the western genre could no longer be ideologically mined and reworked? Manifest destiny was actually fueled—at least to the young and upwardly mobile in this series—by the need to make art.

The Dispersal and Proliferation of Audiences

Several lessons emerge from this example. Exhibitionist histories and televisual embellishments suggest that producers no longer presuppose the "mainstream" as the key to good audience numbers. There are either enough people with sufficient educational and artistic capital who are receptive to this intellectual surplus—those with degrees in the liberal arts and humanities—or those who do watch are financially valuable enough to make up for their numeric inferiority vis-à-vis the mainstream. The whole presentational exercise in *Northern Exposure* aims to cultivate and to teach the viewer that he or she is distinctive. The demarcation of viewer specificity rules this kind of television. Moreover, the logical fact that more people watch *Northern Exposure* than have mastered the great books canon, suggests again that *there is no singular or average viewer or way of viewing*. While prestige televisuality cultivates distinction, it also survives only if it does not alienate other viewers. Televisual excess, then, exploits not just stylistic embellishment and intellectual surplus, but also involves the loading up of different audience appeals within the same program. Even distinctive shows, then, also seem to have something for everyone: melodrama, pathos, and pleasure of storytelling, not just sophistry or embellishment. Televisual shows, therefore—at least the ones that weather the new season cancelation rituals of mid-fall—are rarely monotexts. While the stylistic demands on audience that define televisuality are, in a philosophical sense, necessary, they are typically not sufficient, or exclusive, properties. If they were, the presence of televisuality on the channel spectrum would be marginal or fleeting at best.

Recent statistics on people who watch television outside of the home bear out the revolutionary extent of this diversification of viewers—and the proliferation of viewing contexts that accompanies this diversification. By 1993, more than 28 million adults each week watched some television outside of their home, in locations such as the workplace, college facilities, hotels, motels, restaurants, and bars. Such out-of-home viewing is not even reflected in the television ratings. Nielsen Media Research has concluded that "young and active demographic groups, important to advertisers and traditionally thought to be light TV viewers, do a significant portion of their viewing in unmeasured proportions."[11] Even as the nuclear family no longer offers stable footing for ratings research given widespread changes in the domestic setting, many audiences are now no longer even defined by the home or the family. Multiple sets within the home and viewing markets outside of the home have attracted substantial corporate interest. Whittle

Communication's Channel 1 has been successful since 1990 at placing commercial-educational television within public school systems. Their lure? Cash-strapped public schools jump at the chance to acquire new audiovisual gadgets—TV sets, VCRs, and satellite dishes—while advertisers get two minutes of each twelve-minute educational newscast to hawk their products to a captive audience. As a result of these cola, candy, and acne ads Whittle Communications makes over $600,000 per day, by "mak[ing] sure that [children] are never out of reach of a TV show—or of TV advertising." Other corporate incursions into place-based media include Turner Broadcasting's two-year-old Checkout Channel for supermarket lines, the Airport Channel for travel-based television, the Health Club Network for exercise mavens, and a variety of networks in development for fast food franchises. HBO has branched out into what it calls the Visitor Information Network, which "provides in-room services" for hotel-housed guests—including promotions for both the host city, its attractions, and fine restaurants. Trucker Television's American Transportation Television Network delivers at truck stops across America "news, road conditions, and [according to a press release], lifestyle programming."[12] From trucker hash-houses on the interstate, to four-star hotels, to supermarkets, to captive inner-city schoolrooms, programming—and the modes by which viewers watch it—is being further splintered and individuated.

The conversion to narrowcasting *within* network and cable programming during the 1980s—a formative influence in the success of televisual shows like *Moonlighting* and *Northern Exposure*—has now shifted *outside* the traditional channel spectrum. As long as money can be made from captive or place-specific audiences, each subculture or viewing group is a potential site for more individuated lifestyle programming. Even politics have evidenced this corporate snowballing toward the supposed programming needs of individuals. The growing tension between the press and Bill Clinton was summed up in his curt reply to angry reporters: "You know why I can stiff you on the press conferences? Because Larry King liberated me by giving me to the American people directly." "We're not pleased" replied the president of the White House Correspondents Association.[13] By removing the network news divisions from their traditional gatekeeping chores, Clinton had touched one of journalism's most sensitive nerves and was participating in the more general process—or mythology—of de-mediating television. By March 23, 1993, Clinton had given interviews to MTV and to local reporters around the country but none to the press corp. Clinton's supposed media savvy—typified by his wearing shades and wailing sax on *Arsenio Hall*—was not just televisual facility. It also demonstrated Clinton's mastery of the icon-driven world of narrowcasting.[14] This growing sense of de-mediation in television, is more apparent than real, however. The broadcasting pyramid, an institution formed and legitimized by delegation and monopoly, is indeed being leveled with the incursion of many new players. Consider, however, the corporations behind the new individual friendly and ever more accessible networks—HBO and Turner Broadcasting in place-based media; MTV and its heavyweight Viacom in the new press populism. Capital is not being dispersed as rapidly as each of the new networks is being stylistically and demographically individuated. The splintering audience, then, is really a shell game—many of the same corporate players diversify across niches for multiplying audiences aware of their individual differences.

The Corporate Cult of Diversity

If any mythology grabbed the industry's imagination in 1992–1993 it was the possibility that a mind-boggling 500 channel utopia loomed just off the technological horizon. Implicit in many prognostications was the sense that Reagan was right: a boundless free market would indeed save America, even as it provided individual channels for every interest. Gleeful assessments argued that this degree of diversification would both improve programming and solve systemic political and racial problems. A former adviser to the FCC and the White House, and board member of the National Association of Broadcasters (NAB) commented, "There will be a plethora of niche [cable] networks *responsive to the needs of specific cultural groups* within our multicultural society." In addition to providing "ownership opportunities" for minorities, "these culturally specific niche networks will require management teams that are sensitive and responsive to the needs of their target audience."[15] Whether this was just wishful thinking or self-fulfilling prophecy by those with vested interests in the expanding status quo, the industry had clearly begun to retheorize the very idea of a target market. Since the mid-1980s, the cable industry had distinguished itself by honing special interest boutiques for discriminating viewers. First rights to syndicate many televisual shows—*China Beach, Beauty and the Beast, thirtysomething, Moonlighting*, for example—were grabbed by the Lifetime network, making it the self-styled "woman's network." Broad divisions by gender would soon not be enough.

Many other networks began to use viewer diversity and multiculturalism as a corporate call to arm. Consider in this regard industry appeals by long-term cable survivor USA Network.

> USA Brings You Targeted Support for the Changing Face of Your Market. Every day, marketing cable seems to get more complicated. Ethnic groups are large and growing and each has its own special interests. What appeals to one group may not appeal to another. *Now USA Network brings you the "Response Plus" Ethnic Marketing Kit*. It has what affiliates need to target acquisition and retention efforts to specific ethnic groups—African Americans, Hispanic Americans, and Asian Americans. The kit contains TV, radio, direct mail, newspaper, and guide ads that focus on USA programming with proven ethnic appeal.[16]

In the new world of cable with its cult of diversification, ethnic and racial specificity displaced the home as a prized programming acquisition. Although critical theory typically presupposes that television covers over racial difference and homogenizes ethnic specificity, industry practice now proves otherwise. TV does indeed theorize race. It also brings to bear a theory of essential ethnic differences even as it talks of targeting specific ethnic groups. By 1993, segregation and racial determinism were acceptable parts both of progressive social thought and free-enterprise communications marketing. Waves from cable's new cult of diversity swept back to the aging networks as well. Even venerable CBS fantasized in 1993 that it was niche programming. Executives rationalized to the press that because of the network's class consciousness Tom Arnold's new sitcom would work at CBS: "The blue-collar sensibility has always worked well for CBS, from *The Honeymooners*, to *All in the Family*."[17] Even if such niche claims were preposterous—

Television has always taken interactivity to the bank—with contests, purchases, and 800 numbers. *I Witness Video* is little different from *Star Search* except that winners (citizen producers) get to be entombed in an on-screen videowall. (ABC, ATT, ABC)

given the generic and historical variety of programming on the majors—the networks were now desperately invoking the very same buzzwords of diversity that had fueled the competitive emergence of cable.

The Question of Interactivity

Critical theory, from Hans Enzensberger to John Fiske, has generally presupposed that increased audience activity would help viewers overcome the mind-numbing debilities or subordinations encouraged by television viewing. By interacting, reading against the grain, and reconfiguring program texts on the viewer's or fan's terms, audiences could counter the hegemonic power of mainstream television and its privileged meanings. The myth of interactivity has been held up as a key to this oppositional liberation. This counterstrategy of viewer activity is now undercut by the industry, however, since the consciousness of interactivity now frequently comprises even television's most basic appeals to viewers. At the halftime of the NBA semi-final game between New York and Chicago in 1993, for example, NBC heralded themselves as "having made history." They then stroked the viewing audience by intoning, "Thank you for making NBC number one" during the May sweeps.[18] Clearly, part of the viewer's game was watching with a consciousness that the act of watching was also simultaneously a kind of ratings vote, a participation in the no longer secret process of head-to-head network competition. Sweeps are now seasonally discussed everywhere from *TV Guide* to *En-*

tertainment Tonight. Interactivity, then, is a self-conscious appeal made by the networks to viewers. Although academics have looked at the radical potential offered by the remote control, networks now reward viewers for voting with their remote control devices. Under this guise, active viewers "get what they want" from the networks.

At the same time that networks publicly applaud viewer activity and choice, they counterprogram to ensure audience share against the new and volatile viewing practices. Television executives at Warner Bros., for example, took on the audience's penchant for remote surfing by producing and distributing in-house promotional specials on *The Making of Kung Fu, Time Trax,* and *Babylon 5.* The upside for stations was that Warner Bros. offered the specials free of national barter ad time, thereby creating more extended contexts within which their shows could compete.[19] When shows are successful, then, the remote is a populist interactive voting tool. When shows tread water in the programming clutter, however, remote surfing is an audience ritual that elicits vigorous countermeasures from program producers like those at Warner Bros.

Audience activity is therefore very much a part of the public discourse of both network and syndicated television. Such institutions are not afraid to show how programming extends out into the lives of its audience. Even the consumer electronics giants that bring the audience contemporary television make interactivity a part of their public mission. Sony's lavish brochures educate the viewer in the ways of the new television and audio: "Do You Have Sony Style? Introducing *Sony Style* magazine, your personal guide to Sony consumer electronics. Discover over 300 pages of the fun and excitement Sony brings to life. At $4.95, *Sony Style* helps you make the best decision when buying Sony."[20] Not only does Sony make the product, they define the very parameters of lifestyle choices with three hundred pages of products and explanations. With photographic illustrations done in the manner of the angst-ridden fashion poses of Robert Longo's art world paintings, Sony teaches you how to build your lifestyle around media products. Audience activity outside of programs is not something the audience must initiate. Corporate players Sony, Warner Bros., and NBC are more than willing to help the viewer with that interactivity.

Yes, But What Kind of Interactivity?

John List, a man who had murdered his wife, mother, and three children eighteen years earlier, was featured on *America's Most Wanted* in 1989. Eleven days after the show aired, fugitive List was behind bars. Industry figures agreed: "The show works."[21] According to the producers of *America's Most Wanted*—one of the premier showcases of tabloid televisuality—more than 40 percent of the criminals featured are arrested as a result of viewer input. Broadcasters, always defensive about their moral status as both private enterprise and public service, used this activity as a defense: "Reality programming has always taken its share of licks from TV critics, but the two realities we are most concerned with are those of missing children and of the proven ability of *America's Most Wanted*—read television—to get results: 231 fugitives captured with the help of the show's profiles."[22] Well before America was wired for electronic and digital interactivity, reality programming had gotten the viewer off of the couch and on the phone. With the success of tabloid cop shows, interactivity—a market-proven embellishment from the well-oiled and lucrative cable ghettos of Christian television—spread across the channel spectrum. The activated televisual viewer was now neither the proverbial housewife nor the distracted consumer of academic theory, but a vigilante—a televisual bounty hunter energized by patriotic appeals to American morality, law and order.

Interactivity, then, was not a cybernetic product, but a way for programs to seal a relationship with viewers. Even if one did not actually call the 800 number to report suspicious neighbors, one knew that the community of American law enforcement was always there at the touch of a dial. One could always call out from the home in a crisis—a comforting thought given the possibility that one of the other reality shows, *Cops* or *American Detective*, might break into your living room at any moment. Even as it crowded the screen with dense but low-tech combinations of image, text, and pricing information, the Home Shopping Network (HSN) had earlier proven the prototype for interactivity viable on two levels. First, multiple, simultaneous channels of communication linked producer and viewer—cabled video and telephone lines. Second, commodities were exchanged for viewer cash—cubic zirconium for Master Card. Even in more exotic technical variations these two conditions—multiple and simultaneous electronic communication and financial exchange—typically work to constrain digital interactivity. The interactive cybernetic future of the 500-channel environment looks suspiciously like a marketplace, pure and simple.

At the annual National Cable Television Association convention in June 1993, three kinds of interactive systems were showcased: systems that functioned as program guides to help viewers navigate through the chaos of 500-channel systems; systems designed to offer information and products to viewers; and interactive videogame systems. Of the three different system types, therefore, only half of one (information) had the potential to provide viewers with new forms of service. The others simply functioned to support existing cable, entertainment, and financial services. Interactivity, apparently, means suturing oneself via menus to *existing channels* or to interactive games, which are not interactive at all. Interactive providers merely download Sega game programs onto a PC at your local cable company. But why would anyone want to put their Sega or Nintendo activities at the mercy of a much bigger and more volatile system down at the understaffed cable company? Why not just plug Sega Genesis in at home? Or, consider the absolute absurdity of ordering a pizza—the recurrent example for many developers—through the TV Answer system. Promotional hype described the nature of interactivity in this system when the viewer utilizes a Hewlett Packard terminal: "When the customer takes an action, such as ordering a pizza or answering a poll question posed on the local news, the terminal sends a radio signal via satellite to a central office in Reston, Virginia. The central office processes the subscribers request or response."[23] They then contact your local pizzeria to initiate delivery. Of course this facile cross-country satellite link and requisition is technically impressive, but it ignores one fundamental issue. Why not just phone Domino's Pizza down at the corner? It's faster. It's cheaper. It's easier. The telephone, furthermore, is an elegantly simple and efficient technology. With cable and its cohort of developers at the helm, interactivity means consumerism at its worst. The visual displays and the complicated points of viewer interaction clearly fit the lineage of excessive videographic televisuality—but the visionary inter- activity paradigm comes across more like fast food retailing than on-line libera- tion.

Televisual Activity

Although a more in-depth study of the audience for televisuality is beyond the scope of this book, the examples we have looked at so far suggest three correctives. First, although academics may wish it otherwise, an active audience is not neces- sarily any better than audiences that are written off as passive. Viewer activity, after all, has supported the aims of broadcasters and advertisers from the start, and theorists of fan subcultures forget that fan activities bring with them specific socioeconomic interests and political commitments. Textual or fan activities can still serve either dominant or progressive ends. Second, a large audience is not necessarily a better audience, indicator, or basis for television sampling or analy- sis. Given the penchant for niche narrowcasting in the 1980s, and the corporate cult of diversification in cable of the 1990s, the best way to understand Ameri- can television and its audiences is to look closely at its ever-narrower demographic slices and the ways that programmers strategize those slices. The intellectual sur- plus that characterizes prestige televisual programs is symptomatic of a much broader programming agenda, one that cultivates and rewards distinction in eth- nic, racial, and class terms. With the flight of programmers away from the main- stream, stylistic individuation has become more paramount than ever, maybe even more so than content. To white adolescents, for example, stylistic individuation by African Americans has now become a dominant repository and clearinghouse for radicality on primetime.[24] Third and finally, textual counter-readings and interactivity are not necessarily more valuable than actual or practical politics. It is naive to think that textual appropriations, reworkings, and interactivity are guar- antors of progressivity, since the same tactics fuel the frenzied development of new digital and cable interactive systems. In the same way that earlier theorists had overestimated the radical potential of new technologies, later theorists have overestimated the political value and radical potential of textual process, electronic facility, and bricolage. White supremacists and pedophiles are more at home on the Internet than on the networks.

It is worth considering, finally, why many primetime televisual shows also be- came cult shows that attracted fan followings. *Beauty and the Beast, The X-Files, Quantum Leap, Star Trek,* and *Max Headroom* all initiated fan activity not sim- ply because they were visual, but because they also utilized self-contained and volatile narrative and fantasy worlds, imaginary constructs more typical of sci- ence fiction. Their preoccupation with alternative worlds—a defining focus of virtual reality—justified and allowed for extreme narrative and visual gambits and acute narrative variations. Like sci-fi, televisuality developed a system/genre of alternative worlds that tolerated and expected both visual flourishes—special ef- fects, graphics, acute cinematography and editing—and narrative embellish- ments—time travel, diegetic masquerades, and out-of-body experiences. Such forms, simultaneously embellished and open, invite viewer conjecture.

Other televisual shows that started with realistic settings—like *Moonlighting,* *thirtysomething,* and *Northern Exposure*—eventually *became* highly conscious alternative worlds after extended runs in primetime. As stardom transcended characterization, and Hollywood gossip and entertainment discourse transcended the meager confines of plot, viewers came to expect stylistic volatility because of the shows' highly visible pretense and personalities. Even *Seinfeld,* once it became a household word, began to leave the realistic burdens of the sitcom and to play against the very confines of the genre. The celebrated episodes in 1992–1993 when Seinfeld pitched a "show about nothing" to network executives at NBC and traveled to Hollywood were less exercises in self-reflexivity than evocations of the *Twilight Zone.* So too were episodes of the *Jackie Thomas Show,* whose plots emulated gossip about the series, and about Thomas's relationship to *Roseanne* and ABC. In these cases, the show itself became a frame—both a stage and an altered state—that real stars entered, embellished, and showcased. Stardom and gossip defeat the dramatic obligation or need for narrative coherence and make each episode a conscious exercise of artifice and embellishment—a litmus test of the star's televisual prowess at traversing culture categories.

Critics would do well to look at the industry's current theorization of the viewer. In the 500-channel universe, the viewer is frequently positioned as livestock, as active but bucolic. "Just as a network needs to look like a network to the *itinerant grazer . . .* [so a local station] need[s] to *tap into local idiosyncrasies,*" says Hollywood emigrant and programming guru Brandon Tartikoff.[25] Such a view suggests both the value of developing a distinctive audience and the central role in programming survival of televisual idiosyncrasies. Such cynical attributions about the viewer, however, may be good for a few insider laughs at the annual National Association of Television Program Executives (NATPE) convention, but anyone who has ever herded cattle or hogs knows that such subjects are neither stupid nor lacking in initiative. They frequently turn on their masters.

Having addressed the issue of audience by considering the relationship between intellectual surplus and televisual embellishment, and by surveying the ways that industrially sanctioned audience activity works to preempt oppositional readings and interactivity, a more in-depth study will tie together two of the recurrent themes of this book. The industrial apparatus and the stylistic constructions of liveness together create a problematic place for the televisual viewer. It is not enough, then, to interview fans, or to see how texts position the viewer. The industry also creates physical conditions that simultaneously constrain, allow for, and reinforce a certain type of viewing. Neither viewer surveys nor shot sequences adequately explain the televisual viewer's place in this industrial apparatus, nor can they suggest the degree of investment the industry makes in *its* participatory fantasies.

Case Study: Audience and Mode of Production
The Politics of Taped and Remote Liveness

A lot of folks don't like to be videotaped.
—Lt. Sergio Robleto, LAPD, during murder investigation[26]

I have a vision . . . I have a vision . . . I have a vision . . . te-le-vision.
—Bono, during U2's *Zoo TV* tour

Toyota repackages the populist home video revolution with academy film leader and gangsta suburbanites. (Toyota)

Two innocent bystanders carrying camcorders are murdered by a crowd of between eleven and forty people as they videotape customized cars on Crenshaw Boulevard in Los Angeles.[27] Eight months after the L.A. uprising, the assailants mistake the victims for police informants. Like that chrome? You're a dead man—if you have a camcorder. Although the crucial evidence—the videotape from the victim's camcorder—disappears, police investigators hope to match distinctive patterns on the soles of the assailants' designer athletic shoes to patterns in the scars and flesh wounds on the bodies of the dead camcorder operators. Recently discovered home videotapes show Rodney King at home, on a couch watching television as the Simi Valley verdicts are handed down. He curses and second guesses the D.A.'s tactics on camera, even as he continuously massages his nerve-damaged forehead. Nine months after the verdicts, and in anticipation of King's upcoming civil rights trial, close acquaintances have leaked edited versions of this "secret" tape to the news media. KNBC calls their reworking and showcasing of the camcorder tape on January 26, 1992, an exclusive. Toyota airs a long series of commercial spots on national television called "Toyota Home Video." Crass appropriation of the personal by the public or mutual exploitation? Happy, jumping buyers, perform and produce the footage for Toyota. Each spot is numbered for the viewer and set off in a black graphic box as videotape, complete with academy film leader. This is, to borrow Nam June Paik's classic comment in *Global Groove*, "participation television"—yet it is done by and for a multinational corporation. Anxiety about the massive foreign trade imbalance with Japan? No way. We have our tape.

Live-Remote and Videotaped Interventions

Tape is everywhere, and everyone seems conscious of its presence. For all those who thought that television was defined by the flow think again. Increasingly, television has come to be associated more with something you can hold, push into an appliance, and physically move around with a controller. With the widespread dissemination of the tape and the VCR, the very status of the television and its image has been altered. Not only has the conceptual frame through which we view television changed, but the relationship between the public and private spheres that characterized television vis à vis the world in the past has been altered. The pages that follow focus on two questions, one is technological and historical; the other is political. First, in what ways does television's current flaunting of videotape as a material object relate to a simultaneous historical trend evident since the early 1980s, that is, to television's increasing penchant for stylistic excess? Second, in what ways does this stylistic exhibitionism, centered around videotape, change the idealized terms of the viewer-television relationship? Television, since 1980, has hailed the viewer differently, and its technological mode of production is partly to blame. Participation is as much on the minds of broadcasters as it is on the minds of academics and oppositional producers. Given this, how can viewers engage television's participatory fantasies in constructive or subversive ways?

Three Generations of the Revolution

There is an historical basis for these questions, linked to the "failure" of one media revolution and to the supposed success of another. For guerrilla video activists and video artists in the 1960s, tape was the means by which a movement of individuals could wrest control of the dominant media or at least sabotage its mindless workings with alternative imagery. Finally able to control the means of production also meant changing the terms of the relationship between the mass and the personal. TVTV and Antfarm and others staged hit and run actions in public, but utilized often quirky and very personal points of view to debase their enemy.[28] With a few exceptions, the reception of their works was limited to small screenings and audiences. Although histories of radical video are usually collections of apologetics for individual artists or practices, many utilize a shared historifying device—the cliche that the visionary video revolution ran out of steam and that many of its members were coopted into the very mainstream that they disdained.[29] Yet it is shortsighted to romanticize the radical moment of guerrilla television in this way.

From a contemporary historiographic perspective, much more was at work in this confrontation and failure than shifting personal perspectives and political climate. Guerrilla television was, in fact, a clear example of a technological use that preceded and predated the terms by which it could be effectively developed and widely adopted. To use Brian Winston's terms, although the "scientific competence" and technical "prototypes" needed for change were available, there was no "supervening necessity" at work here that could drive this new technology into a widespread and pervasive form.[30] Rather than fall back on great man theories that view individuals like Nam June Paik as visionaries who cause new technologies and uses,[31] it is useful to see and describe the first videotape revolution in a broader cultural and technological perspective. Tape, like many television technologies, was developed first for military usage, including tactical air combat. Artists and activists, however, also quickly saw and demonstrated its potential. Yet a persuasive enough model for the use of tape had not been developed effectively enough to insure that video's *portable* forms would be adopted by broadcasters in the late 1960s. In short, although the scientific competence and working prototypes were available during that time, no persuasive or dominant intellectual paradigm existed in broadcasting to force the widespread use or adoption of small-format portable tape.

Unless half-inch alternative tapes were particularly sensational, broadcasters simply refused to air them or to emulate their methods.[32] The conceptual framing paradigm necessary for historical change was not in place. Broadcasters typically rejected small-format tapes on the basis of two recurrent myths: they were either not technically up to broadcast standards, or they were not in the public interest. Both claims were clearly problematic and somewhat duplicitous.[33] The tapes, for example, could be electronically modified and enhanced for broadcast, and many tapes had clear public and social motivations. As arbiters of fairness, however, broadcasters acted according to their own definition of the public good. They also simultaneously upheld what David Bordwell, Kristin Thompson, and Janet Staiger consider to be one of the three explanations for technological and stylistic change in classical Hollywood: the industry's "adherence to standards of quality."[34] Such an ideal is also a business strategy that grows out of a broader ideology of progress. Standards of technical quality then, according to these theorists, also function to maximize corporate profits.

Why was this new technology, the model of portable videotape, not enthusiastically embraced by the industry? Television economists describe those various forces that affect and encourage the adoption of new technologies as externalities. Positive externalities include such forces as increasing economies of scale, the timing of introduction, and the number of users.[35] At this point in history, there was really no financial incentive for television to shift to the new technology. First of all, the FCC had, many years earlier, after much contention, settled on a very different standard, one designed for color rather than black-and-white broadcasting: NTSC rather than EIAJ. Secondly, the EIAJ half-inch tape format used by independents also never really had any influential adopters or important corporate sponsors—two other typically important externalities. In addition, as economists Bruce Owen and Steven Wildman point out, financially secure corporations have no reason to risk changes in production technology and personnel, and this weight of tradition creates an inertia that works against technological change.[36] Half-inch videotape was not rejected simply because the broadcasters were conservative capitalists, although that designation may indeed have been accurate. Rather, there was simply no positive externality in place to encourage the kind of portability that came with tape. Film was proving more than adequate for

Competing liveness modes. The live remote microwave hookup to a central station "nerve center." TVTV cooperative's portable-tape guerrilla ambush of network heavies dispersed across Republican convention floor in *Four More Years*, 1972. (KABC [left], Electronic Arts Intermix [right])

news, and television had settled on a very viable alternative to portable tape, the live hookup. Meanwhile, Sony kept developing new generations of nonbroadcast videotape formats.

A second videotape revolution also failed to achieve its anticipated ends. In the 1970s Sony finally introduced a workable tape-in-a-cassette format.[37] The three-quarter–inch Umatic format was intended to form the basis for a home video revolution, but it never did. The relatively large size and limited length of three-quarter–inch tape failed to attract consumers in wide numbers. The format did, however, finally find a home as the backbone of industrial video production—a corporate arena where it continues to dominate. From the early to late 1970s, however, the nightly news continued to be shot on film. After much resistance from unions and conservative industry practice, three-quarter–inch video finally eclipsed newsfilm in the late 1970s. Systemwide economies also collapsed along with film. Labs closed in all of the major cities. Union membership dwindled. Economies of scale came to be a factor in this shift, for the per-unit costs of video equipment dropped as the volume of production went up. The adoption of three-quarter–inch tape by broadcasters was a self-fulfilling prophecy. Once the industry locked in on this industrial format a bandwagon effect rippled through the industry. Once television gave in to portable tape, the transition was swift. Personal and activist video was still not a widespread reality. Yet one of the key negative factors used to resist portable videotape in the earlier period—broadcast standards—was now commonly being downplayed by broadcasters. That is, the industry ignored the very terms by which it closed the gate on independent tapemakers in the 1960s and early 1970s. Stations began to regularly broadcast a low-band format that was clearly not designed for broadcast and that was only marginally up to broadcast standards.[38] The official logic surrounding portable tape had begun to change. Tape entered the newsrooms, but it did so on broadcasters' terms. What was designed as a personal videotaping system also turned out to be the dominant form in an arena that was neither personal nor journalistic, neither public nor private. The three-quarter–inch format ultimately became the workhorse of the nonbroadcast corporate sphere.

The revolution—if one can still call it that—finally came in videotape's third generation. Half-inch and 8mm tape saw extensive usage on television in the 1980s. Celebrated in reality shows like *America's Funniest Home Videos, Rescue 911*, and *I-Witness Video* and spectacularized on laser guided smart bombs in the Gulf War, anything now seemed to be fair game for the camcorder.[39] Videotape footage showed up in sitcoms, feature films, and in those genres resilient enough to survive the collapse of the expensive primetime dramatic series.[40] By 1989–1990 television had a ravenous appetite for videotaped footage and seemed to welcome consumer-producers with open arms. Not only were tapes now made by nonprofessionals with little training, but stations showed off and celebrated these populist origins. The viewer was being celebrated as one of television's producers.

Mechanized Extensions of the Camera Eye

What had historically prepared the ground and opened the access gates for this third generation of tape users? There was, apparently, a clearly accepted industrial logic to small-format videotape by the late 1980s, and a conceptual framework that helped legitimize its usage by broadcasters. The answers to such a question cannot be addressed by focusing on the convergence of positive externalities alone. Negative externalities also played a major role in delaying the revolution. Specifically, broadcasters had developed their own powerful paradigm, one that appeared to satisfy the same need for remote portability and liveness that small-format tape offered. The increasingly complicated live remote could do the same sorts of things for broadcasters as tape, but it could do them on the industry's terms.

Portable videotape was not just a discovery or invention—it was a different way of making and conceptualizing television. It was also an institutional threat in that it undercut the kinds of commitments that the industry had already made—commitments to expensive pedestal cameras, to microwave technologies, satellite hook ups, and to a highly specialized labor system that worked as a clear extension of the studio, not as an alternative to it. Videotape's nemesis during its first two failed generations was the live, multicamera remote. The remote was a production mode that gave the audience the liveness that historians claim was at the center of broadcasting from the start. Remotes also seemed able to transform even mundane events coverage into memorable spectacles. The remote mode came replete with a wide range of multiple visual perspectives, an array of keyed graphics and image-text combinations, and—unlike portable tape—*immediate editorial interpretation*. Better yet, the remote's overwhelming technology brought to situations and events a highly visible and physical presence that simultaneously allowed broadcasters to cover the event and perform in it. ABC's *Monday Night Football* perfectly exemplifies this ability to transform an actual event into a cultural phenomenon about television. Now a national institution in its own right, *Monday Night Football* is really more about the celebrity announcers and elaborate orchestration of live and layered television imagery than it is about football. When this network remote covers, the nation watches. Remote productions presented complicated events, but also typically made broadcasters highly visible players in those events.

There were persuasive reasons for network investments in the live remote paradigm. The remote was, notably, one of the few genres on television that was made by and for the networks. As network productions, the remotes did not carry the financial insecurities and development risks that one associated with many other independently produced genres: sitcoms, dramas, game shows. Nor, more importantly, were networks allowed to share in the potentially lucrative syndication profits of such genres. The live remote, therefore, cut out both the production company middlemen and the high program licensing fees that came with them. In addition to allowing most of the ad revenues to stay in-house, the arrangement also provided a way for the live-remote program to showcase and market the network itself. The live-remote was self-serving in a way that portable tape-based programs could never be.

Television legitimized and promoted itself with remotes in a number of important ways. First, the pedestal, tripod, or vehicle-mounted cameras became literal extensions of a central studio, roving but disembodied eyes of the producers. This was Michel Foucault's "panopticon" in its finest form.[41] The multicamera configuration, with microwave or cable hook ups for simultaneous feeds, kept the editorial power to select and sequence imagery in the hands of a very few. Remotes, then, were really manifestations of the very same structure of power that had dominated the studio for decades in the form of the three-camera live studio style. The difference now was that the mode was being used to dominate and control the world outside of the studio as well. This production mode brought with it a power structure that was elegant and efficient at the same time. It created an extremely vertical organizational hierarchy: with many technicians and staffers but very few creative directors. The remote and three-camera style, for example, designated its camera people "operators" rather than "cinematographers" or "videographers." There would be no directors of photography in this system, and—unlike the world of alternative tape—no real cooperative decision making. The driving force behind this network of visual extensions were the executive types in control rooms, behind glass, and in concrete buildings—fortifications appropriate for the kind of power and status that network broadcasters had obtained.

Skycam in the Orange Bowl, Helmetcam in the Bud Bowl, the Monkeycam on *Late Night with David Letterman*, mobile sports and news vehicles, laser-guided video bombs and Tomahawk cruise missiles in the Gulf War—all of these can be seen as a manifestation of the remote paradigm as well. As physical extensions of the studio, all drew attention to the immense visual and editorial power of broadcasters. Owen and Wildman point out that in the competition between alternatives, different production technologies can achieve or fulfill the same ends or objectives.[42] In this case, the live coverage expected by audiences in the post-Vietnam era could have been realized either with the portable tape or remote paradigms. It is really no coincidence that broadcasters chose the latter. The choice and dominance of the remote over tape in the 1970s exemplifies what Bordwell, Thompson, and Staiger describe as "trended change."[43] Unlike portable tape, the remote was not a threat to the broadcaster's status quo. The live remote allowed television significant continuities, rather than discontinuities. Production style, management structure, and labor all could remain essentially the same. Broadcasters also underscored their own legitimacy and technological prowess by emphasizing the remote. No matter how infinite and spread-out the eyes of the remote paradigm may have seemed, television used the mode to centralize power and reduce diversity. This choice makes sense in a society that licenses a few broadcasters by delegation. Broadcasters chose and emphasized a liveness mode that justified and legitimized their existing corporate structure and investments—its heavy capitalization and centralization, its equipment and concrete. The extensive and prominent eyes of the remote allowed broadcasters to have it both ways: television appeared to move out into the world, even as it fortified its own vantage point.

Cultural Capital as a Supervening Necessity

While this notion of the live remote as trended change, explains how portable tape remained underutilized in broadcast television through the 1970s, it does not account for why the remote paradigm eventually gave way to the portable tape paradigm in the 1980s. The current preoccupation with videotape cannot be tied to one single cause. A range of factors, economic, technological, and conceptual, have converged to make tape a favored mode of production and reception in recent years. Not the least of these factors was the growth of what sociologist Pierre Bourdieu would describe as "cultural capital"—an aesthetic and experiential knowledge of the looks and possibilities of the television image.[44] Tape became an apt symbol for two driving factors in what academics have termed postmodernism: consumption and cultural distinction.

Tape teaches consumption. By 1986 more dollars were made by feature film producers through video and television than through theatrical exhibition. More often than not people experience film only as video. Tape is what one typically watches when one watches "film." Furthermore, starting in 1980, CNN and others began teaching audiences about the endless possibilities of tape. Disasters like the space shuttle *Challenger* explosion were played like endless möbius strips. Such tapes were perpetually redefined when broadcasters encrusted the footage with graphic subtitles, boxes, dates, names. The same thing happened with other newsworthy disasters. The home video shot by tourists during the San Francisco earthquake in 1989, and George Halliday's Rodney King beating footage in 1991, both lived extensive global lives as tape-based news events. Broadcasters even cloned populist production organizations like CNN's Newshounds, which were really thinly veiled bottom line, national surveillance projects that made semiofficial networks out of camcorder-toting citizens. The camcorder became the inexpensive farm system for cost-conscious major league broadcasters.

The music video genre was also instrumental in preparing the conceptual ground for a new kind of audience consciousness and format. It did so in two ways. From the start, the jukebox approach of sequencing videos on MTV did much to characterize the genre as a collection of individual units or objects. First, the fact that you could also buy these videotape-objects at music and record stores underscored the artifactual nature of the video. As video, television was being conceptualized differently. Second, videotaped footage became an almost obligatory element in the collage style that came to be known as music video. Even though such works were invariably shot on film, this footage announced itself as video—by its scan lines, pixels, and/or electronic artifacts. The self-reflexivity of most music videos also taught the viewer-fan the role that video played in the production process. The apparatus that produced tape was frequently the content of the video. Music videos, therefore, were simultaneously: 1) primers of production method, 2) products of videotape, and 3) consumer operations that caused and dispersed ancillary videotape forms. Everyone could do it, or so some ads suggested. Radio stations in Los Angeles ran promotional campaigns asking fans to submit their own music video interpretations of hit songs, that is, their own independent home-video productions, in order to win tickets to concerts during which the winning videos would be played.[45] Consumers were now, apparently, well aware of videotape as a production element. Videotape was something that they could both consume and produce.

Tape also teaches distinction. An important stylistic convention in film emerged during the third videotape generation. Feature films increasingly found ways to utilize videotape footage as filmed footage in feature films. From *Being There* to *Videodrome*, large screen images of video began to populate the screens of cineplexes. Films like *Down and Out in Beverly Hills* used home videotaping as an element in the narrative. The viewers of this film were shown the son's videotapes as an example of video art. Camcorder art was apparently now a common juvenile activity found in the domestic sphere. *JFK* obsessively reminded viewers of the wide-ranging status of its original footage, a dense combination of 35mm film, home movies, and videotape.[46] The viewer was expected to shift and interpret the historical and fictional status within scenes on the basis of their recognition of stylistic and technical marks and artifacts. Watching the film was an investigation of the very nature of film footage, as source material was drawn and quartered for the viewer. Viewers were expected to enter the investigation by and about footage that flagged itself as a material substance and as physical evidence. MTV did a widely televised special on Kevin Costner when *Dances with Wolves* was released.[47] Even though the special was shot on film, the cinematography was done camcorder style, with in-your-face focal lengths and frenetic and drifting handheld camera moves. The background for the program was comprised of a video-wall of electronic monitors and of mural-size simulations of the movie and of Costner's videotaped visage. Natural-man Costner was enveloped by the glow of electronic monitors and videotaped shuttle effects. The earthbound world of *Dances* was celebrated, ironically, in an artificial space of electronic video. What the viewer was given, then, was an experience more like a video-installation at the Whitney Biennial than a typical making-of documentary. This kind of showcasing via videotape, was pervasive. *Sex, Lies, and Videotape* was merely the first movie to hype the guise on its marquee.

Formal Permutations of Videotaped Footage

This chapter has examined several sets of oppositions relating to production practice: the competition between two different production paradigms; the relationship between two different video revolutions; and interventions between the public and private spheres. Since the 1960s portable tape and the live-remote offered alternative routes by which broadcast television could approximate the same goal: liveness and immediacy. Significantly, broadcasters ignored the possibilities of portable tape in favor of their own version of liveness, the remote. This de facto decision to favor one mode over the other resulted from a combination of economic, stylistic, and technological conditions, but it also had clear political implications. The section that follows looks closely at a range of historical productions that utilize videotape, and asks questions about the condition surrounding those productions and their implications. Why were the possibilities of portable tape ignored during what I will term the first video revolution? How did broadcasters benefit by investing instead in the live remote? How have historical conditions changed the way television values and invests in tape?

The tables have been turned in the competition between live paradigms. Contemporary television now seems obsessed with the nature and stylistic possibilities of portable tape. Videotape and its artifacts are prominently displayed across the broadcast spectrum and in many different formal permutations. What, finally, are the cultural ramifications of the latest video revolution, when mainstream television appropriates the oppositional production modes developed by radical media groups. Participation television no longer seems like the unproblematic savior that it once did. Participation, in fact, is a heavy-handed focus in many mainstream programs. It is also part and parcel of television's comprehensive extension into the private sphere. Tape is one key to participation and it is everywhere.

Three Prototypes: Tape, Remote and the Lab

Portable tape, the live remote, and the video lab had all demonstrated powerful, but very different, ways to celebrate liveness and its artifacts by the late 1960s and early 1970s. One of the best examples of alternative media still available through distribution is *Four More Years*. Produced by TVTV at the Republican Convention in 1972, the half-inch tape bore all the hallmarks of radical tapemaking: it was a cooperatively produced, hit-and-run personal intervention into the public and political arena. The anonymity, ironic humor, and pedestrian demeanor of the producers allowed them to covertly work the cracks of the well-oiled political machine. The coop's low-tech approach, on the other hand—existing light shooting, handheld camerawork, unedited conversational takes, and extreme portability—beckoned politicos and media personages alike to let down their guard. One cannot imagine that any of TVTV's unsuspecting victims would have liked the final product. These video activists were, in the final analysis, wolves in sheep's clothing. Tricia Nixon acts the gracious confidante; weary patriarch Cronkite philosophizes; Mike Wallace gets on-board the interviews like a team player—that is, until his political alarm goes off at a question about "advocacy journalism." Only Roger Mudd seems to recognize the real threat from these illegitimate guerrilla types. While Wallace patronizingly lectures the long-hairs on the evils of advocacy, Mudd plays the silent type—a network heavy in the shadows—and merely blows cigarette smoke into the questioner's eyes. With portable tape in these hands, everyone is graciously allowed to hang themself. A prototype for the kind of reportage that MTV would discover in the 1992 presidential conventions, this guerrilla action by TVTV was both a resounding fulfillment of the media insurgency war plans outlined by Abbie Hoffman in *Steal This Book*, and a hip, sacrilegious, and down-to-earth approach that was hailed, even by many mainstream critics twenty years later, as the best and most insightful convention coverage.

Yet, if the networks saw the power in TVTV's approach, there was absolutely no evidence of it. In fact, despite all of TVTV's on-camera successes, hardly anyone saw the production, at least in terms of network numbers. The weak link for the alternative tape makers, then, was that they lacked a viable and effective distribution system. If mainstream audiences saw the tape it was only as an afterthought months later on a handful of PBS affiliate stations or later yet at museums. As part of an underground distribution system, however, the lessons of such tapes were clear: guerrilla activism could be waged and waged provocatively with the new portable tape tools. In one emblematic scene, the TVTV recordists are challenged, as the tape rolls, by a convention security guard who cannot believe that their press cards are legitimate. After examining the press badges the official grudgingly retreats, and the on-camera soundman breaks into an on-camera blues-harmonica version of the "Republican Convention Drag"—all without missing a beat. This was more than just in-your-face "journalistic" coverage. With these kinds of giddy excesses and ecstatic outbreaks, the massively engineered public political spectacle—coproduced by the networks and the Republican party—became a personal space for unruly performance art. *Four More Years* was less agitprop, than a kind of decentralized contagion, one that infiltrated and exposed political faultlines and duplicities alike.

Although the networks continued choreographing newsfilm stories with wall-to-wall verbal narration as their stock in trade on the nightly news, producers outside of the networks began to see the power of the new highly portable video tools. One of the best examples of how the mainstream attempted to engage taped portability—while at the same time redefining it for mainstream ends—came in the video documentary *The Making of a Live Television Show*.[48] The tape is more than just a one-hour documentary about the making of the Emmy Awards show in 1971. It was also a veritable paean to the spectacular powers of the live remote. Produced at the same time that TVTV was working, this was broadcast television's version of the new and revolutionary half-inch portable tape format. Like *Four More Years, The Making Of* foregrounded low tech. Its low-resolution images were black and white, handheld, prone to blooming when pointed at lighting hot spots, and susceptible to sync break-up when the deck changed positions too rapidly during a walking shot. The perceptual effect of both the TVTV and Emmy tapes was verité: both were live, portable, and apparently spontaneous.

Yet *The Making Of*'s low-tech eyewitness look was merely a disguise: its half-inch portapacks were used to cover and laud one of broadcast television's unique and proprietary abilities: the production of a big budget, high-tech live television show. The type of language and metaphors used in the narration are significant. First of all, the narration was dramatically and omnisciently read by none other than Mr. Guerrilla Video himself, Orson Welles. Not only was this the voice of god, but it was also *the* voice of aesthetic authority, one that said, in effect, "What you are about to witness—is great television art." In somber and earthshaking baritone, the narration repeatedly describes producing the Emmy show in military terms: as an intense and logistically demanding "assault," as "D-Day." When on-camera directors and choreographers are not masquerading in military terms, the narration emphasizes the biological nature of the production unit in Hollywood. As the viewer sees the miles of arterial cabling and conduit snaking behind the scenes at the Emmys, the Welles-god makes his diagnosis: "This is the heartbeat . . . this the nerve center" of the production. By casting the whole project in physiological terms, the episode biologically echoes and reinforces the military analogy: this is a well-oiled and cohesive unit, comprised of highly skilled, interactive components/staffers, with a unique, powerful and all-seeing "center" of consciousness/authority.

The control room of the remote—a darkened arena filled with banks of monitors, cigarette smoke, and directorial generals who frenetically snap their fingers to marshal cuts—is the fortified neural center in a process; a live television cortex that cannot be replicated anywhere else on earth. The irony of this making of is far from subtle. An inherently low-tech, populist, and *decentralized* media format is appropriated here to construct a blatantly deferent homage to the unique and highly *centralized* abilities of Hollywood broadcasters. Half-inch tape provided the producers with dynamism, but was really only used to legitimize television's cult of professionalism. Low-tech is exploited, paradoxically, to worship high-tech, to reinforce television's long-standing and proprietary mythos of space and time conquering liveness.

A third prototype for the use of liveness artifacts, developed during this same period, came in the form of the TV lab. The networks had, for some time, utilized elaborate control rooms where source materials were synthesized, mixed and assembled for broadcast. Yet the idea of a site where the actual *process* of synthesizing such images would become the *content* of the production, originated at the very margins of network television, in the television labs at WGBH in Boston, KQED in San Francisco, and WNET in New York. Low-tech videotape artists were awarded access to these PBS experimental labs to produce visionary TV in the late 1960s and early 1970s. *Global Groove*, produced by Nam June Paik at WNET's TV lab in 1974, is emblematic of how the labs simultaneously engaged and manipulated both live and taped video. Widely publicized at the time as one of the first and most important videotape makers, Paik had been shooting and manipulating half-inch tape since Sony introduced its portapak in 1966. Characteristic of the lab mode, Paik brought with him to the studio for *Global Groove* a large collection of taped footage to electronically cook in the lab. Unlike the nerve center of network television's remote, the lab model was pitched to viewers as a private performance site, as an expressive televisual space for important artists.

The TV lab prototype in *Global Groove*, long before CNN. Nam June Paik hauls in everything but the kitchen sink for videographic cooking at WNET. (Electronic Arts Intermix)

Paik stirred a rich variety of taped imagery into the lab's videographic soup: tapes of mantra-chanting Allen Ginsberg, anecdotal philosophizing by John Cage, countercultural theater, softcore fan dancing, time-lapse photography, and various and anonymous fragments of street footage. To this taped source material, he added electronic and luminescent markings. Using both the station's production switchers and analog video synthesizers, the looks that resulted ranged from subtle patinas and silhouetting to the psychedelic layering, matting, and keying of multiple images. Traditional Asian performers were mixed with neodadaist Fluxus performers; Charlotte Moorman on the video cello was intercut with pop dancers rendered in day-glo video. Paik's obsessions were clearly intervening the public airwaves. Yet his interventions were contained, and his private musings ghettoized. A wraparound, complete with on-camera host and apologist, situated the half-hour show for viewers clearly within a high-culture frame. So did the lengthy list of arts and foundation grants credits that followed at the end of the tape. For viewers, then, *Global Groove* was framed as an important art world ghetto, even as it demonstrated how television could eventually shift to a more synthetic model of video production. CNN, MTV, and others would utilize the very same kind of frenetic electronic synthesis years later. The lab at WNET was a prototype, therefore, of how TV could have its cake (videotape represented in a live studio) and eat it too (through endless graphication and stylization). When the lab prototype is used today, all that has been typically jettisoned is WNET's artistic ghetto. Like the Emmy show's redefinition of portable tape to fit network expectations, Ted Turner's redefinition of the lab model at CNN headquarters in Atlanta was, in a logistical sense, also a kind of militarist transformation. Global discipline rules CNN coverage, not an expressionist id. In the surviving variants of the portable tape and lab prototypes on television today, opposition, expressionism, and activism are now essentially absent categories.

The "Successful" Video Revolution

Camcorders seemed to be everywhere by the late 1980s, and network TV cashed in. *America's Funniest Home Videos* premiered in 1990 on ABC alongside Lynch's *Twin Peaks*. Yet, unlike Lynch's offering, *America's Funniest* survived and achieved multiyear hit status with programmers and audiences alike. Although television itself had been a public target for hit-and-run interventions by tape activists in the first revolution, the tables here were turned. Shows like *America's Funniest* were instead public interventions into the private sphere. Producers went after the audience by creating program forms that drew attention to the audience, both by isolating and heavily stratifying viewer positions. Consider the many layers of audience involvement and participation—indeed, the many variants of audience—on a typical episode of *America's Funniest*. There are, for example, the families recorded on submitted videotape; a fake family playing cards on stage; host Bob Saget performing with homelike flats; an in-studio audience of families; and the real audience, comprised of those families viewing the show at home. The audience, then, textually multiplies on this series like Hydra's offspring. Numerous and overt mediating devices set off and distinguish these variant audiences for the viewer. These presentational boxes include the family's prominently displayed camcorder; the show's image delivered to the viewer's home and framed on a TV set; and various technical mediations in postproduction that constantly reframe the videos with graphics. What has been created, then, in each thirty-minute episode of *America's Funniest Home Videos*, is an explicit visual diagram—first, of multiple TV families and second, of technical divisions or barriers between those families.

  Given this overt visual and narrative stratification of constructed family positions, the show posits audience activity, as an ability to negotiate numerous levels and navigate familial barriers. The goal: to move through mazelike levels of

Endlessly replicated surrogate families and their new appliance on *America's Funniest Home Videos.* (ABC)

participation in order to attain the position of award winner—that is, the spot-lighted audience at center-stage. The very structure of the show underscores audience participation in a number of ways: viewers are allowed to select the channel, to attend the show, to act in competition, to compete as finalists, and to buy advertised products. The show repeatedly makes clear, in addition, that the participatory exchange is two way. The producers celebrate this quid pro quo by announcing their generosity: they graciously send the weekly program into your home; provide a live and entertaining show for the studio audience; give expensive VCRs and camcorders to all of the finalists; and give $10,000 and $100,000 cash prizes to the grand winners. *America's Funniest Home Videos* suggests that it is television by, about, and for the viewer. As an artificial but mutual back-scratching ritual, such shows not only utilize the technical tools of the first failed revolution, they also exploit part of its ideology—the aura of populism that used to drive outsider media.

This celebration of participation and ritual of exchange covers over an array of important institutional forces: the hidden role of commercial sponsorship, broadcast licensing and advertising fees, and an extensive range of gatekeeping chores performed by the production company. The home-video program genre actively effaces its corporate and institutionalized infrastructure even as it promises participation TV. Camcorder prizes stand in for advertisers. Cash awards and the host's stand-up rhetoric stand in for the production company. These home video program producers define themselves, then, as specific consumable items rather than the profit maximizing corporations that they are. Of course, the same could be said of all commercial television. The difference here, of course, is that ABC is not just providing entertainment that the audience can lose itself in. Rather, the show constantly constructs and showcases the viewer in relationship to the ritual of audience participation. Consider for example, the spots for RCA's consumer video products aired during *America's Funniest Home Videos* in 1990. The ads proclaimed that the "Pro-Edit VHS Camcorder" produced the very spot that was being broadcast to the audience. The not so subtle lesson? You too can produce broadcast/network quality video production—and you can do it at home! This pedagogical connection to the home is made complete when the live Bob Saget offers the very same unit advertised in the spots to the show's finalists. As if to underscore what is already very obvious from watching the show: "You too can be a producer—and we might even give you the equipment and cash to pull off your broadcasting aspirations."

Although the very idea of picking up a camera and VCR as a weapon has always played a part in alternative media, mass market television, at least in the home video genre, does not ostensibly resist such interventions. Instead, it appropriates resistant modes by adopting and overdetermining those same activities. In this case the audience and the private home are highly articulated, codified, and interactive parts of the program. ABC encourages and rewards the consumer revolution. Even as the higher and more cinematic televisual modes stroke niche individualism, the earlier primetime home video shows resurrect television's historic domestic ideology in order to overhaul and sedate the oppositional potential implicit in decentralized portability. By doing this, the producers suggest that they are somehow on the outside even as they quietly structure and secure for themselves a position of insider privilege.

Teaching a Supervening Necessity

The eventual adoption by mass-market television of both the technology and the populism of alternative media raises again the question about why portable tape was essentially ignored in the late 1960s. Although the working prototypes and scientific competence were easily available at that time, there was a lack of what Brian Winston would term a "supervening necessity"—a conceptual motivation

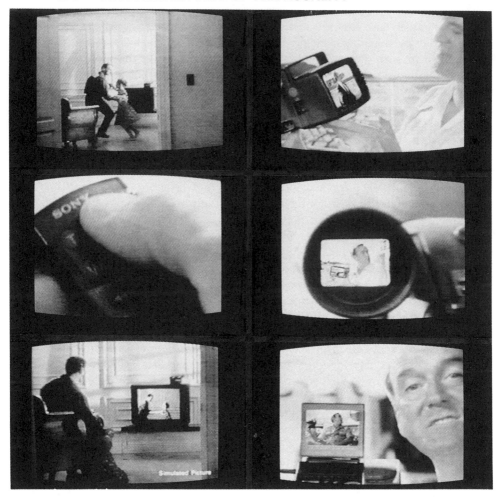

Teaching a supervening necessity. Sony reconfigures the single-parent family around the sensitive male. Magnavox creates carnivorous video that eats video, with Python's John Cleese, in its vacationing videographic hall of mirrors. (Sony, Magnavox)

for shifting to the new prototypes. Advertisements during the later video revolution, by contrast, worked to naturalize certain functions and to promote other motivations and logics for videotape. Advertisements about videotape, in both print and broadcast in the 1980s, clearly provided a teaching function—an industrial hand-holding operation for viewers—that helped create a supervening necessity for portable tape. The corporate benefits of mass hand-holding were twofold and complementary. Consumer giants moved more product, and the networks—facing declining viewership and a level of competition absent earlier—showcased the one producer who promised to force viewers to return: the viewer.

If RCA demonstrated that you too can make broadcast quality productions during spots in *America's Funniest Home Videos*, Sony pitched its product, the

Handycam, in 1993 as a way to save intimate moments. Of course this sounds merely like an electronic update of earlier "Kodak moments." Yet Sony placed this ability within a newly configured, and much more progressive, single-parent family. The sensitive male in Sony's spots is also able to interiorize and preserve those special moments for all to see by using the Handycam. Bathed in warm baroque lighting, father and daughter skillfully dance before the all-seeing Handycam—a device that, like the new male, is simultaneously smart, hip, and sensitive. Magnavox took another approach in their 1993 video ads. Their technology is lauded as "smart, very smart." The Magnavox spots combine the technical wizardry of multiple keys and mattes, with the Pythonesque humor of an ever-multiplying John Cleese. These ads clearly tout video products as a form that endlessly consumes other video forms. Less a hall of mirrors than a way of grabbing the world for personal collection, Cleese's cruise vacation here is electronically mummified on video as a Safarilike ritual. Befuddled and confused at the end of this endless consumption of video images, vacationer Cleese acknowledges that he will only be convinced that he had a good time if it is recorded on tape. The videotaped artifact, for both Magnavox and Cleese, is finally the ultimate way to guarantee and verify human experience.

The characterization and depiction of personal needs and public rituals in video in spots like these—modeled as home movies (RCA), parent-child bonding (SONY), and vacationing (Magnavox)—suggested that videotaping could fulfill an essential and broadbased function in American culture. Alongside the many other public uses of videotape in feature films and television programs in the 1980s and 1990s, ads also created a special need and logical place for video in the personal and domestic sphere. In fact, television programs and ads repeatedly characterize this construction of video's personal function as analogous to, rather than separate from, the public sphere of television. Television shows and video technology alike benefited from this supervening conflation of the public and private spheres. In this way, even new technologies were naturalized as an extension of two of mass culture's most central continuities: the family and the home.

Synthesizing Prototypes, Recuperating Tape

U2's ZooTV tour in 1992 was considered by many the penultimate step in the evolution of videographic excess—the *Gesamtkunstwerk* of our era. It was an international rock-and-roll tour; it was a pay cable special; it was a national syndication broadcast; it was music videos, CDs, talk shows, and, more importantly, an excessive performance of video style in the televisual age. It was an audio extravaganza and a television production spectacle at the same time. Camcorders, satellite feeds, frenetic electronic switching, voice-overs, collaging—you name it, ZooTV had it. In fact, the spectacle used all the hallmarks of radical and oppositional videotape. Yet the net effect of the ZooTV spectacle on the audience was in many ways problematic.[49]

ZooTV was significant because it synthesized both the portable videotape and live-remote production modes from the first video revolution, even as it simultaneously combined them into a different kind of hybrid electronic overload. Several of ZooTV's tactics are worth looking at more closely. Camcorders that

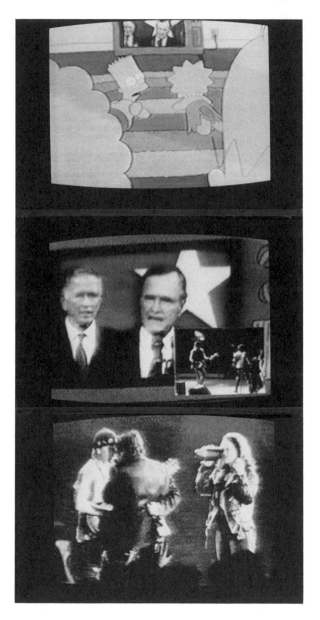

The televisual spectacle celebrates an aesthetic of theft. President Bush hits the Simpsons as "no Waltons"; Bart Simpson hits Bush back in primetime shortly thereafter; and U2 steals the whole ball game in its off-satellite onstage mixage. Camcorders are everywhere. But who gets to use the cameras, and what can be shot? (Fox)

promised extreme portability and spontaneity are here patched live into the show's main production switcher. Their purpose was not to produce tape outside of the center of power, but to focus attention in on the center. Production cranes and camera mounts of the live-remote cameras, on the other hand, also extended inward rather than outward. At one point megastar Bono aggressively grabs and yanks these cameras into his mouth and crotch. These operatorless camera extensions are positioned, then, as infatuated with and possessed by the body at the center of the stage rather than by the world outside. The question of who actually gets to do the videotaping in this spectacle is perhaps the best indication, however, of the relative power of the oppositional paradigm. Television images were everywhere—on stage and in the sky—but they were not in the hands of unauthorized fans. Camcorders were in fact zealously confiscated at the gate during the tour, making it clear that this democratization of television would only happen in the hands of the chosen few. Inspite of the apparent videographic chaos on stage, therefore, and the multiplicity of large-screen images that dominated the background, ZooTV ultimately provided a very controlled and constrained form of video output. The U2 juggernaut marketed cloned specials on broadcast TV and sent multiple music videos into rotation on cable. Participation, however, turned out to be merely fodder in a ritual of consumption that had video at its center. Much like the Magnavox series of "smart TV" ads, ZooTV was a frenzied videographic ritual that ecstatically taught us how to consume more TV.

At the same time that the ZooTV tour was winding down, television equipment giant Phillips began airing an expensive series of ads promoting their new DCC recording format.[50] These spots were shot in featurelike letterbox, with cinematic images that evoked German director Wim Wenders's feature film, *Until the End of the World*. This stylistic connection is important, for U2 was strongly identified with Wenders's project throughout and after its production. The videographic-cybernetic world that *Until the End of the World* presented to film viewers was also simultaneously celebrated by ZooTV and consumer electronics giant Phillips. Like Wenders's extensive international coproduction, Phillips represented a transnational European media company noted for, among other things, its pioneering development high definition television (HDTV). DCC was in some ways Phillip's HDTV high-resolution equivalent in audio but was, significantly, both portable and tape-based. Given the corporate-aesthetic archaeology of the "clear sound" advertising campaign, then, consider how these spots both taught the viewer about digital tape and naturalized specific consumer practices.

Phillip's "clear sound"spot campaign focused on a disruptive technical and aesthetic force that threatened to destabilize a global, capitalist media corporation. Emulating Foucault's panopticon, Phillip's corporate Big Brother was shown obsessively monitoring the global expanse for any and all unauthorized media technologies or transmissions. Under the paranoid shadow of this techno-totalitarianism—and apparently ignorant of its own global market share—Phillips presented itself as the radical threat to this conservative status quo. Phillips defined *its own* media technology as cutting edge and outside, as a guerrilla action that would send shockwaves through business interests and the fattened mass media alike. A closer look at one of the sixty-second primetime spots, however, demonstrates exactly how Phillips defined oppositionality. The three examples of DCC-induced liberation fantasized by the spot, depict in turn: a moment of monastic ecstasy; a flourish of operatic sentimentality; and a display of working-class excess, as a waitress erupts in a frenzied dance in a public restaurant. Radicality, according to Phillips anyway, is defined first as emotion, and second, as aesthetic emotion. More important, all three of these idealized altered states were induced by prerecorded music. The conceptual frame erected around this new tape technology— that is, the supervening necessity that will encourage the adoption of DCC—is apparently its ability to play *prerecorded and manufactured* media. There is no

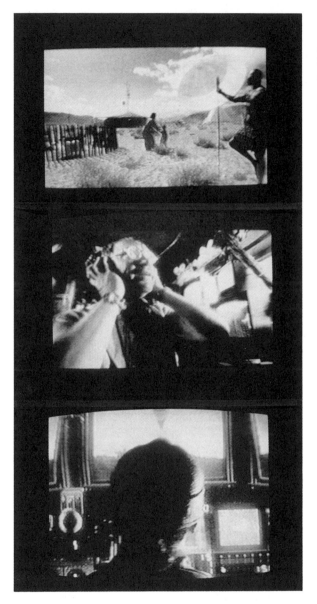

Phillips (Wenders–Bono–U2–ZooTV) announces its Clear Sound digital revolution against the global Big Brother, panopticon. (Phillips)

conceptual space allowed or even available here for alternative uses, even though tape, by its very nature, presupposes decentralized use.

Although the era and industries are different, the process by which oppositionality is appropriated and exploited bear a number of similarities. In the first video revolution, television controlled the definition of portable tape by avoiding it. By investing in an alternative paradigm, the live remote, television ignored both the threat and promise of tape, and legitimized its own ends and investments. In the era of digital sound and HDTV, on the other hand, the dominant industry *immediately* announces and defines for the audience the logical uses for its new technologies, through stylish teaching and hand-holding operations on and off primetime. The supervening necessity and logic for a technology now come pre-packaged with the technology.

Complicating matters further still is the fact that the very question of radicality has been transformed into a marketing strategy. Oppositionality is here but a shifting sign, whose connection to actual social practice or audience power is clearly arbitrary. Modes of viewing are not simply things that technology does to us or causes. Nor does television change when new alternatives are available. Rather, television seems to change, but only on its own terms. *The question of audience can no longer be isolated from the issue of television's industrial base, since television now creates and sanctions specific rituals for viewers, ones inextricably tied to new video production and home entertainment technologies.* In the case of portable video in the 1980s, there were both aesthetic and economic benefits to showcasing videotaped footage. On one level, the process embraced and hailed the new audience as active and free-thinking, even as it reduced practical politics and oppositional activity to mere semiotics, to questions of style and lifestyle.

10 Televisual Economy
Recessionary Aesthetics

Fueling the frustration among CBS brass and affiliate station managers was the realization that the . . . [network] had few promising mid-season replacements for ailing shows already on the schedule. Unrealistically high hopes seem to rest on the return of *Beauty and the Beast* and the debut of *Loose Cannon.*
—Network insiders explaining management shake-up, 1989[1]

In the quest to develop innovative programming without making sky-high deals, one of the things we've found is that the young, fresh voices are the people most capable of creating new and interesting shows.
—Amy Adelson, senior vice president, ABC Productions, 1993[2]

Even as primetime stylishness was moving center-stage, industry conditions were pulling the rug out from under it. All three networks were affected. By 1989, *Beauty and the Beast* was in programming purgatory, and Wall Street was betting on the demise of its network patron CBS. The same season, the press spread the word that NBC had taken a risk in renewing *Quantum Leap.* To the executives who had made the creative decision to renew, the show's subsequent and abysmal ratings in September and October were far from comforting. Series star Scott Bakula was paraded before the press in an attempt to stem the attrition one week before the November sweeps. His hype was also a thinly veiled last gasp: "Our future will depend on how well we do in November. We've developed some *wild* episodes"(italics mine).[3] These included segments in which the thirty-five-year-old actor played a retarded child and sang in a musical. So this was the televisual formula during times of crisis: answer programming threats blow-for-blow. Perched on the cancelation precipice, the structure and looks of televisual shows got "wilder," not more sedate. Televisual shows that did otherwise, like *Beauty and The Beast,* when it shifted its appeal to a broader *male* audience, faced an even swifter demise: they lost their most loyal female demographic.

Perhaps the most dramatic trial of the primetime televisual aesthetic took place at ABC. The network went out of its way in the months leading up to the fall 1990 season to till the critical ground for its latest aesthetic showcase: *Cop Rock.* Unlike anything before it, *Cop Rock* was a highly touted prestige offering by auteur producer Steven Bochco. More than just an extension of his earlier hits, *Hill Street Blues* and *L.A. Law, Cop Rock* was a hyperactive generic exercise that collaged fragments from the ensemble cop show, the film musical, and music video. Every scene in the series was heavily stylized, self-conscious, and overproduced. Feature-film style action, wild automobile stunts, Steadicam flourishes, smoke, directional lighting, sensitive courtroom dialogue, and black rap music—written by white writers—were all part of the visual assemblage.[4] Each episode commenced by showing the actors sitting-in during the show's recording session; groovin' along with pop celebrity Randy Newman and his backup as they did the theme song. The televisual and teleaural apparati were knowingly flashed for the viewer. You, I, Stephen Bochco, and the American public were all expected to dig it.

Cop Rock went down in flames in the fall of 1990. Pulled from the programming wreckage, Bochco disappeared from the headlines for months following the show's collapse. Sensing a fundamental change, critics now quickly turned on the very figure who, several months earlier, could do no wrong. *Cop Rock* and Bochco evoked for television what *Heaven's Gate* and Michael Cimino had become for United Artists and the film industry almost a decade earlier: a sacrificial marquee figure, and an omen for programmers who had cut their teeth on big production values and aesthetic and financial risk.

The primetime market had changed. At the very moment that mainstream television was celebrating high style, American business was celebrating its cult of deregulation and hostile corporate takeovers. In January 1986 Capital Cities tookover ABC. Nine months later GE swallowed NBC, and Laurence Tisch's boardroom rescue in October 1987 was actually a corporate coup. Each action carried with it a marked shift in the way that the business of network television would be conducted. Shortly after, ABC and CBS began cutting their personnel and programming costs; thousands of employees were laid off. The bottom line became a corporate obsession; mean and lean the central principle of network management. The great irony of this self-styled downsizing is that it occurred even as all three networks reported record profits. Although audience share had continuously dropped along with profits and predictions for over five years, advertising revenues went through the roof in 1989.[5] Fox was finally profitable, and MCA and Paramount were talking realistically about the possibility of a five-network economy. In retrospect, these business changes would have a negative and long-range impact on quality televisual programming. Network television is relatively slow to change, given its program development pipeline and conservative dependence on proven precedence. Downsizing was merely a way for the new owners to cash in their chips and functioned as a kind of temporary holding action. The network stylistic slide appeared to continue as well.

A New Signature: Austerity

Last year we were the innovators. This year, I don't think we have the edge we once had.
—Robert Iger, president of ABC Entertainment, 1991[6]

What we have learned about the audience is that they are very literal. They don't want to be fooled or spoofed. They want to know for sure what they are watching.
—Howard Stringer, president of CBS Broadcast Group, 1991[7]

High style was no longer necessarily the badge of accomplishment that it once was. During the 1990–1991 programming season, the United States underwent its worst recession since 1982. Ad revenues, the traditional guarantor of expensive, quality programming, began to drop precipitously in 1991. Even as the fourth network, Fox, had finally managed to survive and hold its own, industry experts were predicting dire consequences for the networks. CBS, the weakest of the three majors, was widely thought to be headed for sale or massive reorganization.[8] Cable and pay television continued to gain market share against the very networks that forged the mythology of stylish self-consciousness. The Mike Tyson–Evander Holyfield title fight, a *single* nonnetwork broadcast on pay per view in the spring of 1991, earned $55 million dollars for its cable promoters—more than the *entire* first quarter revenues of national network CBS. The playing field for broadcasters was no longer safe or secure. Over one-third of ABC's primetime schedule was losing money in 1990–1991.[9] After increasing the fees they paid for some primetime shows to record amounts in order to cover escalating production costs in 1990, CBS and NBC were forced to quietly negotiate rollbacks with program suppliers to lower levels in the face of decreased network viewership and advertising revenues.[10] As a result of these factors, the corporate axe began to fall, with aesthetic as well as economic consequences. Although the logic of the televisual spin proved resilient, the cost of high-style exhibitionism—of designer and marquee television, of event-status programming—was under attack. This chapter aims to consider the nature of those changes and to question the dire interpretations and forecasts given by network presidents Robert Iger and Howard Stringer. Is innovation over? Are viewers tired of subtlety, irony, and high production values, as the executives suggest? Or is the industry and the practice of televisuality merely being organized and orchestrated on a different scale and economic plane?

Several dramatic shifts in programming occurred in the months between spring 1990 and summer 1991. Televisuality and stylistic consciousness were pervasive during the onset of this period. Industry voice *Broadcasting* magazine touted ABC's *Twin Peaks* as a "critical *and ratings* success" (italics mine), important because the excitement it created, along with *Doogie Howser, M.D.* and *America's Funniest Home Videos*, "has ABC looking like the network of the future."[11] The *Village Voice*, on the other hand, used the new television as a basis for a special issue entitled "Rad TV." Television was now the hip and stylish embodiment of an art that "cocooned" and "soothed the savage yup."[12] Others positioned television as a form of vanguard poststructuralist theory: "If NBC, CBS, and ABC are the history of television, then Fox is the slyest deconstructionist around."[13] Aesthetic consciousness was not, however, simply the province of New Yorkers. The Emmy Award nominations of 1989–1990 indicated the broader importance that style had taken by 1990: an overwhelming number went to the newest and most offbeat shows and many went to Fox, as the industry began to recognize the network's self-consciously innovative shows. *The Tracey Ullman Show, The Simpsons*, and other nontraditional Fox offerings netted a total of twenty-six nominations.[14] Significantly, it was auteur David Lynch's strange and gothic-styled ABC series that garnered the most nominations. Lynch's brooding, primetime series *Twin Peaks* received fourteen total nominations, causing ABC executive vice president Ted Harbert to remark of primetime programming, "It's a clear sign that people are looking for smart fresh programming."[15] Whereas earlier shows in the mid-1980s, like *Max Headroom* and *Pee Wee's Playhouse,* had aggressively established new boundaries for television style, *Twin Peaks* demonstrated that televisuality was now an economic as well as aesthetic factor in programming. The strategic placement and promotion of *Twin Peaks* in primetime spring programming was a major factor in the financial reascendance of ABC as a programming force. By the summer and early fall of 1990, apparently, televisuality was clearly not a special aesthetic ghetto in broadcasting. Style became a part of the network's public face and identity. Everyone, it seems, got on the bandwagon—originality and innovation were *very* important. ABC's Iger predicted that the new programming for fall 1990–1991 would demonstrate "some risk-taking, perhaps more than there [had] been in the past."[16]

Television's growing penchant for stylistic exhibitionism, however, immediately ran into a long succession of obstacles and setbacks. *Twin Peaks* was moved around the weekly schedule in a desperate attempt to increase viewership. Network executives later acknowledged that this bouncing around cost them valuable viewership and revenues.[17] By spring 1991, *Twin Peaks* was canceled. It was not, however, alone in its ignoble demise.[18] Prestige hour-long dramas, with high production values and self-consciously innovative ends, ended en masse. *Dallas*, the premier primetime soap of the 1980s, tied up its loose ends in May with a final two-hour special structured around numerous dream sequences and out-of-body travel by J.R. Amid much speculation and indecision, *thirtysomething* was finally canceled, in spite of its great critical acclaim and numerous Emmy Awards. The equally stylish *China Beach*, which had become one of network television's most explicit exercises in dramatic experimentation, aired its series-ending finale in the late spring.

Highly visual primetime programs were not alone in bearing the brunt of the new austerity moves. Game shows, long thought to be the reliable revenue mills of daytime programming, also were canceled. *Wheel of Fortune, the* game show of the 1980s—the highly visual, spatial, and gestural spectacle that made Vanna White one of the best known celebrities in the nation—was canceled by NBC in September of 1991, along with the network's only other surviving game show, *Classic Concentration*. ABC had canceled its remaining game show, *The Match Game*, earlier in the summer. Once again, production costs were blamed. One production executive claimed that "the economics [are] that networks can do reality shows in-house, with their own production staffs, and do them cheaper than they can buy game shows."[19] Network news divisions also faced drastic cuts. Having lost significant viewership to CNN during the Gulf War spectacle, the major networks acknowledged defeat by further cutting back on news bureaus, coverage, and allocated resources. One more arena in television competition was given over to cable. Production was expensive and the cuts were widespread. Many argued, in addition, that the era of the serious dramatic series—the genre that helped create and perpetuate a look of visual excess in shows like *Dynasty, Hill Street Blues, The Equalizer, Beauty and the Beast*, and *Moonlighting*—was over. Cheaper forms of programming were slotted to take the dramatic series' place. In the face of network television's new penchant for sitcoms and cheaper forms, Iger put the best face possible the network's growing deaestheticization: "A new signature has been written for ABC."[20]

Periodizing history can be an idealist trap, since culture rarely conforms to the neat packages and categories of analysis. Nevertheless, 1991–1992 saw dramatic changes take place within a broader process that had begun much more gradually in the early 1980s. It is possible, as I have argued, to see the rise of televisuality and stylistic consciousness as a specific strategy that dominant television used to compete in the new and increasingly deregulated market of the 1980s. With the veritable control of high-end production at their disposal, good access to film origination in Hollywood, and relatively clean affiliate franchises that secured their potential earnings in distribution, network television was a relatively low-risk enterprise in the 1970s. Cable television, satellite, and pay television changed all of that. In 1980, 91 percent of the television audience watched network television. By 1991, the figure had fallen to 63 percent—so low in fact that that the expenditures frequently outpaced revenues. In 1991, for instance, the production costs of a typical comedy went up 8.2 percent and the cost of a primetime drama 5.1 percent; this despite the fact that advertising revenues increased by only 3.5 percent and were scheduled to decrease in 1991–1992.[21] When money goes out faster than it is taken in, only the most visionary CEO will continue down the programming path that leads to such an equation. The victims, in short, were precisely those programs with high production values, unorthodox scripts, and complicated forms of audience address—that is, the very genres that had perpetuated and promoted televisuality.

In retrospect, televisuality can be seen as mass-market network television's attempt to do what it does best in the face of growing competition. In much the same way that Hollywood individuated and distinguished itself from upstart television in the 1950s (that is, by offering unique *stylistic* experiences: wide screen, technicolor, 3-D viewing), network television had taken upon itself an intensive program of innovation and stylistic development during the 1980s. In the deregulated world of conglomerate media corporations, survival depended upon developing new and different products and new looks. Within contemporary television, for instance, major film studios like Warner Bros. and pay-cable franchises could now be simultaneously owned by the same corporate giant, Time-Warner. With the direct linkage between film producers and cable, then, the television networks, in short, have been increasingly cut out of one important loop in the television market. Blockbuster films, in some cases, can now only be viewed on cable— with all of the revenues going to a single non-network conglomerate. Even when faced with stiff competition and breakthrough innovations by its competitors (for example, the frenetic avant-garde visuality of MTV, the graphic environment of CNN), the networks have always had, up until this point, enough centralized capital, production resources, and distribution clout to appropriate and exploit those breakthroughs for their own ends. They countered MTV with *Miami Vice, Friday Night Videos,* and *Max Headroom* and CNN with *very* flashy magazine news shows like *West 57th Street, Primetime Live,* and *20/20.* The networks countered all of the other less powerful and local forms of television (that is, cheaper program forms that originated on video), with their unique and expensive film-style look from Hollywood in *Crime Stories, Dynasty,* and others. The scramble for stylishness, then, evolved both to compete with and outperform specific formal breakthroughs, as well as to establish a distinguished aesthetic aura of institutional quality that could not be manufactured elsewhere.

Programming Changes

If anything promised to displace televisuality as a dominant aesthetic tendency during the 1990–1991 programming year, it was surely reality programming—a category that included shows like *Cops, Rescue 911, Unsolved Mysteries, Hard Copy,* and *America's Most Wanted.* Several factors helped make reality programming a central concern of programmers. First, reality shows (like sitcoms) are much cheaper to make than primetime dramas, and so offered the networks a chance to salvage their financial statements and corporate earnings. Some of the more popular reality programs only cost between $200,000 and $300,000 per half-hour to produce. This is but a fraction of the production costs for primetime dramas and action adventure programs that typically cost between $900,000 and $1 million per episode to produce.[22] Second, reality programs had crashed the programming scene unexpectedly and had done so in a way that indicated massive popularity. Ironically, it was a program that premiered on ABC around the same time as *Twin Peaks,* that suggested the breakthrough quality of reality forms: *America's Funniest Home Videos* was much less expensive than Lynch's work and turned out to have a more widespread and lasting audience. It has survived and spawned numerous clones, even as *Twin Peaks* died. What could be cheaper, after all, than airing home video footage as a primetime program? The networks, after all, do not have to pay producers, union crews, or independent production houses, and truckloads of home video tapes flood into the studio from the national populace.

Two other reality phenomena promised to challenge television's preoccupation with stylishness. In January and February of 1991, the United States and its allies carried out their planned attack in Kuwait upon Iraqi forces. Endless replays of low-resolution, grainy, and chaotic footage filled the airwaves and cable systems for weeks. Gun-mounted camera footage on supersonic fighter bombers was replayed to an international audience, even though viewers could seldom see anything other than sighting devices and calibration marks. As viewers watched amorphous, electronically smeared, and monochromatic imagery, Pentagon briefers carried on overly detailed and comprehensive descriptions of what the audience was *supposed* to see. The authority of the narrated word dominated feeble apparitions that viewers imagined as both carnage and and evidence of technical superiority. The exquisite imagery of *China Beach* and *thirtysomething* gave way during this period to an endless succession of gray blurs. In the wee hours of the morning, the promise and aesthetic threat of this kind of reality programming was even greater. For hours on end after midnight, CNN would randomly play incoming satellite footage, before editing and without narration. The practice of airing random and unedited footage from remote locations continued during the Soviet coup and overthrow of Mikhail Gorbachev on August 25–26, 1991. No one watching could really know what was happening when molotov cocktails went off. Indecipherable shouts were mixed with vague images; flames illuminated crowded figures in the night. Were people dying? What was happening and to whom? For the American viewer, reality was coming home to roost in the form of pure and ambient imagery and sound—all without context and interpretation.

One final factor suggests the emerging promise of the reality pose. Rodney

King's brutal beating by Los Angeles police officers in 1991 was also recorded on home video. The outcries of injustice that followed the tape's emergence were unequaled. Racial rallies and protests followed. The Los Angeles police commission was overhauled, threats of violence were made, and the heretofore entrenched and untouchable white police chief was ultimately forced to promise his resignation. The tape was played many times on almost every station in the country. The actual quality of the tape—as we will see in the final chapter of this book—bears the same kinds of marks that operated in other reality disruptions: the Gulf War and *America's Funniest Home Videos*. The Rodney King beating imagery was monochromatic, high contrast, jerky, and low resolution. The viewer's realization and horror about what is happening in the tape occurs over time as the viewer works to make out details in the chaotic scene. What subsequently happened to this footage on television, however, bears examination.

Each of these highly celebrated reality displays promised to undermine the conventional ways that viewers value highly styled televisual programming. The final appropriation of these reality threats by television, however, suggests that the apparent antithesis between ruptures of reality and displays of style is not so complete. Invariably, rampant and amorphous images like these—footage that at first seems uncontrollable, given its origination outside of the institutions of television—eventually succumb to the very stylizing processes discussed earlier. Television, apparently, does not allow such images to remain disruptive or real for long. Chaotic gun-mounted camera and battle footage was wrenched out of context and made a part of many excessive stylized forms: condensed reports, graphic openings, and orchestral celebrations of American military superiority. CNN immediately made icons of the war part of its sophisticated graphic environment emanating out of Atlanta. Even if random live images were not culled to illustrate reports, they could find themselves ground up in the graphic environment of postproduction for use as highly visual wallpaper. The potentially disruptive and incoherent qualities of the tapes on *America's Funniest Home Videos* were also constrained and overhauled. Significantly, the first thing one notices about the program's stylistic operation is that any ambient sync sound is stripped off. Cheap music and obvious sound effects are added in their place. The unruly nature of reality is, then, anchored within the show's stylizing logic and formal control. Images are heavily edited and captioned. The end result of this aggressive repackaging is introduced by an upbeat emcee, and then projected before a live studio audience. If there was ever a program form defined by the overhaul of ontological realities into a videographic and visual spectacle, this was it.[23]

In all three cases—CNN's Gulf War, ABC's *America's Funniest Home Videos*, and the Rodney King footage—the disruptive potential of uncontrolled reality was suppressed through a wide range of intensive stylistic operations. Televisuality does not just work this kind of footage over as an afterthought or response. Rather, it clearly has an ideological appetite for such footage. In much the same way that I described the televisual documentary as a kind of ideological safari in chapter 8, television also *prospers and empowers itself by collecting and pictorializing dangerous images*. Television consumes and embellishes itself in much the same way that Delacroix's romantic paintings of the early nineteenth century turned the threat of race, distant cultures, and nature into spectacular and exotic furniture—

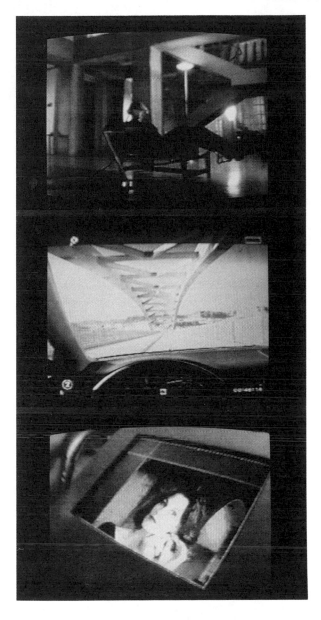

Reality programming? Not in the world of television advertising. Honda hit the fall 1993 season with a highly stylized, high-tech, thirty-second virtual-reality domestic drama to showcase the new Accord. Far from being hidden in this revolutionary interactivity, Honda's gender roles were little more than those of *The Donna Reed Show*. (Honda)

vistas surrounded and displayed within baroque frames.[24] The ideology of the exotic also transforms disruptive and dangerous low-resolution images on television into furniture. Such pictures either literally become a part of the newscasters high-tech set or fill the viewer's television box like encased wild-game trophics complete with graphics that inscribe broadcast ownership. In this way, by considering television's appetite for the exotic and dangerous, the semiotic and videographic processes used by the industry that I described in chapters 3 and 6 can be viewed as a kind of high-tech taxidermy. Television bleeds the life out of its unruly game and flaunts a stylized model of it for the viewer. The stylistic repackaging of reality like this is not, then, simply a form of mannerism. It is also an important tactic of appropriation by which television legitimizes and empowers itself.

Reality programming does not, as executives Stringer and Iger imply, mean the demise of televisual exhibitionism and stylistic self-consciousness. Reality programming is merely the working out of televisual and mannerist impulses on a different level. CBS may fold, and the networks may shift to less expensive program forms, but other institutions and programmers will take their place, will continue to pursue and perform style. Commercial advertising, the stylistic avant-garde of television, shows no signs of de-aestheticizing. The boundaries and frames that constrain style continue to disintegrate: Saatchi and Saatchi's Eveready Battery rabbit continues to invade commercial spot after simulated commercial spot; single ads are now broken up into multiple parts and scattered over commercial breaks; mini soap operas are created as continuing serial spots for Taster's Choice coffee; long-form infomercials evaporate completely the distinction between program and advertisement. With the choice of over one hundred channels now coming to some cable markets, and five hundred on the horizon in others, consumers face immense viewing choices. ABC cites this factor, and the resulting tendency of viewers to endlessly zap channels with remote controls, as the reason for the demise of serious aesthetic forms on television and for the *viewer's new lack of commitment*.[25] But this excuse is both shortsighted and revealing. If one views the emergence of excessive visuality as a counterstrategy by the industry, as I have argued, then stylistic extremes will continue as a way to fight viewer disinterest, channel grazing, and infinite choice—in short, will continue as a way of attracting and intensifying the spectator's visual gaze. The new economic and corporate realities of television, then, should not be seen as the basis for an aesthetic reversion to reality, to information, or to the glance. Rather, the current changes in television suggest that even more value will be placed on artifice and embellishment. Such things give programmers power.

The Changing Industrial Footprint of Televisuality

By 1993, after nearly four years of recession that left only one network consistently profitable, network heavyweights sounded more like survivalists than visionaries: "It's not business as usual anymore. We have got to find ways to *recreate this business so that it will survive* into the next decade," declared Ted Harbert, ABC programming executive.[26] Networks moaned about their costly multiprogram commitments to "heavy-hitter producers James Brooks and Steven Bochco." Executives were rewarded for developing reality shows and sitcoms, like *Home Improvement* and *Roseanne*, shows that both originated on video. Yet quality TV was not always seen as a liability and a programming dinosaur. Televisuality was actually being regionalized and franchised on a widescale basis, both because of changes in technology and in production economics.

Networks found ways, for example, to continue producing distinctive cutting-edge programming by shifting to a younger and less expensive production force. Amy Adelson, senior vice president of creative affairs at ABC Productions justified the programming and aesthetic value of *inexperience:* "In the quest to develop innovative programming without making sky-high deals, one of the things we've found is that the young, fresh voices are the people most capable of creating new and interesting shows."[27] Adelson's comments conveniently cover over the fact that youth and inexperience, while agents of innovation, are also cheap. Transforming the cutting edge, then, also became a kind of ageism, a justification for pushing aside the expensive and middle-aged deadwood in the industry. Although personnel shifts in the above-the-line ranks (producers, writers, and directors) could be justified from an aesthetic point of view, for below-the-line crafts people (cinematographers, editors, production coordinators, art directors, set decorators, carpenters, and others), the new personnel shift looked more like union busting.

As competition increased, and the viability of original syndicated programming spread, budgets would no longer apparently support the expensive production tastes of primetime network programming. Complaining about the impossible logistical hassles of shooting and permits in Los Angeles, producers fled to nonunion states for production, to North Carolina, Georgia, Oregon, and to Vancouver. Translation: union crews, no matter how good, were now simply too expensive. Hollywood producers, along with their ever more underemployed writers and directors, went to the Directors Guild and the Writers Guild, to ask for dispensations to shoot syndicated programming outside Southern California. Sensing the seriousness of the economic situation, the guilds quietly began to consent. The Screen Actors Guild, painfully aware of the explosion of nonunion production, both in the form of low-budget syndicated programming and feature projects, no longer held the ax above its underemployed membership and set up a system of lower contractual arrangements for such work. By 1993 the notions that the unions in Hollywood were dead, and that shooting in Hollywood was no longer practical, were oft-repeated explanations passed on in the confines of producer's offices.

Typical of post-recession changes were shows like *Baywatch* and *Acupulco H.E.A.T.* Both were distributed nationally and internationally, *H.E.A.T.* was produced by an inexpensive nonunion shoot, and both proved to be cost-effective syndicated cash cows. While *Baywatch*—known to some insiders as "Babewatch"—was really a throwback to the T and A of *Charlie's Angels*, its clone, *Acuplulco H.E.A.T.*, showed how inexpensive syndicated production could exploit the concept-crunching punch of the televisual pitch mentality. Internationally coproduced by Europeans and North Americans, shot with nonunion Mexican labor, managed by a skeleton crew of Hollywood guild members, and shot in exotic locales south of the border, *Acapulco H.E.A.T.* was intended as a hybrid cross between *Baywatch* and James Bond. The show's writing was far from profound, but the series had the elements thought to guarantee cross-cultural international success—visual flash, violence, and skin. Its national airing in this country and abroad proved that exhibitionism could survive even under the severe economic constraints of recessionary Hollywood. Unemployed Writers Guild screenwriters missed the point, then, when they moaned that the assassin of quality was reality programming. Syndicated and reality programs were part of a much bigger and more influential picture: a marketing appetite driven by the snowballing, cost-conscious, and competitive myth of the 500-channel environment. GM was not alone in sending jobs south of the border; Hollywood did the same to keep its programming fires burning.

Counter-programming, a marketing strategy that helps fuel stylistic individuation, has also taken on increased importance in the face of heightened competition. Despite a number of notable and costly victims of counter-programming—*Miami Vice*(NBC) lost to *Dallas*(CBS) in 1986-87, and *Seinfeld*(NBC) lost to *Home Improvement*(ABC) in 1991–1992—Fox demonstrated that counter-programming was the continued key to its network attack in the wide-open environment of cabled television. Having successfully used *The Simpsons* to counter-program the nationally dominant *Cosby Show* in the late 1980s, Fox premised their entire 1993–1994 season on a systematic strategy of stylistic counter-programming against many of television's dominant shows. Film director Robert Townsend's new comedy-variety show *Townsend Television* was thrown against *60 Minutes*, *America's Most Wanted* against *Roseanne*, and *In Living Color* against *Seinfeld*. All three series took on the network heavies with relatively inexpensive televisual genres—race-based comedy-variety or voyeuristic reality shows—and all three celebrated outsider or excessive attitudes.[28] Fox hoped to strip away the reality freaks who watched *60 Minutes* by using their own voyeuristic series— *America's Most Wanted*—to abduct CBS's lucrative audience for its Sunday strip.

Part of the new openness and volatility in the programming environment was also due to a fundamental breakdown of traditional program production categories and relationships. Government regulatory changes to fin-syn rules and the 1990 termination of the judicial consent agreement—policies that had capped the number of program hours that networks could produce—now meant that the big three could tap into the once-restricted business of producing and owning the very programs that they broadcast.[29] Program production, historically the source of massive syndication revenues for independent program producers, now also provided profits for the traditional networks. ABC, CBS, and NBC could now sell more of their own wares to syndication. The percent of in-house produced news and entertainment programming at the major networks rose from 12 percent in 1989 to 27 percent in 1993. The networks were, in essence, fighting for a toehold in their own market.[30] The incestuous nature of this kind of "competition" distressed outside production company bidders, who were not lucky enough to be located in offices a couple of doors away from chief network programmers. The networks were accused of having "become the umpire in their own game."[31] Yet the scramble by any and all to actually *make* the programs indicated two much more fundamental changes. First, the static and stable boundaries that used to exist between the networks and producers were being abolished, and with them a conservative kind of quality control. Second, the afterlife of network series—now seen as potentially endless—was being recognized as the real financial game in town. The goal of a network broadcast no longer solely governs the origin and design of even network programs. Ancillary, individuated markets are very much on everyone's minds, including the networks'.

Changes in technology—in digital compositing, videographics, and high definition television (HDTV)—have also helped perpetuate and reincarnate televisual exhibitionism. As high-end production tools drop in price, individual stations, cable, and syndicated networks now buy and produce complicated electronic looks that were once the sole domain of expensive specialty boutiques. Jan Phillips, executive producer of Telezign, notes that "Stations are a little more constrained as they see budgets shrink, especially in the broadcast area." This shrinkage, along with the increased power and affordability of compositing equipment, caused a number of broadcasters to start creating their own promos.[32] Fox in 1993, for example, was the first network to buy a Hal digital compositer and produce its own promos, rather than booking expensive outside production houses. Further regionalizing and decentralizing televisuality are one-stop services that provide, not high-end graphics equipment, but custom made *visual sequences* for local and affiliate stations. In spring 1993 News in Motion, a division of Knight-Ridder Tribune Graphics, went on the air with a service offering "animated illustrations of major news events." Local news operations can, with this on-line service access flashy styles that *"look just as good as the networks"*[33] (italics mine). Even cash-poor affiliate and local stations in minor markets, then, could now be fed *prepackaged, stylistic flourishes for each late-breaking news story—third-party looks* guaranteed to compete with the big boys. Televisuality, therefore, had become both valuable and cheap enough that even backwater broadcasters—traditionally little more than clearinghouses—could stock their news programming shelves by purchasing preembellished and prepackaged reality products.

Franchising and regionalization are not the only ways that technology has helped reinforce the industrial centrality of style. The future of televisuality is also tied up in the long and costly battle over HDTV. NHK in Japan and Phillips in Europe had viable working prototypes for high definition television in the mid-1980s, an accomplishment that should have given them clear superiority over American developers in launching the new HDTV systems. In the spirit of competition, the FCC established an extended contest in which the three major consortia developing HDTV in the United States could test, showcase, and compete in the race to standardize an American version of the system. In the spring of 1993, after months of testing and extensions, rather than choosing a winner—and thereby creating losers and possibly litigants (!)—the FCC encouraged the competitors to modify their plans so that a joint American HDTV format could be agreed upon. Given the high costs and risks of developing a new system of this scale, the competing consortia gladly agreed to this patent pooling and to the rallying notion of a single American HDTV format.

The international community was profoundly less ecstatic. Trade publications predicted that the new American standard would dominate international markets and eclipse even the more established Japanese and European alternatives. While this type of government supported monopolistic practice has a long history in American television, the *consensus HDTV initiative* stands in stark contrast to over twelve years of communications policy based on the Reagan-Bush ideal of deregulation. It also stands as a potential challenge to the notion of free-market business practice at the center of the Federal Trade Commission (FTC). One only has to recall the patent-pooling trust that stole and nationalized radio from the Marconi Corporation in 1919–1921 and the licensing freeze by the FCC in 1948, which effectively created national communications monopolies in the form of NBC, CBS, and ABC, to grasp the long-term implications of such decisions. The apparent lack of concern by regulators over issues of free-market and monopoly in 1993 had perhaps less to do with a change in political climate than with a recognition that the *scale* of the communications market had changed. The 1920s and 1948 government monopolistic initiatives had an immediate and lucrative impact on the national market and the fate of corporations like AT&T and NBC.

HDTV arose, by contrast, in a television market that has clearly become international in scale and comes after decades in which state-of-the-art television technology had been taken over and controlled by international Japanese and European corporations like Sony, Panasonic, Hitachi, and Bosch. In this context, *consensus HDTV* was less a gesture of monopoly than a forceful reemergence of American nationalism. Domestic free-market constraints make little sense, after all, in an international arena made up of multinational competitors with pervasive but irregular government subsidies. In the face of protests of massive entertainment software piracy and the threat of continued technological and market-share decline, consensus HDTV was a way for the industrially bruised United States to reestablish control over an essential technological and transmission gate, one that would give them more than just high-resolution and stylish images. Consensus HDTV also promised to guarantee American producers, studios, and networks a lock on the very proprietary entertainment product that most transcended nationalist and regional boundaries—expensive primetime programming and feature-film works. American television has always ruled international markets not because of its hardware, but because of its high-production value software. Consensus HDTV, if successful as a unified transmission gate and home-delivery system, would help American television continue its style-driven market dominance.

The Darwinian imperatives of the new television and cable environment mean that sponsors and advertisers have to create better mousetraps if they want to continue attracting profitable audiences. Niche television at first looked like a gold mine for advertisers and programmers. "The smaller the rating, the better the demo," remarked Tom Winner, executive vice president and director of marketing services at CME-KHBB Advertising, about cable's self-fulfilling niche prophesy.[34] Yet, while niche narrowcasting guaranteed sponsors their target audience, the endemic advertising that followed also meant that the niche cable networks and their ads had little appeal for viewers not interested in their focus: golf, fitness, or war documentaries. Audience growth was limited by narrowcast boundaries, which meant that advertisers and programmers had to find better ways to build distinctive audiences.

Changing market and economic conditions also helped perpetuate and hybridize televisuality. That is, stylistic exhibitionism was also influenced by the breakdown of traditional advertising relationships and categories: by the horizontalization of sponsorship, and by the creation of à la carte agency economics. When Michael Ovitz signed Creative Artists Agency's (CAA) major deal for advertising with Coca-Cola, the advertising industry was clearly threatened by what it saw as Hollywood's incursion. The CAA deal, however, was threatening not because it was a takeover, but because it represented a significant structural change in the ways that campaigns were designed and managed. The CAA model broke up the dominant ritual of single long-term contracts—arrangements that delegated comprehensive multimedia ad campaigns to single large agencies—in favor of what was termed an à la carte approach. *À la carte promised more freedom, since it allowed clients to simultaneously utilize multiple agencies.* Following this structural leveling and diversification, other ad agencies, like Wieden and Kennedy, signed deals with Nike to produce original sports programming as well as ads.[35] Clients and advertisers were, in effect, now producing their own programs, rather than depending on the purchase of time from networks or sponsorship of production companies. Institutional distinctions were further breaking down.

Two precedents proved the programming power evident when agencies and programmers cut across and ignore traditional categories. Ted Turner had organized the international Goodwill Games in the early 1980s. Turner's private-enterprise Olympic clone provided both the sporting event and its televised coverage, even as it culturally validated and legitimized the Turner conglomerate as a global media player. The presidential campaign of 1992 also challenged the verticality of traditional advertising and showed that successful promotion in the new electronic media environment required simultaneous facility on many levels of televisual communication: from MTV to talk shows to call-in shows to network spots to the long-form infomercials of Ross Perot. This comprehensiveness and media diversity proved problematic to the big agencies that had ruled Madison Avenue since the 1950s. The diversity of advertising tastes, and appetite for stylistic looks, threatened to outrun the stable of personalities housed and controlled by many agencies. The Michael Ovitz/CAA/Coke deal essentially cut out the constraining agency middleman and provided instead an extensive range of personalities rather than long-term allegiance to a single corporation. Stung by the implications of this deal, ad execs defensively protested that "CAA hasn't cornered the market on creativity."[36] Yet the issue of agency diversification was clearly in the wind as clients were now simultaneously utilizing specialist boutiques along with numerous media buyers to promote their products. The business structure that supported television production for decades was clearly changing. The ideological myth of limitless creative choice began to rule television advertising even as it did programming theory.

The same lesson was not lost on Hollywood heavy-hitter Barry Diller, who left the prestige and aura of Los Angeles studio feature-film production in 1992 to man the helm of one of the lowest televisual forms: the QVC home shopping cable channel. The omnipresent, electronic world of QVC had it all—new technology, low overhead, rote interactivity, and an accounts receivable department that dwarfed anything in the film industry. Furthermore, even as Diller was "giving America what it wanted," he had cut out the very institutional middlemen who were said to have strapped television and the entertainment industry for so many decades: labor, the guilds, screenwriters, costly program development, ratings rituals, and sweeps weeks. The system was elegant. Diller and QVC were not manufacturing entertainment programs to attract money; they were simply manufacturing money.

The Ancillary Afterlife

A recurrent theme of this book has been that televisual exhibitionism is driven as much by industrial conditions and economic crisis as it is by changing cultural and aesthetic tastes. Two trends suggest that television shows will have to work even harder to raise their distinctive silhouettes above the programming clutter. First, the fall 1993 schedule included the greatest number of new shows ever

The endless ancillary afterlife. Pee-Wee's mid-1980s HBO specials rose from the ashes for broadcast on KNBC in 1993. Gun-camera Gulf War footage became a veritable military-documentary industry thanks in part to the Discovery Channel. *I'll Fly Away* was lifted from network purgatory by PBS for their more discriminating niche in 1993–1994. (KNBC, Discovery Channel, PBS)

introduced in a season by the major networks, forty. "Last season we tracked thirty-six new programs beginning in the fall," network ratings analyst David Poltrack stated. The fact that "at the end of the season only eleven remained," suggests that a massive failure rate not only defines but also fuels program development. According to Poltrack, viewers today are much less inclined to view a new show. In 1983 fully-one third of all viewers would try a typical new show during the first two months of the season. In 1992 the number fell to 18 percent; and in 1993, the number fell to 16 percent. Even reality programs declined by fall 1993 to 10 percent of the schedule.[37] All of this suggests that the ideologies of newness and program individuation—pervasive in programming practice, production discourse, and advertising—have had a deadening effect on audiences. Yet this overload of individuation is taken by the industry not as opportunity for surrender, but as the very reason for increased televisual exhibitionism.

Consider in this regard Brandon Tartikoff's call to arms to local programmers in the face of the niche obsessed and demographically splintered 500-channel competitive environment. The former programming president at NBC and Paramount warns, "As the channels grow around you, your local production needs to grow as well." Although you don't need to "out-Hollywood Hollywood . . . *your signature shows and productions may need to multiply by the same factor as channel capacity does.*" [38] According to this vision, although the scale and televisual modes used by affiliates may shift from the prestige guises of Hollywood—that is, away from cinematic modes, loss-leader marquees, and epic narratives—survival demands, nevertheless, that stylization increase in geometrical proportion to competition. The distinctive signature look, according to this programming equation, is even more important now as the lifeblood of economic survival.

In some ways, the 500-channel mythos has become a kind of heavenly niche afterlife, both to bottom-line programmers and televisual evangelists like Tartikoff. This 500-channel telos has become both the punishment for programming's hellish competition in this life, and the future savior for those who play their development cards right. The mythos may either be wishful thinking or a self-fulfilling corporate prophecy, but the preconditions for this utopia are, in some ways, already in place. Cable has taught television that programs never die; they just keep resurfacing in the niche afterlife. *Moonlighting* and *China Beach* live on as women's shows on Lifetime. *Young Guns* is resuscitated on Pat Robertson's new right Family Channel. *Miami Vice* keeps strutting on USA Network. *Twin Peaks* assumes its place within the aesthetic pantheon of European cinema and independent feature films on cable's auteurist boutique, Bravo. Gulf War coverage will also never die; footage of endless bombing runs keep replicating on Discovery Channel series like *Wings Over the Gulf.* When these signature acquisitions lose their distinctive auras, the boutiques will simply send them on to other niches—to Nickelodeon and to syndicated superstations. There they will join *Donna Reed* and *Mr. Ed* in a timeless ancillary utopia.

Although programmers, development people, and advertisers now clearly grasp the implications of this ancillary afterlife, television scholars need to rethink how we do history and analysis. It is, for example, clearly shortsighted to categorize the shows discussed in this book solely as a 1980s programming or postmodern phenomenon, especially given the fact that their televisual excesses will continue to hybridize and reincarnate indefinitely. If the 500-channel telos means anything, it means that television history can no longer be conceived of as a sequential linear or temporal process. Rather, simultaneity and endless televisual redefinition will replace periodization as *the* defining quality of television history. Is *Get Smart*, for example, a show of the 1960s or—given its continual transmission on Viacom's Nickelodeon—a show of the 1990s? It is, arguably, both, and television historiography would do well to acknowledge this dual status. The pervasive quality of simultaneity, then, will make the scholar's analytical task as difficult as that of the poor viewer who is faced with endless choice. Cable futurists argue that such

Televisuality lives on in the 1990s. A long-term loss-leader role for Spielberg's *seaQuest dsv* at NBC combined the cinematic mode with expressionistic set design and funky "Video Toaster" graphics. *Sea Monkeys, Beakman's World,* and *Mighty Morphin Power Rangers* extended Pee-Wee's hyperactivity and *horror-vacui* into updated markets. (NBC, CBS, Fox)

viewers will have to navigate through the torrent of future cable choices with the aid of smart, interactive, PC-driven menus rather than with a meager one-choice-at-a-time remote.

If one scans the cable channels in 1994, one will undoubtedly find television that is cheap, or cheap-looking: infomercials, C-Span, and talk. If one talks to network producers in 1994, one can find those who still consider writers to be the basis for program development.[39] Yet these factors clearly coexist with other fundamental mythologies: that television profits by hailing a discriminating viewer; that marquee programs can anchor and protect mundane programming in evening strips; that signature programming attracts Emmy nominations; that critical acclaim boosts network and affiliate morale; and that stylistic individuation brings with it a built-in promotional angle. At least in its present trajectory, then, American television continues to privilege both low-risk videographic embellishment and high-risk loss-leader programming. Consider how Frank McConnell, senior vice president of Warner Bros. Television rationalizes the ever-present risk of programming failure as an external compromise: "A bad timeslot is the worst thing that can happen to you as a creative person. It's a thing over which you have very little control. It sounds corny, *but the most important thing for us is being proud of a show because it's good.*"[40] Pride, purity, and corporate quality. Legion criticisms and flippant write offs to the contrary, the primetime television mill imagines that very much of what it makes—or wants to make—is special and distinctive signature programming. Televisuality, clearly, is less a landmark of aesthetic accomplishment than it is a linchpin of corporate psychology, an organized frame of mind that keeps the industry's programming machine churning.

11 Televisual Politics
Negotiating Race in the L.A. Rebellion

Ethnic diversity isn't just a fact of life, it's a fashion theme for spring.
—Mary Rourke, *Los Angeles Times*[1]

Here I am in my own backyard—and I'm covering a war.
—KABC television reporter Linda Moore, April 3, 1992

It has become an apparent truism that justice was defeated in Simi Valley because the evidence of aggression was stylized and overworked. Why did the prosecution fail to establish that Rodney King was a victim? Because the tape of the beating was played in slow motion, freeze-framed, talked over, interpreted, distributed, and—most damaging of all—closely analyzed.[2] To critics, the visual evidence had been tampered with through a highly publicized ritual of deconstruction and overanalysis.[3] Satiated with redundant images of King's beating, the audience-jury was thought to have been deadened as well to the reality of violence.

I hope to address several problematic issues surrounding this event and the supposed analysis-induced miscarriage of justice that accompanied it. First, the ever-expanding coverage of the King trial, the L.A. rebellion, and its aftermath suggest that stylization and deconstruction are not aberrancies limited to televised legal proceedings, but are fundamental modes of mainstream televisual representation as well. In the way that difference is represented, and the other is packaged, these modes have clear political implications, but their effects depend ultimately on the context and functions given those modes. The Rodney King–L.A. rebellion phenomenon was less a crisis of style than a televisual form of crisis management.

Second, television coverage of the L.A. rebellion challenges some assumptions about the gendered nature of television and its depiction of the other. The ideas of television as feminine and mass culture as modernism's other have helped establish the field of contemporary television study in important ways.[4] Landmark work by Tania Modleski demonstrated how television narratives are organized around the daily rhythms and pleasures that culture sanctions for women.[5] Beverle Houston feminized the "endless consumption" of television by stripping the new critical theory of its phallocentric bias and arguing that the medium's defining properties were oral and maternal pleasures.[6] Extrapolations from these perspectives, however, have tended to underestimate fundamental masculinist aspects of the television apparatus, traits that have been operative in both the industry and its programming for many years. Even Lynne Joyrich's important demonstration of the rise of "hypermasculinity" as a reaction to increased feminization, reinforces the notion that television's norm has somehow been feminine.[7] The linkage of television, distraction, and the feminine—typically set in binarist opposition to the spectatorial modes and desires of film—has come at a cost.[8] The L.A. rebellion and its aftermath suggest that the television apparatus is *also* clearly and problematically masculinist, especially in the ways that it performs style and fetishizes production technology.[9]

When sergeant Stacey Koons cast Rodney King as "Mandingo," in a book manuscript that hit the press well before his trial, and anchor Paul Moyer and his Channel 7 Eyewitness News reporters targeted window-breakers as "hoodlums" and "thugs," television viewers witnessed classic ways by which the ethnic other is rendered marginal and alien.[10] But there was more at work the first evening that Los Angeles burned, April 29, 1992, than the invocation of these kinds of static stereotypes.[11] The anchors did occasionally apologize for their impulsive outbursts, as they did after calling on-camera window-breakers "creeps."[12] More than just attempts to reestablish news studio decorum, such outbursts and apologies betrayed the fact that the news readers were also simply taken-aback spectators. The same reporters then elaborated sociological justifications for their racial designations. Repeated reference was made to the fact that the senseless violence was being perpetuated by "people with nothing better to do" than to look for "an excuse to go out and trash buildings and start fires."[13] A persistent but not so subtle othering process was now in gear: there was no motive or reason for the violence; the hoodlums were simply inactive people with enough idle time on their hands to entertain themselves by creating a disturbance. With this kind of officially concerned verbal discourse on television, fifty years of Los Angeles social, political, and economic history—indeed the very notion of causality or context in any form—simply vanished.

Although the mainstream has always created a center on the inside of culture with this kind of verbal discourse, television during the rebellion quickly brought to bear a large number of other, nonverbal tactics for containing the dangerous other. As much as any other recent phenomenon in television, the L.A. rebellion demonstrated the fundamental role that the ecstatic performance of style plays in constructing and managing the other. Once the conflict started, desperate news institutions appeared unable to keep up with the unfolding threat. Without their normal scripts and teleprompters, news-reading anchors and reporters merely free associated. The sometimes incoherent verbiage that followed quickly indicted those who free-associated, as did their lack of actual knowledge about the communities in Los Angeles.[14]

The *televisual* apparatus, by contrast—construed by broadcast corporations as hard-wired, automatic, and omnipresent—quickly and immediately engaged with the chaos outside and presented it as knowable, understandable, and containable. In rebellion coverage—in both local and national manifestations—style was clearly more than icing on the cake. It was a fundamental way the other was managed and packaged, and it fulfilled an important role in television's crisis management. Coverage of the L.A. rebellion is important because the social uprising caused a simultaneous crisis in television's system of representation as well. Faced with a social rupture of this magnitude, television's privileged journalistic and narrative systems of representation were outrun by a furious performance of electronic televisual style. The coverage also stands as a challenge to a number of recurrent and privileged theoretical notions—especially those that link television inextricably with liveness, the glance, and the dehistorified present.

The analysis that follows uses two tripartite models—one historical, the other ideological—to understand television's crisis coverage of the L.A. rebellion. From an ideological perspective, crisis coverage convulsed with three recurrent control

The borrowed and stolen icon, encrusted with graphics and legally reinscribed with time-code. The icon, anchored as political book and graphic postcard. (Fox, KCBS, KCAL)

fantasies: hyperactive embellishments of masculinity, race, and autotechnologies were all thrown into the fray to establish and maintain television's command presence.[15] The ideological effects of this linkage between race, gender and technology during crisis provides one key to the politics of televisuality. From an historical perspective, on the other hand, television representations of Rodney King and the L.A. rebellion underwent three different phases and transformations as well, each with its own favored televisual tactics. Television's strategies of containment, furthermore, changed in each period, even as the threat of the dangerously racial other was redefined, managed and naturalized. The term "Televisual Mill" describes aptly the first phase, a period that stretched from Rodney King's videotaped beating in March 1991 to the Simi Valley trial in April 1992.

The Televisual Mill
Rorschach, *Vanitas,* Stigmata

The great irony of the beating footage seen round the world was that people were reacting *perceptually* to very little. The handheld, low-resolution, monochromatic tape footage was essentially electronic noise. The form was amorphous and vague—a kind of Rorschach test that could be infinitely read into. But unlike the Abscam sting footage over a decade earlier, the King footage was not just legal evidence, it was also a visual schematic ripe for constant and immediate commercial and journalistic redefinition. Like a palimpsest, the grainy video slate would be scraped and erased, encrusted and reused in a seemingly infinite number of ways. Once on satellite, even its owner and originator, George Halliday, lost control, which lead to suits and countersuits, threats and counterthreats about the video's use and exploitation.

An icon was being mass produced, and its dispersion throughout culture seemed self-perpetuating. In fact, the mass production of the beating icon in the period between the actual beating and the Simi Valley trial clearly suggested that television is a kind of televisual mill—endlessly grinding out different stylistic permutations from privileged and charged visual fragments. The low resolution and amorphous source slate became, in many manifestations, highly stylized and visually complicated program openings, mural-size screens in newsrooms, and graphically constructed and flying visual artifacts. The beating icon had value as both an anchor for journalistic discourse and as a cutting-edge component in newsroom interior design and station marketing flash. The beating icon was continuously borrowed, stolen, and encrusted with graphics. NBC, CBS, CNN, Fox, and others fought to insert their own logos over Halliday's date graphics marks. Dispersed by satellite, the noisy image became a commodified "projective test"—a tabula rasa for mass market mental projections and readings.

The legal system continued and complicated the mass media's obsession with fixing the icon. In one of many courtroom variations, an alternate time-code system was keyed over Halliday's original time information in order to re-secure its time. Electronically keyed style arrows pointed out details impossible to discern with the naked eye—and probably absent on the tape in any credible sense of the word. The obsession with fixation, however, did not just occur by time-place insertions and graphic anchors. In another case, the dispersed icon was objectified in book form for Amnesty International's legal case against the police system. Television stations regularly created graphic postcards out of the fragment for intros and previews. Other appropriators included fraternal and lobbying organizations who took partisan stances either for or against the police department. A legal hermeneutic was emerging that placed Rodney King's image at its center. The rhetoric of the trial itself, centered around videotaped images and numerous mockups and derivations of images. Live television coverage of the trial positioned the audience both as jurors and as police officers. In over the shoulder close-ups, the audience was allowed to see from Stacey Koons's perspective, even as he

Televised trial and visual analysis from Koon's spatial point of view. Use of force expert dominates and anchors the image in voice-over Karioke. (Ventura County Court)

struggled to make sense of the visual fragments and details. King's absence in court, by contrast, prevented any of this narrative camera language from fixing *his* point-of-view. On another day of the trial, a use of force expert dominated and anchored the video image in a kind of voice-over *Karioke.* Because the image the expert analyzes is so open and lacking in detail, the officer's performance evoked a Rorschach test, a projective exercise the openness of which allows subjects to invent and fix their own meanings. Continuous live coverage of the trial clearly made visual analysis and evidence the audience's business. Courtroom witnesses provided the audience with a surrogate model for participation, one that made visual anchoring and fixation the crucial viewing task.

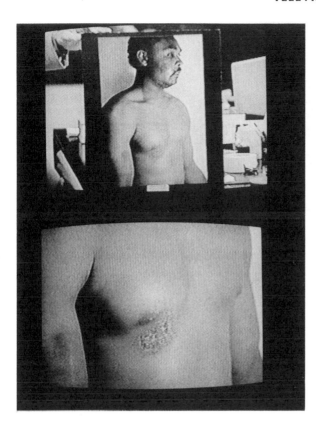

King's sacrificial body transformed into stigmata. (Ventura County Court)

A daily obsession with the credibility of the victim's flesh produced visible evidence that transformed King into a post-religious, mass-cultural *stigmata*.[16] Cloned icons of this sacrificial subject were dispersed throughout the medium and wounds of the martyr were broadcast from the courtroom into the living room on a nightly basis. The original, open and amorphous video evidence also initiated and replicated higher resolution color images. Highly resolved 16 by 20-inch color photographic prints—a medium of superior ontological value when compared to video—were used to fix the authenticity of King's scars and flesh-wounds. The legal and televised issue of credibility, then, hinged on a battle between three competing discourses: surveillance video, Catholic iconography, and medical diagnostics. The very authority of the image and the authenticity of King's body were at stake in the competition between these institutionalized but divergent image-discourses.

In the days before and during the trial, King was also transfixed in fake print form, embellished on videographic marble surfaces, and arranged tastefully in still-life fashion, complete with baton and badge. As evocations of the past and constructions meant to suggest nostalgia, digital arrangements like these evoked the *vanitas* of earlier European still-life paintings. In this tradition, painters self-consciously arranged and posed objects for viewers and patrons as symbolic moral messages: on temporality, on mortality, on the very meaning of life. The video *vanitas*, repeated nightly on national network feeds before and after each unfolding

King's beating evidence transformed into moral *vanitas*. (Fox)

story, inscribed, prepackaged, and reduced the Rodney King story to the status of a singular tragic moral.

Crisis Management
Televisual Autopilot

Televisual Autopilot describes the second and most intense period of local rebellion coverage—a phase of unstable crisis management—from the moment the verdicts were made public in late April through the weeks of armed conflict that followed in May. If the pretrial televisual mill ritualized electronic, mass-production moralizing as benign, then televisual autopilot betrayed another one of broadcasting's control fantasies: the idea and ideal that crisis coverage means hard-wired omnipresence and automated response. Even though reporters and anchors frequently seemed clueless about the chaos on the ground, that fact that television could see the unfolding spectacle, and could see it everywhere, justified and underscored the very authority of broadcasters to speak. Live coverage has always been the trump card of broadcasters, and crisis coverage on a massive technological scale legitimized two of television's most persistent mythologies: its cult of technical superiority (the result of years of heavy capital investment in concrete, satellite, and microwave technologies) and its cult of journalismo (the superiority of delegated professional reportage over democratic or populist media). While the earlier milling of Rodney King packaged him in rote moral guises—the stigmata and the *vanitas*, symbols that neatly fit the long tradition of tragic African-American victimization—crisis televisuality attacked the now exploding threat of race through higher-tech ecstasies of masculinity and militarism.

Once the Simi Valley verdicts were handed down, community violence erupted in response to tentative police actions. A botched arrest attempt in the neighborhood near Florence and Normandy instigated a hasty and disorderly police retreat. The visual chaos of early unsanctioned camcorder footage taken by residents during this confrontation was clearly reminiscent of the King beating imagery. Like Halliday's earlier beating footage, these images were handheld, sketchy, and unresolved glimpses of violence. Other unsanctioned footage evoked images of the apocalypse. In camcorder footage later snapped up by concerned tabloid *A*

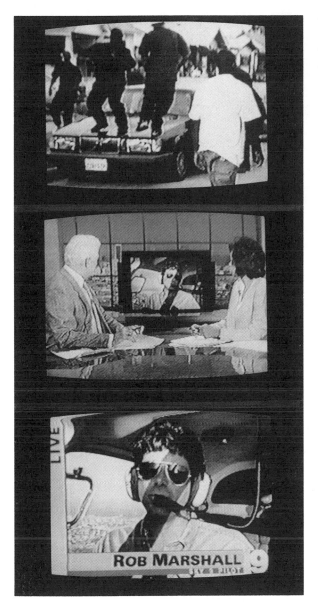

The visual chaos of unsanctioned camcorder footage. Bunkered news center and forward air-control stacked-up over "flashpoints." Chopper pilots as mechanized eyes and surrogate interview subjects. (KCAL, KCAL)

Current Affair, a cleric, Bible in hand, calls for divine protection above a recently slain victim. The street footage from the start of the crisis, like the King video before it, had the blurry and bewildering openness of a projective target.

This volatile openness was short-lived. Within minutes, the official news media called in the choppers—an immediate response that kicked the televisual apparatus into autopilot. Choppers were dispatched from bunkered, observation tower–like television-studio control rooms. Refueled and in the air around the clock, swarms of chopper cameras eventually stacked up above flash-points, and acted as literal and symbolic extensions of the bunkered and centralized stations

Olympian dialogue between station tower and skycam. The vertical hierarchy and an obsession with militarization during the crisis conflated Los Angeles with the Gulf War and the arsonists with crazed Iraqis. (KTTV, CBS, NBC)

below. The pilots become mechanized eyes of the spectacle. Since there were, apparently, no human subjects below to interview, much of the coverage in the next few days involved interviewing the pilots, who were essentially the station's own mechanized eyes, not rebellion subjects or victims. Television's relative lack of contact with actual human participants in the rebellion also lead to a crisis of representation. Paired graphic boxes were setup to show dialogue between station tower and skycam. Anchors in front of control room vistas and mechanized camera eyes reflected on the meaning of the world cleanly above the misery spewing forth below them. KTLA Channel 5 proudly announced to the viewer, in their skycam dialogue boxes, that KTLA represented Hollywood in the violent spectacle that was unfolding below. *These* participants were not, then, just anonymous reporters. They were concerned industry players.

During crisis coverage, the obsession with stylistic fixation took many other less Olympian forms, including the metaphorization of the event in dramatic military terms. While the apparatus celebrated its automated technological abilities during the period of crisis coverage, television also actively worked to militarize itself. By drawing battle lines in explicit on-camera graphics, television conflated the multiplicity of participants and motives into a paradigm of the polar pitched battle, with an us and a them, with victims and assailants, good guys and bad guys. The decisiveness with which television militarized the situation is not so shocking if one considers the long history of strategic counterinsurgency plans espoused and formalized by the LAPD. Mike Davis has documented how militarization and Vietnamization were systematic and highly publicized parts of the LAPD during the 1980s. He describes the LAPD's "search-and-destroy missions" in South Central, the Vietnamization of large urban neighborhood assaults like "Operation Hammer," the construal of African-American housing projects as "strategic hamlets," and LAPD's overall strategic goal of pacifying Los Angeles.[17]

The rebellion Rorschach, and the media's militarist response, then, were not subjectivist at all. Television reporters were not simply grabbing impulsive connections from a journalistic and individualist id—they were spewing forth *projections from a systematically manufactured social unconscious.* So too, the tactics of live coverage were not merely "determined" by the technologies of chopper-borne and up-linked video, nor were they as automatic and hard wired as broadcasters imagined. They too were fabricated from a long tradition of logistic control and surveillance, honed from ritual aerial freeway chases on the nightly news and from years of spotlit aerial search-and-arrest missions in South Central. Although many on the streets celebrated the fact that LAPD "was finally getting its ass kicked," few recognized that the conflagration and its coverage were actually self-fulfilling prophesies that followed years of civic and militarist strategic planning.

Repeated on-camera references likened the aerial spectacle to the endless oil fires in the Gulf War, as smoke billowed to the horizon from countless fires below. This reporting tactic did not just create a visual analogy, but forcefully inscribed the unfolding events within an existing, and easy to grasp, militarist framework. Through this televisual staging, the arsonists below assumed the unenviable symbolic role of crazed Iraqis. Significantly, the ploy suggested that the victims, like the oil reserves in the gulf, were positioned as the physical (rather than human) resources of Los Angeles: what was at stake was American property. The credibility problem created by calling these views "close-ups"—as NBC news did in their graphics—underscored the sense that the only thing the viewer was actually close to was the viewpoint of an individual pilot high up in the air. The aerial eyes were posed, then, as sensitive professionals, musing on the tragedy and lost physical resources of Los Angeles, as individuals filled with remorse and traumatized by declining property values.

CBS went further and hyped Florence and Normandy as "Ground Zero"—thereby invoking the specter of fifty years of paranoia over global nuclear destruction. This militarist and geographical fixation centered the conflict in a local

The specter of nuclear warfare. Florence and Normandy as Ground Zero. CBS's digital Betacam locks-in on target for air superiority in Strangelove-like smart-bomb video opening. (CBS)

skirmish rather than within the social system of Los Angeles or a broader urban context. The Reginald Denny beating footage was recognized on the very day that it aired as a mirror image of the King beating. Reporters immediately and repeatedly characterized this incident in polar terms—as black retribution against the King beating and the white verdicts. The new Denny icon was further stylized and embellished by *48 hours* as part of their high-impact opening. A computer-generated digital betacam locks in on the target, suggesting "air superiority" —a term with resonance dating back to Vietnam. The pilotless betacam flies into the target at Florence and Normandy. By visually riding this Strangelove-like doomsday apparatus down to ground zero, CBS unwittingly identified its line of sight with a different kind of retribution, one made popular in smart Gulf War video-bombs. The network news heavyweights at CBS thereby positioned themselves both as a form of air superiority and as journalistic, but partisan, retribution against Denny's injustice. The network's militarist self-consciousness and high-tech air superiority did George Halliday's unwitting video-crusader-partisan mode one better, with one important exception: CBS clearly positioned itself on the other side—the official side—of the assault.

Having established visual sight in L.A. rebellion coverage as the preferred mode of observation, the complicated semiotic potential of the televisual apparatus shifted into high gear. Seemingly endless configurations and embellishments were spun out to manage, treat, and stylistically encrust rebellion imagery. ABC, for

The "good" mayor Bradley weeps over his fallen people in 1930s-style montage, while the ever-sensitive Ted Kopple uses burned out buildings as apocalyptic wallpaper. (ABC)

example, worked to fix the chaos with a personal center. Evocative of a 1930s-style montage the good mayor Bradley is stylistically forced to weep over his fallen people and city. This form of fixation attempts to center the chaos by giving it a human and emotive core. Without a handle, the stations grasp for ways to personify an out-of-control social story. In another case, *Nightline*'s ever-caring Ted Kopple used burned-out buildings as apocalyptic wallpaper. This visual tactic made his special both relevant and centered around an accessible and knowing subjectivity, his own. In search of a televisual Kurtz, Kopple is quick to travel up-river in South Central's heart of darkness. President Bush and Governor Wilson will soon do the same thing.

Although gnostic scholars like Neil Postman argue that television does otherwise, the rebellion footage immediately and continuously contextualized the event.[18] On KCBS, floating and transparent digital supergraphics were fused with fast-breaking images in order to fix the time and place of each incident leading up to the outbreak. In the days following the outbreak, television was in fact obsessed with temporal fixation. During the crisis, television worked hard to contextualize the rebellion in other ways as well. In one of many such instances, the itchy fingers of KABC's station director/switcher pull in and combine two live images in a split-screen wipe. On the split screen that results, King's lawyer continues to translate and paraphrase to the press King's now-famous plea for Los Angelenos to "just get along." Even as the audience hears and views these pleas

Televisuality's obsession with temporal fixation and context. An either-or moral binarism ruled, as when video was turned into electronic film negative. (KCBS)

screen-right, the left side of the screen simultaneously shows armed Army National Guard troops storming off the back of a military truck in South Central. More than just an effort to show the audience what was going on at the same time that King made his plea for peace, the left screen also was the station's defacto warning about what would happen if the audience does not follow King's plea. Simultaneous and overlapping pictures, therefore, continuously editorialized the crisis. Visual diptychs like this one littered coverage with moral warnings.

A logic of binarism ruled rebellion coverage. Moral preachments were not just products of journalistic storytelling or free association by news anchors. Binarist judgments also occurred in the mundane ways that the conflict was visually pictured. In one case, the footage of the failed arrest at Florence and Normandy was freeze-framed, then electronically inverted as a photographic negative. The videographic negative, obviously, was an imported visual paradigm that dramatized and assigned a dark side to the struggle. At the same time, however, other visual paradigms and stylistic tactics implicitly encouraged the viewer not to worry. For example, television coverage during the rebellion attempted to describe the situation and the LAPD as a book, with sequential chapters and a conclusion. This graphic mode not only imagined an orderly transition from out-going Chief Daryl Gates (widely perceived as one cause of the conflict) to incoming Chief Willie Williams, the device also brought with it an artificial expectation of order. The digital transition used here brought with it confidence that one chapter will inevitably follow another, and that the Los Angeles conflict—like all bound books—will have a clean beginning and ending.

The foreign view: L.A. as Dresden-like firebombing on Germany's ZDF. The domestic view: rebellion prefixed as a book, with neat chapters and a conclusion. (ITN, ZDF, KCBS)

There was a pervasive sense during the first few days of the rebellion that television was struggling to gain control of the situation. One immediate response to this flood of fragmented and disorienting information was the shift by broadcasters to multiscreen video displays. Across the channels, viewers confronted double, triple, and quadruple screen configurations. The multiscreen device was useful for staging surrogate interviews between anchors and reporters as on Fox; staged video-conferenced fights between supposed ideologues on *Larry King Live* and CNN; and simple mischief in the newsroom as in the case of one of Edward Olmos's many in-studio interviews. Olmos was repeatedly set up as the voice of community calm and constructive action in many on-camera appearances during the crisis. In the live studio at KCBS, however, Olmos instead decried the systemic injustice of the economic and social system in Los Angeles. The uncomfortably surprised anchors—clearly hoping for a more docile mouthpiece, one who would simply encourage calm—became openly skeptical of Olmos's political critique. Increasingly impatient with the diatribe, the station pulled in a fourth box of imagery unbeknownst to Olmos, one that showed a fiery conflagration outside of the studio. While Olmos earnestly continued his critical discourse off-screen left to the now-absent anchors, his message was overwhelmed by the station's visual statement. Again, the political and social complications of the situation had been channeled into a polar dichotomy: calm or destruction. Olmos the political theorist has been silenced and contained, *framed as an expressionist picture in the newsroom's digital trophy case.*

Loading up the screen: boxed spectacles as diversionary televisual defense against unruly in-studio guests. (Fox, KCBS, CNN)

Not only did television try to cover the crisis in as many different visual guises as it could, it was extremely conscious of how the world was reacting to its visual coverage and to the violence. American viewers were shown what was termed the "foreign" view of L.A.: gun-toting Korean vigilantes on Britain's ITN, and the Dresden-like firebombing imagery displayed on Germany's ZDF. American television was very conscious, then, of its own televisual performance and emphasized this awareness and facility. By reappropriating and showcasing even foreign, and potentially critical, readings of the rebellion, American television showed that it could play the game of textual one-upmanship as well.

Immediate visual historification of rebellion as race riot. Watts-era pleas for "passover" are replicated in 1992. Ahistorical? Television constructed a concrete then and now in desperate, crisis catch-up mode. Yet instant-history ignored the visual fact that looters were Caucasian and Latino. (KCBS, KTTV)

Television also immediately historicized the rebellion as a race riot. Clearly challenging those academic theories that characterize television as ahistorical and obsessed only with presentness and simultaneity, crisis coverage, in fact, made historification a regular televisual strategy. Contemporary images from South Central were repeatedly paired with similar icons from the Watts riots. In one shot, the 1960s pleas for a "passover" of "negro-owned" properties are replicated in 1992. Televisuality may deal with the wrong history, but academics are misguided if they ignore the fact that historification is very much a central preoccupation of the televisual apparatus. This parade and facility with history was embraced by stations in the catch-up mode and helped them make sense of the apparent anarchy. Finding and constructing a concrete and connected then and now, helped the stations visually reduce the conflict to race riot status—even as on-camera reporters verbally equivocated on whether or not the rebellion was induced by race.[19] This televisual and historical fixing and framing around race glossed over fundamental economic and class-based aspects of the rebellion. Television's inability to deal with both race and class exposed the leaky nature of the televisual apparatus. Stylized historification also conveniently ignored the newsroom's own evidence that many of the looters were Latino and Caucasian. What did not fit television's historicized and binary racial model, then, was simply ignored. The visual evidence showed that the looters in many cases were not African Americans at all, especially as the rebellion spread to Hollywood and to the Pico-Union district. Partly as a result of technical changes in recording and storage technology,

Airlifted Chicago news anchor performs high-speed, journalistic drive-by shooting. Network reporter sets up live, on-camera tag-team wrestling match between Korean and African-American in the ruins. Gates protests that Kopple hasn't even read the LAPD written document during their televisual summit. (WBBM, CBS, ABC)

historification is now a very easy ritual that television routinely uses. The possibilities for historification now seem tantalizingly limitless, for most major stations have amassed large amounts of archival material from years of news production. Proprietary access to this raw material provides endless possibilities for historical reworking and hybridization. Constructing a history along binarist or reductive lines is dangerous however, as this example shows, for images can belie the theses that news editors force upon them.

If digital graphics, file footage, and electronic stylization all worked to stylize and historically fix the onslaught outside—in a kind of televisual holding action—

then dropping reporters onto the scene, after the fact, could help reestablish and legitimize the station's authority to say anything about what was going on. In one case, anchor Bill Curtis of WBBM-Channel 2, the CBS affiliate in Chicago, was airlifted and inserted into the Los Angeles war zone. Curtis performed one of television's favored visual guises—the journalistic drive-by shooting. Rent-a-car wheel in one hand, microphone in the other, Curtis speeds past rows of burning buildings in order to explain to viewers what is "really" going on in Los Angeles. A broken windshield later in the day served as proof of the extremes to which this reporter would go to bring back the inside story. "Inside" must, in this case, have referred to the inside of the reporter's car.

After the smoke cleared, another reporter covered his look back on the riot in a method more akin to tag-team wrestling. On-camera with a distraught and ruined Korean shop owner, the network reporter invited local African Americans to "come on down" and enter the dialogue about the riots even as the camera rolled. This done, the reporter crossed his arms in a defensive posture, even as the interracial argument that he arranged was set into ever more hostile motion. Before airing this reporter-staged, racial grudge match, however, the segment producer/editor tastefully cut away to another scene before any overt violence actually ensued. This was, after all, CBS—good network journalism—not tabloid sensationalism.

In like manner, ABC's *Nightline* set up a heated on-camera exchange between Ted Kopple and Chief Daryl Gates. The event was pitched as a kind of televisual summitry between opponents of comparable power. After presenting an edited, point-by-point indictment of Gates and the LAPD, Kopple allows Gates to appear on the set at show's end to even up the score. Far from an exercise in self-flagellation by Kopple, Gates is left with the impossible task of having to refute the dramatic visual evidence the audience has just seen. Behind Gates is a bank of monitors, a constant reminder to the audience that the rebellion violence stands as undeniable evidence against Gates. Gates convincingly accuses Kopple of not having even read a lengthy written document which shows that the LAPD's *did* have plans in place for dealing with urban riots (these plans were criticized by Kopple in his edited segment). Gates's angry jab at Kopple notwithstanding, ABC clearly masters this televisual arena, for the chief is allowed only a token verbal rebuttal, even as he frantically waves a dense but unread written document. Kopple takes his ceremonial lashes, but Gates clearly loses. Such is the image-driven world of televisual summitry.

Containment: The Big Response
Primetime's Transcendent Polis

Containment, the third and final stage, refers to the more ostensibly serious, response of the industry, and includes that period when rebellion-specific entertainment programs were showcased by the networks as season premieres in August and September of 1992. After the flames were out, and local television had exhausted its angles on crisis coverage, orgies of technology and militarism lost both their anchoring power and command presence. Industry heavyweights now began to weigh in and formulated what might be termed the big response. Hollywood,

Within days, rebellion footage is fair game for May political candidates. (Gil Garcetti Campaign, Barbara Boxer Campaign, Gray Davis Campaign)

cable, politicians, and network entertainment programmers all became active players in what appeared to be a more reasoned and planned strategy for reconciling the effects of the rebellion.

Reconciliation had, in fact, started within days of the violence, when the rebellion footage—like King's before it—became fair game for any and all takers. The political primaries in California in May made extensive use of the footage on both the right and the left. The district attorney attempted to novelize the riots with an epic textual opening. Future senator Barbara Boxer symbolically positioned herself as a lone and isolated photojournalist in front of the apocalypse.

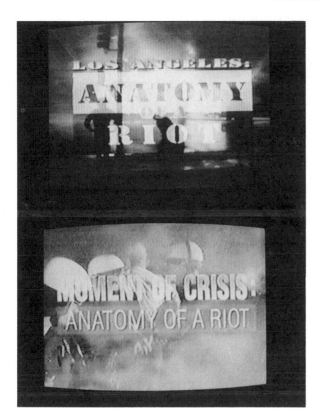

With time, favored metaphors shift from militarist to anatomical analogies. Biology and anatomy position threat in diseaselike terms. (A&E, ABC)

Senate aspirant Gray Davis, on the other hand, visually cast himself in contrast to all that was wrong with the state of California—that is, as antithetical to the decay symbolized by the hybrid incumbency of Gates, Wilson, and Bradley. Keys, mattes, and graphics stitched candidates into the fabric of the rebellion, even though most had little contact with the situation while the buildings actually burned. The historical causes of the riots, then, were now not nearly as valuable as the appropriation and exploitation of the rebellion as a historical cause. Like a loose historical cannon careening on deck, the rebellion was available to any who felt inclined to grab on. With the rebellion now overhauled and viewed after the fact as a *cause* rather than as an *effect*, any candidate could pitch him or herself as the answer and not as a cause, which many of them were.

The favored metaphor for framing the event also shifted from the confrontational and binarist modes operative during the crisis period to an obsession with anatomy during the industry's big response. It is worth considering why viewing Los Angeles as a "body" served the needs of the media during a time of social crisis and its aftermath, that is, when society was trying hardest to contain and naturalize its threats. The body, implied by the L.A. rebellion special entitled *An Anatomy of a Riot*, both unified and effaced difference, since the figure postures culture as a single organism. The metaphor also functioned as a more organic and ostensibly benign variant of the melting pot myth; a synthesizing paradigm erected on the myth of cultural effacement. Biology and anatomy, in addition, are also

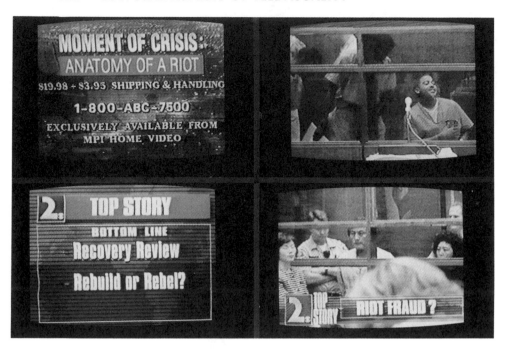

Commodification. "Traumatized by the horror?—call ABC's 800 number." TV used an obsessively polar logic and grabbed callous and awkward images of Denny's jailed assailants. (ABC, KCBS, KCBS)

useful in positioning social threats in diseaselike terms. While the rebellion can be looked at as a kind of social disease and symptom, the notion of sickness and disease also places the threat on the *outside* of the social body. Hence, the looters, rioters, and demonstrators—infections frequently conflated and identified as people of color—are pictured as threats to the equilibrium and health of the system. Anatomical figuration, then, repeatedly legitimized and reinforced television's favored inside/outside dichotomy as a response to contagion.

The big response also contained the threat by commoditizing it. At one point, the spectacle was sold by ABC directly to viewers on VHS for only $19.95. Anxiety and trauma about the horror and hate that you have just seen? Call ABC's 800 number and have your credit card available. During the big response, coverage unabashedly became commodity. Even in the more extensive phase of reconciliation after the rebellion, television thought in polar terms. One of television's visual pretenses was the promise that it could both think and interpret for the audience. Less than mindless, television actively paraded interpretation and analysis before the viewer's eyes. Infinite follow ups to the rebellion fixated binary conceptual schemas. KCBS's options for response? There were really only two choices: "recovery review—rebuild or rebel." Binarism was now being institutionalized, especially in the service of rhyming headlines or couplets. No raw and decontextualized image leaked out of these televisual icons. Viewers were warned (how) to think.

Following the uprising, the courts were jammed with rebellion-related cases,

The big response: Doogie, Bill, and George in televisual Mt. Rushmore. (ABC, NBC, NBC)

giving television ample visual material to process throughout the months that followed. The images that resulted frequently discredited ethnic participants of *any* type. On one evening, Reginald Denny's African-American assailants were caught by the cameras callously laughing behind the courtroom's barred holding area. KCBS left no doubt as to their guilt when it isolated the laughter. Korean shop owners were also forcefully discredited when KCBS broke its top story about how some merchants had set fires to their own stores. With this tortured visual attribution, even the groups most victimized by the rebellion were discredited by postrebellion association: the Koreans were also arsonists. NBC did a story months

L.A. Law and *Doogie Howser* do the camcorder ritual. Legal trajectory of Simi trial piggybacked onto *L.A. Law*'s episodic arc. Large-screen TVs now pervasive on each set of *L.A. Law* for the first time. (ABC, ABC)

later about the hard-working blacks who were "also" victimized by the riots. In this segment, a self-made African-American businessman sits alone on his property. Looking back on images of the conflagration, he provides a sad and tragic ending to NBC's followup. NBC's moral? The rebellion was shortsighted and self-defeating. The dangerous other here gave way to the domesticated and contained other—to a model of middle-class reflection and property-owning patience.

Episodic television, however, also enacted and relished the big response. By September, Doogie Howser had joined President Bush and network patriarch Bill Cosby as a member of the televisual Mt. Rushmore. Even as classical dramatic form seemed dwarfed by the social threat, this triumvirate regressed to one of the most basic forms of one-to-one communication. Each left his narrative world, took courage, faced the massive national audience in close-up and direct address, and appealed for calm and cooperation. Even prestige entertainment programming, then, gave in to direct address—to point-to-point unmediated communication with the audience—when the threat was great enough. Primetime used eye-to-eye contact when deck-clearing consciousness-raising was in order.

But the King beating, and the L.A. rebellion also provided rich iconic templates that could be narratively used again and again in primetime. Visual configurations operated as master structuring allegories in ways once attributed only to narrative archetypes. *L.A. Law* and *Doogie Howser, M.D.,* did the camcorder ritual as central parts of their programs. When the industry did not feel the need

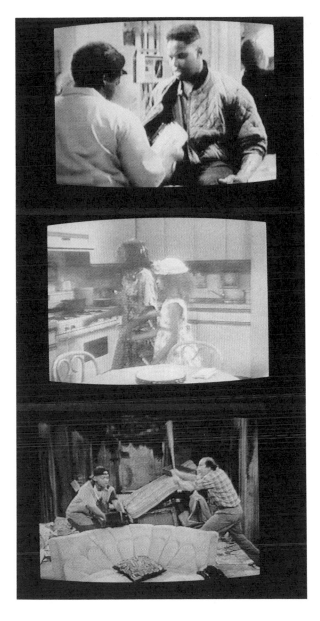

Doogie's African-American medical aide is allowed to "teach" African-American patient not to steal consumer electronics. The rebellion was an excuse to create South Central backstory in *Fresh Prince of Bel-Air*. Retrospective textuality provided the show with enough guilt to connect with its lower-class roots. (ABC, NBC)

to directly address the audience with messages about containment, they acted out and performed pointed socio-political morals. Doogie's medical aid, for example, sensitively disarmed a young African-American patient of his recently stolen consumer electronics contraband. *Doogie's* master narrator, then, tastefully allowed another African-American character to teach this lesson to the audience and to the shop-lifting African-American patient who had stolen "to get back at the system." This sort of heavy-handed social teaching is clearly no longer restricted to Mr. Rogers's preschool province.

Episodic television danced around the threat in a number of other "sensitive"

L.A. Law's sensitive characters gaze at burning city like grieving family members. Sister Souljah is heavy-hitting import on *A Different World*. Primetime masqueraded with recognizable news photo templates during the big response. (ABC, NBC)

ways as well. An episode of *Fresh Prince of Bel-Air* used a sepia-toned flashback scene to re-create an entirely new history for the show. This revised L.A. rebellion–based premise for the show revealed that the sitcom family now had its economic roots and origins in South Central. This form of crisis-induced textual regeneration provided the family the quantity of guilt necessary to enable them to reconnect with their lower-class roots. It also created comic opportunities whereby the Prince demonstrated just how easy it is for an African American to be mistaken for a looter. The bat-wielding Latino who attacks him in this episode actually turns out to be an old family friend. Retrospective textuality like this was in full force during the industry's big response.

The L.A. rebellion provided a rich visual menu from which episodic television could order up icons in September. Photographic icons of mass arrests and urban wreckage taken directly from the pages of the *Los Angeles Times*, were reenacted in both *Fresh Prince* and *L.A. Law*. Primetime also imported rebellion-related scenic imagery and wallpapered dramatic scenes with visual evidence of the conflagration. In *L.A. Law*, the cast's sensitive characters gaze at the burning metropolis and emote, less like Nero, than like a saddened family member grieving a loss. Sister Souljah made a much-publicized appearance on *It's a Different World*. Her unabashed on-camera critique of the King beating trial—footage of which plays behind her in the electronics showroom—was set up in the press as a special, insider's point of view. Having gone toe-to-toe with presidential candidate Bill Clinton over the racial implications of the rebellion weeks earlier, she

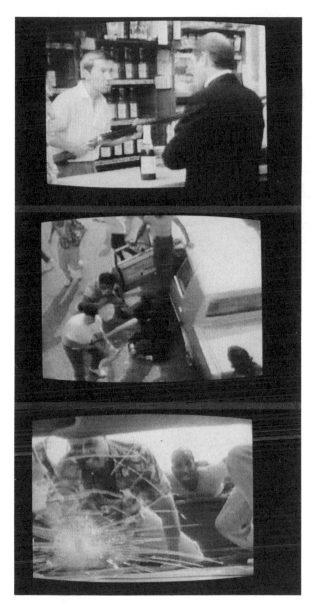

Primetime's repertoire of available L.A. rebellion icons? Gun-crazed Korean shop owners fire 12-gauge shotguns into crazed African-American mob leaping through plate-glass store windows. The Denny icon was also rigidified into resusable cinematic motif in primetime, in a way that invariably placed the audience inside the glass, with surging African-Americans outside. (ABC)

was imported by the series producers to elevate the moral air and relevance of *It's a Different World*. Souljah doubled as a police brutality critic and as a high-profile advertising ploy.

L.A. Law, meanwhile, fused the legal trajectory of the televised Simi trial with its own episodic arc. This *textual piggybacking* heightened the constructed liveness of the show even though the trial footage seen in the episode clearly predated production of the primetime drama. Absent during the previous seasons, large screen TVs now popped up everywhere on the *L.A. Law* sets. Like the family and audience at home, the *L.A. Law* workplace family was now inside, looking out

Empathizing with what it means to be Homer (that is "black") "like me." (ABC)

on the world through the television window.[20] Like middle-class home-owners, they too would be forced to pull together in the face of the uncontrolled onslaught taking place around their property. The *L.A. Law* episode used the full repertoire of L.A. rebellion icons. A senior law partner on the way to his wedding, for example, confronts his worst nightmare—a gun-crazed Asian, a Korean shop owner. The shopkeeper soon fires a double-barrel twelve-gauge at African Americans who are wildly leaping through his plate-glass windows. This racial stereotype of the Asian as strange and incomprehensible is as acute as *L.A. Law*'s depiction of the crazed looters. The show also rigidified the Denny beating spectacle into a reusable, high-angle scenic vantage point. In their variation on a theme, *L.A. Law* reinscribed one of *its* characters as the victim in the Denny aerial-beating-guise. The *L.A. Law* characters and the audience are both placed on the inside of the shattered glass, with the surging African Americans on the outside. This repeated spatial rigidification of inside versus outside not only conflated the audience's point of view with that of the lawyers', but, more importantly and ironically, set up both the characters and the audience as the victims of the violence in Los Angeles. The L.A. rebellion provided drama's essential fuel: dangerous antagonists (the people of color coming at "us" from out there) and innocent victims (the white, disoriented characters and audience at home).

The bullet-in-the-breast motif. Franchising victimization: "We are all erotic Rodney Kings." (ABC)

Each of the subplots in the episode also masqueraded with visual disguises taken from the earlier conflict. In one scene a marshall acts as courtroom VCR operator. *L.A. Law* mimicked the Simi trial, then, but with some problematic differences. The victim in primetime was no longer a black man, was not Rodney King, but was a Homer Simpson impersonator employed by a theme park. The defense dramatizes the wrongfulness of his firing by having the entire jury don Homer Simpson's headgear. Arnie must get the jury inside of Homer's skin, apparently, in order for them to truly understand victimization. The lesson is clear. The law firm wins. Now, apparently, the jury members can empathize with what it means to be Homer (that is, "black") "like me." One troubling implication of this plot line is that the Rodney King beating has been reduced to the level of peril that one finds at family-oriented theme parks. Furthermore, Rodney King is construed as the essential postmodernist mask—as a Matt Groening clone—his beating an empty foil available to one and all.

But perhaps the strangest of the many King/L.A. rebellion primetime analogies was the bullet-in-the-breast motif—a highly privileged example of retrospective textuality. The previous season of *L.A. Law* ended with a cliffhanger wherein a blonde lawyer was gunned down outside a courtroom. By September, however, the event had been redefined as a symbolic part of the L.A. rebellion. The female victim's anxiety about the violence, her scar, and the social implications of her career are only textually resolved in this episode by a man's kiss, through the man's oral acceptance of her scarred breasts. Once again, primetime television was obsessed with creating opportunities and victim-substitutes from the L.A. rebellion. Here the affluent white woman bears the burdens for society's sins. Here the sin can only be absolved by an erotic act. The jumping assailants depicted in *L.A. Law*'s earlier crisis-management scenes give way here to a much bigger picture: one that makes the substitute victim *erotic*.

The King–L.A. uprising conflict provided a rich visual score that was performed in many variations. In primetime's *It's a Different World*, on the traffic-clogged streets of L.A. and in the courtroom, the police baton-beating of King became a widely recognized ritual that was repeated again and again by those with very different motives. The gestural performance of the baton blows was open to infinite appropriation. This subsequent popularity raises the question of why such televisual scores proved so attractive and pervasive to programmers. In one telling instance, an NBC segment proclaimed in hindsight that all crises can be summed up in single images. The segment directly compared the Rodney King and Reginald Denny beating images to the heroic man-against-tank icon from Tienamin Square and to the infamous on-camera execution of a captured Vietcong prisoner during Tet in 1968. This singular thesis, then, was NBC news's retrospective analysis of the very same spring rebellion that had earlier overwhelmed both police and news media alike. Had amnesia set in at the news division? No, the network philosophized, the riots really could be reduced to one image. NBC, ironically, was ignoring the fact that during the period of crisis its own televisual apparatus spun out and proliferated as many different images as possible. By contrast, and after weeks of reflection and analysis, the high journalists of the big response tried to verbally summarize the rebellion's lessons by resorting to clichés of the most reductive sort—that a picture is worth a thousand words.

The progressive potential of this kind of verbal gloss is worth considering. The images that NBC points to as iconic markers could not ultimately be contained by either the South Vietnamese executioner during Tet or by the Chinese government during Tienamin. In fact, the point-blank firing in Vietnam and the marshaling of Chinese tanks—like Stacey Koons's police assault against King—were staged by the powerful as explicit symbols of intimidation against the weak. Yet all eventually leaked from their constraints. The Vietcong execution, in fact, helped catalyze the American antiwar movement. The Chinese tank confrontation, once on satellite, garnered worldwide support from reformist sympathizers. And the Halliday beating footage was utilized by outsider independents and community groups, as well as by Spike Lee and others, to further various causes. Televisuality is indeed a leaky system. Those who describe it as an inherently illusory, hegemonic, and deceptive system fail to see that it is also an instrumental system, one that can be used by the marginal as well as abused by the powerful.

Conclusion: TV's Command Presence

The crisis of the L.A. rebellion, then, was not simply a matter of arson and anarchy outside of the walls of the broadcasters. For television, the crisis was also very much an internal one. Faced with an out-of-control catastrophe in the streets, television marshaled a command presence by invoking three of its most enduring control fantasies. Exhibitionist displays of technology, masculinity, and race-based journalistic authority were all thrown into the fray on a commercial-free, twenty-four-hour basis. These frontline defenses, however, regularly faltered. Disheveled and unprepared local anchors like Paul Moyer—traditional bastions of calm and dispassion even in the face of earthquakes and serial killers—badly ad libbed their way through coverage. Network types like Kopple were caught making on-air statements that were factually wrong. Chief Gates, meanwhile, continued to insist that the LAPD had a systematic plan. Fantasies of information control—inevitably centered on-air around white, male figures—circulated both in law and order and journalistic circles.[21] The burning of Los Angeles—unfolding live before the viewer—made these assertions of authority far from credible.

As white masculinity staked its claim to control, television slotted people of color into a very different, but much more spectacular, role. Rodney King emerged as the latest in Hollywood's long line of tragic, suffering, and sacrificial victims of color, a lineage that includes the young Sidney Poitier, Paul Robeson, and even Rudolph Valentino. Olmos—despite his sometimes forceful efforts to the contrary—was continually stripped of informational and ideological authority during coverage. Television postured him first as an on-camera stand-in for victims, and then as a long-suffering street-sweeper picking up the city's tragic pieces with a push-broom. Cosby turns in pain to the national viewer during his show, but is never quite able to jettison the entertainment aura NBC gives his sitcom patriarchy. Even Mayor Bradley is visually montaged as a tragic figure, as a wounded person verbally unable to account for his community's trauma. The racial bifurcation is clear: *white males struggle with little credibility to maintain information control, while men of color are spectacularized into recognizable roles defined by raw emotion and victimization.* Lynn Spigel has suggested that this racial split—between white male information control and black male visuality and victimization—mirrors television's favored binary split between entertainment and information. Perpetuating the myth of that division, a generic polarity under threat in contemporary programming, seems to be continually at stake in this coverage as well, even as the very myth falls apart before our eyes.[22] Even the sometimes articulate and politically informed direct addresses thrown back at reporters by looters—fragments of which were captured on-camera as stores were being emptied—undermined the simplistic entertainment versus information model being forced upon persons of color as an explanation for the horrific spectacle. The social crisis of the L.A. rebellion, then, was also a crisis of masculinity and race; a threat to the clean ideological distinctions that prop up and feed fundamental program distinctions. High-tech embellishments of the spectacle notwithstanding, the televisual command presence faltered under the weight of its own insecurity and instability.

The Los Angeles rebellion and its coverage proved to be traumatic for a diverse coalition of forces: urban social programs and government policy were indicted; broadcasters (at least temporarily) questioned their image-hungry role in encouraging and fueling the conflagration; and scholars stopped to consider their own methods and politics in light of the crisis. Coverage of the King beating trial in particular created a crisis of conscience in media studies. Much hand-wringing, for example, greeted the fact that police brutality was exonerated in the legal

NBC News muses in retrospect that all crises can be summed up in single images. (NBC)

system through close textual analysis. Since the same kind of textual analysis, now bloodied by legal precedent, has been favored in critical theory, one immediate temptation was to flee to more naive conceptions of extratextual reality. This reaction, however, ignored two important facts: (1) that televisual texts and representations *are* political acts, and (2) that excessive televisuality can and is being used by independent groups to counter the effects of mainstream television practice. Scholarly insecurities about concrete politics offer no credible basis per se for disavowing televisual practice.

For those interested in social change, the apparent power of televisuality to manage crisis challenged the very idea of resistance. Certainly the months that followed the uprising, however, demonstrated that resistance is possible. Young African-American protestors, for example, disrupted traffic in Los Angeles and performed "the Rodney King." The legal system may have legitimized the LAPD's blows, but it could not own the public ritual that soon took on a life of its own. Street artists and taggers also made site-specific artworks in and among the physical wreckage of the rebellion. In one case, a touristlike network news correspondent used a painted guerrilla counterwork as a kind of pop supergraphic to frame his entrance to the rubble. The street artist gets network play; the network journalist gets some hip visual flash. Crisis televisuality has also been used by community activists and organizers in public schools and in media workshops for urban children. Anna Everett, for example, has shown these stylistic modes to be operative in media programs of public elementary schools in Los Angeles county. Beretta Smith has shown that televisuality was performed by community video

Uprising as widely available performance score: in traffic, on primetime, in the courtroom. Site-specific artworks by street artists and taggers were used as pop supergraphics to frame the network correspondent. (NBC, Fox, CBS)

cooperatives to address and redress King's victimization.[23] Stylish, low-tech political videos proliferated in children's video workshops like those organized by KidVidco. The real victims of the conflict in L.A.—urban school children rather than the affluent characters on primetime television—do speak the televisual language and can use it to engage the status quo.

Close analysis of the evidence from the period of crisis management, furthermore, shows a preoccupation with presentational modes that challenge a series of at least five privileged assumptions about television: (1) television's obsession with militarist guises and production performance underscores fundamental masculinist aspects of an apparatus that is typically theorized as feminine; (2) television's penchant for immediate and visual historicization, along with its marshaling of contextualizing file footage, problematize the view that the medium is fundamentally defined by its ahistorical appetite for presentness and immediacy; (3) the favored theoretical paradigm that television is based essentially on the innattentive glance is undermined by the highly embellished and complicated images evident during the period of the rebellion and King's televisual milling; (4) television's pervasive and continuous efforts at fixation—to interpret and nail down the historical meaning of its televised images—undercuts the popular view that the televisual image is by nature both contextless and illogical since it is antithetical to reason; (5) finally, primetime's obsession with rituals of substitute-victimization during the period of containment undercut the highly publicized liberal concern and social conscience of those shows. This retrospective textual performance, whereby dramatic shows inserted their characters as *the* victims of the violence, displaced the real rebellion victims and participants into symbolic positions outside of the showcase. Even in their calls for racial harmony and a sense of oneness primetime stylistically positioned the other back on the margins and outside of the glass. The aesthetic potential and dramatic capital provided by this master visual ritual of victimization simply overwhelmed the industry's more immediate and explicit liberal interest in moralizing, education, and pacification. The industry's big picture response was at root a textual feeding frenzy—a wild performance of surrogate victimization.

What was dangerous in televisuality's crisis management during the rebellion was not its high-tech stylization of the event, but rather the use of those modes to bolster old-style class and racial politics. Even the big response of the industry, after months of concerned reflection, attempted to contain the real threat to the economic status quo with clichés of the most regressive kind. If crisis televisuality made the dangerous other exotic, alien, and stylized, the serious and liberal voice of the industry contained by covering over difference. The industry's explicit and heartfelt verbal appeals to calm, reason, and cooperation were belied by primetime's use of racial stereotypes and marginalizing representations of the most problematic and sometimes bizarre sort.

The fact that television was so quick to racialize the crisis should come as no surprise, given broad-based demographic changes in Los Angeles. For several years the media had tossed around the notion that whites now represented only 37 percent of the city's population, that whites were now a minority. The rosy picture of happy multiculturalism foisted during the Olympic era in 1984 had long since been dwarfed by the sense that Los Angeles was in decline. The decline mythology was invariably linked to the emergence of a new multiethnic monster and to race. Illegal immigration, white flight, gated communities for the affluent, and the collapse of public health and education all became evidence for decline in nightly news coverage. Seen in this context, crisis televisuality was also an act of desperation, a last-gasp exercise in control by upper-middle–class "victims" jostled into a corner defined by race and economic class. The smiling upper-middle–class Anglo faces that dominated newscasts leading up to the rebellion, then, were far from comforting: the air of decline and victimization was not a pretty picture. Desperate times called for desperate measures, and television was more than equipped with electronic and preemptive command and othering measures. A long-cultivated partnership between the televisual apparatus and its institutionalized fortress mentality had come of age.

There is a lesson in television's representation of the L.A. rebellion. The uprising so overwhelmed television's narrative and journalistic systems of containment and recuperation—systems that have claimed the lion's share of critical attention in the past—that the televisual apparatus was exposed for what it has been on many occasions: a stylistic architecture for managing difference, building consensus, and stylistically packaging the dangerous other.

Postscript
Intellectual Culture, Image, and Iconoclasm

The visual is essentially pornographic, which is to say that it has its end in rapt, mindless fascination.
—Frederic Jameson, *Signatures of the Visible*[1]

The visual in any sense of some pure realm . . . is finally non-existent.
—Stephen Heath, *Questions of Cinema*[2]

Heath and Jameson were not the first to vanquish the visual from the pulpit of the intellect. In eighth-century Byzantium the eastern empire was violently purged of sacred and representational imagery by iconoclasts, by medieval media theorists who laid their theological axes to the very roots of the visual regime. Resurrecting and synthesizing ancient Hebraic and Islamic prohibitions against recognizable imagery with medieval theology—a fitting historical task for a desperate city dug into the apocalyptic junction between warring Eastern and Western civilizations—Byzantine iconoclasts cloaked their denigration and destruction of paintings with a double-barreled apologetic, one that feared visual pleasures and preoccupations as idolatrous diversions and material challenges to God's deity. Bloody confrontations ensued when those termed "iconodules," legitimizers of the image, rose up against the emperor's high theory and his iconoclastic purges. Basing their defense of sight on the centrality of God's incarnation in the worldly figure of the Redeemer, the reactionary iconodules retheologized, that is retheorized, how icons and images were legitimate for human contemplation and reflection. The aesthetic-moral debate was finally won by the iconodules, but only after more than a century of violence, assassination, and exile in which many images were lost.[3] Far fewer casualties have come from debates in mass communications. Nevertheless, contemporary theory's predisposition against the image can be usefully described as part of a much more modern iconoclastic controversy. Both controversies have had striking methodological and political consequences. When contemporary theorists decry the deleterious excesses of the mass-produced spectacle and its imagery, they also inevitably promote a kind of intellectual and cultural asceticism that resonates with the earlier period.

The motivations for the iconoclastic purges of emperors Leo III and Theophilus provide an apt allegory and possible explanation for contemporary theory's denigration of the image. "*Iconoclasm was an attempt to provide therapy for a corporate anxiety.* As city after city fell to the enemy, anxiety about the future was verbalized in the idea that the . . . state had drawn down God's wrath by its idolatry in permitting the veneration of icons."[4] Compare this fatalistic justification, this defense of *aesthetic self-flagellation* as a therapeutic and confessional ritual, to a more modern and more theoretical iconoclasm. Consider, that is, the falling cities as a figure for mass-culture's agressive inroads against intellectual and academic culture. The centrality of the theoretical aversion to the image in Western culture is provocatively evoked by international high-theory trendsetter Baudrillard: "There is something secret in appearances. . . . They remain insoluble, indecipherable. *The reverse strategy of the entire modern movement* is the 'liberation' of meaning and the *destruction of appearances*. To be done with appearances is the essential occupation of revolutions."[5] Baudrillard and Theophilus are, then, two sides of the same coin. They are iconoclastic soulmates separated by ten centuries. Both postulate the destruction of appearances as a key to the liberation myth in Western societies.

Seen against the televisual excesses examined thus far, intellectual scopic purges seem to erect ever-higher barriers between two contemporary institutions: the industry and the academy. Where, then, does this divergence leave the study of media? How does this bifurcation affect theory? What points of engagement can there be between a mass consumer culture that defines itself by the image and an intellectual culture that denigrates the image? Having stepped through various televisual forms and modalities and raised broader cultural issues about televisual technologies, audiences, and economics, it is worth turning back on the very conceptual frameworks that make possible the recognition and analysis of televisual practice in the first place. More than just a conceptual dynamic, the academic response to television takes place in and through real-world institutions with certain vested and historical interests. Pedagogy, economic class, and the functional ways that knowlege is reproduced are all essential but ignored components of critical theory. Even theory, then, should be recognized as a cultural process. Once theory is acknowledged as a cultural form—rather than simply an intellectual activity, involving generalization, framing, deconstruction, or analysis—the possibilities of desegregating theory and practice, of making theory instrumental as well as philosophical, come in to sharper focus.

High theory has devalued the image in ways that contemporary television and its audience belie. The concluding pages of this book, then, aim to do two things. First, a survey of the curious intellectual coalition arrayed against the image demonstrates that debilitating institutional constraints still shackle television study in the academy. Second, contemporary televisual practice suggests several possibilities for breaking open the institutional lock set against the academy's visual nemesis. Historiographic, aesthetic, and pedagogical alternatives all promise to desegregate the iconoclastic split between theory and practice.

Conspiracy Theories

Like incessant bells tolling across the intellectual landscape, prophetic critics of the image range far and wide—from social scientists to concerned humanists to radical situationists. A fundamental suspicion of the image, in fact, is at the heart of much in Western academic thought. Its origins are both religious and rationalist. Yet this methodological suspicion is not always overt in media study, or even necessarily intended. It frequently emerges as a by-product of past intellectual commitments even in studies that look to engage very different phenomena. Five such prejudicial paradigms—theoretical frameworks that I term neo-Gnosticism, journalismo, scopophobia, semiologos, and connectivity—wield assumptions so fundamentally predisposed against the image that they shackle stylistic analysis and cultural studies alike. Whenever these kinds of intellectual models permeate discourses about media, style and image remain absent or undertheorized categories. Although profoundly different in method and pretense, all five views are general theoretical trends that denigrate or conspiratorialize the image.

Neo-Gnosticism

By 1992, the worst criticism that academics could make about something was that it had become an image. Reacting to the Rodney King beating trial and Gulf War coverage, scholars began to echo a shared orthodoxy: "the image had triumphed over *reality* and *reason*."[6] This dichotomy between the visual image and its supposed victims—reality and reason—manifests itself in a variety of subsequent, polar oppositions. The visual world was vigorously mapped out in stark opposition to introspection, the external world of images in opposition to the internal world of the mind, and visual immediacy in opposition to complexity and reflection.[7] The implications of these structuring oppositions and polar assumptions were seldom hidden. The television image is consequently irrational, unreal, and without redeeming logic or history. The basic organizing framework recurrent in these diverse accounts, then, is reminiscent of a rather classical moral dichotomy between good and evil, the mind and matter, the image and the world. A Manichean dualism operates here with scholarly roots that date back at least to the first-century Gnostic heresy and to Platonism's denigration of appearances before that. The lies of sight, we are lead to believe, constantly threaten the truth of the mind. True history is available through the word; false history is a construction of the image.

Is there a cost in accepting this moral dualism? The elevation of history and of reality *as concrete and accessible* correctives to the illusory deceptions of the visual image seem strangely nostalgic.[8] This tactic certainly ignores the long tradition of scholarship that indicts the categories of history and the real as themselves problematic. Certainly the visual image is no less or more constructed than the real and the historical. *The image is both real and historical.* It is both a product of history and an agency of history. To ignore this fact is, in one way, an act of surrender. The proclamations that "history does exist, and it is not the story seen" and that the "image cannot triumph indefinitely over the reality that it conceals" ignore the fact that all of history is available only through representations.[9] On what scientific basis is the image banished from the operations of the mind? Can thought occur without it? On a very practical and day-to-day basis, the image has as much realness as the truth that is here described as its victim. People make images; cultures represent themselves with images; political interests contest images; oppositional groups appropriate images. It is a mistake, then, to attack the messenger rather than the sender. The Rodney King and Gulf War crises were political and economic first, centered around the use and abuse of power. The visual-semiotic problem was clearly a manifestation of that condition.

Although we are strongly persuaded by Gnostic accounts to believe otherwise, media images do indeed have logics and these logics can be utilized in progressive discourses.[10] To assume that the culprit (reason-destroying instant global imagery) is a product of the new audio-visual-digital-cybernetic technology is to participate in a debilitating kind of determinism and acquiescence.[11] Using an age-old moral dualism in order to bemoan the absence of introspection, complexity, and verbal discourse (admittedly important functions) in the conflict does little to aid activists and producers who want to engage and actually change the course of things in the new world order—to change the way power is displayed and produced in and by the media. Are introspection and verbal discourse the best tools for activists interested in opposing the corporate fronts of postmodernism and the machinations of global capitalism as these studies imply? Intellectual paradigms that we bring to bear in accounts of the media—like this moral dichotomy recurrent in American communications scholarship in recent decades—also have political implications for alternative media and oppositional practice. Separatism and abstinence are not very productive responses to the new global imagery—at least not if one cares about life outside the academy.

This intellectual binarism—in but one of the scholarly traditions that tend to denigrate the status of the image—short-circuits and prejudges consideration of the nature and cultural importance of style. Partly because of its philosophical commitments and methodological origins in the social sciences, American mass communications occasionally exudes what might be described as a neo-Gnostic penchant for purity. Reminiscent of the Greek construal of matter as both evil and fleeting, the Gnostic revival supposes that emancipation only comes through knowledge and that knowledge is wholly mentalist in nature.[12] Exemplified especially by the work of Neil Postman and Jerry Mander, media imagery is characterized as mind-numbing and duplicitous, an enemy of the intellect.[13] These moral presuppositions probably owe as much to the important insights of Daniel Boorstin on pseudo-events—seen as a defining property in contemporary American culture—as to anyone else.[14] But the moral critique of imagery is also a logical outgrowth of some of the earliest studies in the field. Propaganda and persuasion research, by implication, positioned the mediated visual spectacle as the very Achilles heel of the modern audience.[15] It is worth noting that the fascist spectacle in the 1930s and 1940s was certainly no less overwhelming than CNN and the Pentagon's in 1991. Both effected crises in communications scholarship. The immediate awareness that scholars were as politically impotent as audiences, and certainly as far removed from any real power, underscored and reinforced the academic moral defense against image critique. It is worth noting, however, that this moral rationale often tends to be *imported* to mass communications from scholars outside of empirical and quantitative research: Boorstin from American history, Mander from the commercial world, and Postman from the humanities. Such *data weak and theory rich* fields are often tapped to infuse and energize what some consider to be the *data rich and theory weak* province of mass communications study.

Yet the fact that the field of American mass communications still celebrates its foundation in empiricism and positivism—in the notion that "all knowlege derives . . . from *observational* experience"[16]—suggests that visuality *should* define research to a greater extent than it does. This defining pressure for observation, then, makes ironic the scholar's crisis-induced moral defense against image argument and inclination invoked during times of crisis. Crises, however, inevitably demand bigger answers than social science is typically willing or able to offer. Yet public anxiety breeds urgency and impatiently feeds on macroscopic generalization, rather than delimited and controlled microscopic method. Seen in the context of public crisis, then, the mass-produced electronic-visual spectacle is certainly one of the easiest and most handy moral targets, even for otherwise patient social scientists.

Journalismo

"Bit by bit we are letting pictures steal our mind. The tsunami of images can crowd out our other faculties: judgement and analysis."[17] Apocalyptic preachments like this one suggest that something significant happened when print people took over the television newsroom and when journalists began to dominate academic training in broadcasting. A paradigm took center-stage that initially saw the visual spectacle as unimportant and inconsequential by definition. Journalismo, an iconoclastic presupposition cherished in the commercial world of broadcasting, publicly celebrates the truth-value of facticity. The paradigm is also an outgrowth, whether intended or not, of the *Dragnet* aesthetic: "The facts m'am. Just the facts." Like institutional law enforcement and Joe Friday, journalism wears this discipline on its sleeve. Such a view worked well for years, especially for television news departments until, that is, the very spectacle that television journalists had created turned on them. In 1993 *Dateline NBC* and its network news division were caught with incendiary devices in their journalistic hands. The intended evidence of the videotape notwithstanding, the smoking gun turned out to be NBC's not GM's. The Rodney King beating tape was celebrated even more by the word-and-fact people as proof that imagery was inherently deceptive.[18] Even the popular press now trumpets the growing sense that pictures do lie, and that they do so on a regular basis.

Clearly images, like words, are problematic. But are they mind-dwarfing tsunamis? It is important to examine such totalizing claims and to reconsider the exclusionary implications of journalismo in both television theory and practice. To argue that a "picture conveys more sensation than information" ignores the inextricable relation between sensation and cognition, and the fact that images are also fundamental forms of semiotic information.[19] The journalistic protest that seeing is not always believing disingeniously ignores the fact that journalists created the myth in the first place. To theorize that a picture can't analyze or remember begs the question and elides the common-sense reality that words can't either. Why polarize pictures and images as antithetical to humans and words? Can any thought occur without images, models, or cognitive maps? When journalists decry the dangers of picture-making as antagonistic to thinking, they ignore an alternative that is just as plausible if not more urgent than their own calls for renewed critical thinking: *why not put picture-making—not just reflective thinking—back into the hands of the people*? This democratic and practical alternative, however, would directly challenge another one of the most privileged myths of journalismo, the cult of delegation.

As long as journalists conceive of the world (or posture it for self-serving reasons) as a collection of inert facts bound to confuse the citizenry and see television's task as collecting, selecting, and mustering information in the name of truth, then journalism constructs for itself on television a heightened and privileged place. The cult of information that drives the fourth estate is in fact a product of the eighteenth century, a libertarian god from the world of Locke, Jefferson, and Mill.[20] The purist principles that drive the cult—enlightenment rationalism and delegated gatekeeping untainted by external influence—tend to bleed journalism of sensitivity to both the nuances of the material world and to the excesses of its own method. As an interlocutor inserted between dead data and the audience at home, broadcast journalism frequently becomes a kind of performance art, a showy marshaling of facticity. Gone is any sense of the world as a semiotic process or as an experience or phenomenon accessible to common viewers. In its place culture has constructed a publicly sanctioned cadre of professionals, uniquely trained and sensitive to the overwhelming nuances of facts. Meaning, we are encouraged to believe, can only result from journalistic discourse and argument—that is, from the verbal mediation of static facts, not from the world at large. Journalismo pines for the ordered and rational world before phenomenology.

This is *not* simply my characterization or criticism. Broadcast journalists themselves champion and justify the cult of professionalism. In a public critique of populism that bordered on the outrageous, long-time anchor and venerable news patriarch John Chancellor cleared the network decks of any ambivalence on the subject. As keynote speaker at the Fiftieth Annual Du Pont–Columbia Awards forum in New York, Chancellor explained the dangers that come with the American populace: "Ordinary Folks are not trained to conduct a serious interview with candidates. *It is perilous to turn everything over to the people* of the United States. You shouldn't reduce the dialogue to the lowest common denominator."[21] This was less a reflection on the failures of the recent 1992 presidential election, than a call-to-arms for broadcast journalism. The American people, we are lead to believe, are actually adversaries, the "lowest common denominator" that threatens television's cult of delegation. Whether or not this was arrogance or simply insecurity, such rhetoric suggests that there is a fundamental crisis in the field. It is no coincidence that this renewed emphasis on saving quality (and delegated) broadcast journalism arises at the same time that two other crises move to center-stage: first, the mythology that pictures always lie, and second, the demise of the network news divisions—now commonly diagnosed as being on their deathbeds. The new world of the 500-channel environment and corporate downsizing have placed once-prestigous and serious journalistic divisions on the network backburners. In the face of crisis cutbacks in both financing and prestige, it is no wonder that television's word people strike out at the threats that come from both image and audience.

There is, however, a historical irony in journalismo's self-righteousness. Words did not always dominate TV news. For many years, a cameraman would simply go out and shoot film of something happening, a film editor would cut the images together, and only after the footage was assembled would a newswriter write a script to go with the pictures. The skill was in getting all the news in while captioning the pictures. The process of gathering credible pictures from the field drove the process and words followed. According to Reuven Frank, former two-term president of NBC News, the word people only took over after "the time most

Americans told pollsters television was their prime source of news."[22] According to this authoritative network figure, then, journalistic dominance in network production practice was really a kind of after-the-fact opportunism. TV had already developed a workable way to do the very thing that many saw as its unique ability: showing things as they happened. Only then did the word people take over: as "correspondents, anchormen, as producers and managers—scripts were written first, then pictures were matched to words." Now, however, stories drive the process and pictures follow. If a picture has no words, it is thrown away. If a word has no picture, one is found or fabricated for it. In the new world of television journalism—largely due to the widespread use of videotape and archival libraries—images now are things that assistants find to illustrate written stories. Images, then, are stripped of any determining authority within the logic of this newer industrial practice. No wonder concerned broadcasters rotely criticize the old cliché that pictures never lie as a way to validate stories for the benefit of an audience it still considers naive and susceptible to such myths. Cynicism about the audience, then, very much rules journalismo.

The problem with the *Dateline* staging was not, according to Reuven, that they were chasing pictures, as many other self-righteous journalistic critics were now saying, but that they were chasing scripts. The *Dateline* script had, in fact, been read by both the network president and an NBC lawyer before the broadcast, but it never dawned on anyone to actually look at the tape. For broadcasters then, words have profound legal ramifications, while pictures are merely window dressing. Although isolated figures like Reuven argue that the word, not the image, is to blame, the dominant logic of journalismo continues to point the accusing finger at pictures, not words. Prophetic editorial calls to flee the artifice of the image and return to truth, therefore, miss the point. Given journalismo's long-standing institutional commitments, the platitude that images have no authority is a self-fulfilling and occasionally self-destructive prophecy. Broadcast journalists regularly get burned by their own logic.

Scopophobia: Hermeneutics of Suspicion

The American traditions of neo-Gnostic research and journalistic purity—intellectual offspring of Enlightenment rationalism—are far from unique in their denigration of the image. Many variants of critical theory, especially those with European origins, frequently share a similar sensibility: an overarching suspicion of appearances and a radical distrust of the visible. This attitude is certainly logical if one considers the origins of modern theory. Many critical theorists, even if by osmosis, are heirs and proponents of a persuasive negative hermeneutic, an interpretive orientation broadly sketched out by benefactors Marx and Freud.[23] This modern distrust of appearance has had its consequences and costs. The negative hermeneutic is iconoclastic, not because it is wrong, but because applications of it tend to prematurely conspiratorialize appearances. Analysis reifies and postures. Readings become substitutes for the texts. Observation and common sense become secondary or marginal skills in the heady air of deconstruction, where literary method replaces text and culture. This was not always so, however, for earlier variants of the theories prepackaged interpretive allegories—of incest, of class struggle, and of the imaginary. Yet even these unified allegorical master codes were frequently used to dominate small and feeble texts, films, and artworks, since the surfaces of such works no longer held the key to their broader and underlying meanings. Such things were, after all, only appearances, distractions at best, and mystifications at worst.

One of the best and most influential manifestations of this distrust of the visible was the concept of scopophilia.[24] For Freud, scopophilia was a compensating sexual pathology centered around the eye and the gaze, an "obsessional neurosis" tied to "the desire to see."[25] The very impulse to gaze was overtly described and sexualized by Freud not just as a kind of neurotic sickness, but as a kind of violence, as sadism. This scopophilic model was widely imported to film and television critical theory, even though at root the analytical perspective was clearly prejudicial against image and sight. Before ever applied to a film or video text, however, the desire to see was identified as a medical "symptom."

Rather than neutralizing the concept after stripping it from its medical context, scopophilic applications in contemporary media theory actually intensified the prejudice against sight. We owe it to the French, for example to Jean-Louis Baudry, Christian Metz, and Daniel Dayan, for having turned this peculiar mental sickness of the eye into a universalizing critical method and explanation for political oppression. Baudry described the evolution of image technology in the West as a political process that aimed over many centuries to falsely liberate the eye, a process based on the construction of a politically regressive ideal ego in the cinema spectator.[26] Dayan stated unequivocally that "the 'I,' the 'ego', and the 'subject' *are nothing but images, reflections.*"[27] By arguing that voyeurism was at the base of the institution of cinema, Metz politicized the perversion of sight even further by describing the desire to see in cinema as a political institution, as a "scopic *regime.*"[28] An entire discourse, then, had evolved whereby the eyes, sight, and the visual domain became mechanisms by which dominant capitalist cultures cloned and hegemonized their subjects. A not so subtle methodological loathing of the visual had evolved, sprung from speculations about scopophilia and the gaze.

Given these universalizing claims about the regressive nature of sight and its pleasures as sadistic and neurotic forms of cultural domination, theory left itself little room to reconsider the status of the image. Theorists had prejudged and totalized the scopic assumptions into an influential form of high theory that distrusted the visible. This suspicious attitude might better be designated "scopophobia"—a fitting term since it shares the same root as Freud and Metz's term, "scopophilia." What the tradition of negative hermeneutics presupposes as a pathological desire (-philia), is also evidence of high theory's intellectual repulsion and fear (-phobia). For high theory, the visual spectacle overwhelms and subjugates.

Why have scopophobic and gaze theories been so popular and totalizing in critical theory? Universalizing extrapolations about visual sight as pathological may have been academic reactions to the sheer power and scope of mass culture. After all, the entrenched and dominant mass media have always been better than anyone else at mass producing highly visual productions, stylized imagery, and pleasurable forms of escapism. The dominant media—the implicit enemy to radical theorists—is also unique in owning the *tools and capital needed to manufacture excessive style*. It follows logically, then, that any viable and radical media work or theory should resist the domain dominated by the enemy and controlled by the owners of visual spectacle. Mulvey's gaze theory, a profoundly influential feminist update of scopophilia, was, for example, fueled by an outsider logic. Without access to the stylizing tools and financial capital of dominant culture—a world built of institutionalized, male sexual aggression, according to this view—radical theorists and practitioners were left with one very clear choice: attack the very eye of the beast. Using this kind of counter logic, antistyle verbal discourses present themselves as incisive tools and interpretive weapons that can counter capitalism's hegemonic lock on the visual.

More recent theory disperses this orthodox suspicion of the visible in a variety of ways. While hermeneutic traditions centered around psychoanalysis and Marxian ideology overtly castigated visual pleasure as regressive or conspiratorial, others theorists simply ignored the image by turning their analysis to nonvisual aspects of media. The 1980s, for example, saw some interest shift away from visuality toward the aural realm—to voice-over, to music, and to narrative.[29] In short, there was a renewed interest in media forms that stood as alternatives to mass media's *dominant scopic regime* and its *fetishization of the visual*. This methodological reaction and shift away from visual analysis suggests that scopophobic tendencies still pervade high theory. While other stratas in film-video enunciation are open to reappropriation and counterreading, the visual domain is still frequently construed as one important key to the media's ideological dominance.

Given the actual operations of mass-consumer culture, high theory's vague but pervasive scopophobia and prejudicial treatment of style are at best curious stances. Images are, for instance, perhaps the primary way that viewers consume mass culture—in dress, in fashion, and in appropriated imagery from film, advertising, and television.[30] The pleasures of the visual may be pathological and politically repressive as high theory states, but ignoring those pleasures is surely shortsighted. A reappraisal seems important given the essential role that imagery fulfills in allowing the spectator and media consumer to play and work at mass culture. Since pleasure has been theorized as a progressive site for both counter readings and alternative practice, intellectual rituals choreographed by scopophobic suspicion are clearly myopic. The emergence of cultural studies as a more popular academic substitute for critical theory in recent years surely is a symptom of high theory's iconoclastic lack. Cultural studies, after all, at least considers significant those things that mass culture values. Rather than pathologizing visuality and style out of hand, cultural studies accommodates the visual image and the pleasures of sight as social practices.

Semiologos: The Saussurian Ghost

There is a fourth iconoclastic ghost in the machine of high theory, one that continues to cast a long shadow on critical analysis. It comes in the form of language, and hails linguist Saussure and theorist Metz as authorizing fathers. Even poststructuralist media theorists who distance themselves from the naive formalism of early semiology persist in repackaging Saussure's outrageous linguistic glosses. Stephen Heath remarks, "*The visual* in any sense of some purely visual realm . . . *is finally non-existent* . . . the visual is always in a production of meaning run through and through by language" (italics mine).[31] Poststructural semiology is, from this account, not just prejudicial against the status of the image, it denies even the existence of the visual. Arguments that cinema is not a language are by now common, yet the actual use of linguistic paradigms in various guises within media theory belies the claim. Film and television theorists continue to invoke the sciences and practices of language for keys to media texts and spectatorship. The universal claim by Heath that "language remains a constant and crucial point of reflection, a junction problem for thinking cinema today," not only overestimates the relationship between film and language, but also stands as a symptom for film semiology and theory in general.[32] That is, once the field started with the model of language at its base in the 1960s, it spent the next two decades trying to dance around the presupposition, trying to modify how the moving image is (or is like) a language or discourse.[33]

The initial privilege given Saussurian semiology over semiotics in media study was a direct result of Metz's comprehensive and influential elaboration of cinema's "science of signs" in the 1960s. No comparable project had yet elaborated the Peircian roots of American semiotics for media.[34] But Metz would change, so that by 1975 his "second semiotics" was a synthesis of psychoanalysis and linguistics. Now, semiotics for Metz investigated the psychodynamics of the film text and film spectatorship.[35] During the 1970s theory was clearly dissatisfied with reductive semiological models and broke out into several new directions.[36] Transforming influences came in the form of narrative and discourse analysis, methods that followed from the suggestive cultural examples of Roland Barthes and from the theorizations of Emile Benveniste.[37] Other post-Metzian scholars associated with *Screen* in England, found in psychoanalysis a nonlinguistic justification for their obsession with language.[38] In one telling passage, Freud's tension between word and image is invoked: "Thus Freud, at every moment that he operates the distinction between 'thing' and 'word', the antithesis between 'visual' and 'verbal', everywhere finds language."[39] Language dominates the image in both psychoanalysis and semiology. Such accounts also characterize the image as more primitive than language, even though it is more "advanced than the olfactory, tactile, and gustatory." Not only does this view create a graded and antiquated hierarchy of the senses, it also polarizes the visual as an enemy of language. Within this context, preposterous claims that "the visual is finally non-existent" seem less incredible. Even when *Screen* theorists acknowleged that there was a visual dynamic in film, the acknowledgment was couched in another language-based term, "inner speech."[40] By attempting to account for the shifting, slipping, polysemy of signification, and by espousing Freud's worst iconoclastic impulses, poststructuralist semiology merely exchanged one verbal metaphor for another.[41]

This critique of the semiological ghost in media theory may seem passé to some.[42] The emergence of semiotics on a transdisciplinary scale, however, has served to put into proper perspective the accomplishments and severe limitations of any media theory that shares or revises the French and Saussurian semiological assumption.[43] The American semiotic tradition, by contrast, can be characterized as a form of sensate and cognitive symptomology, as an analytical process directed at textual symptoms and signals given off by objects in a wide range of social and biological contexts, from legal proceedings to genetics.[44] Given the extensiveness of semiosis, it is clear that the process is not limited to humans, even though language and linguistic semiosis are.[45] This fact was lost on Saussure, who claimed without hesitation that, "Language, better than anything else offers a basis for understanding the semiological problem. . . . Language is the most important of all these [sign] systems."[46] Many decades after Saussure's fleeting and speculative elaboration of the field, other theorists continue to repeat his homage to the master code of language. Even basic definitions of the semiotic framework in mass-media textbooks, for instance, persist in falsely claiming that semiotics means "to take linguistics as a model and apply linguistic concepts to other phenomenon."[47]

While language is limited to humans, humans clearly are not limited to language as a means of signification. The convergence of several factors make this nonverbal alternative in semiotics plausible. Thomas A. Sebeok, building on the work of Gregory Bateson, François Jacob, and Stephen Gould, articulates a number of these scientific perceptions in his essay on the phylogenesis of communications.[48] The ability to imagine possible worlds through the process of modeling; to make tools that could make other tools; in short, to be involved in abstraction—these semiotic functions preceded but were not displaced by the advent of linguistic communication.[49] Semiotics, then, offers a much more useful model for understanding the visual in general and contemporary television in particular than does poststructuralist semiology—a model that presupposes not language, but the ability to "model possible worlds" as the defining property of signification. As we have discovered, the construction of alternative worlds has become a defining factor both in primetime televisuality and in the development of virtual reality technologies.

The physical and biological perspectives that support the semiotic framework, including research in nonhuman semiosis, suggest a number of important correctives. All of nonhuman semiosis, for example, and almost all of human communication, *is* nonverbal: chemical, gestural, motor, visual. Nonverbal semiosis was not displaced by the evolution and acquisition of language, but rather flourished and coexisted along side it. Furthermore, at least from a semiotic perspective, the world fundamentally exists in the nervous system, in the body. If one considers semiosis a process of modeling, rather than a languagelike construction that produces meaning, then both humans and the universe can be construed as aggregates of signs rather than the product of imposed master codes. Finally, any effective semiotic analysis must account for a constant interaction of minimal unintentional signs—even in semiotic systems, like television, that have a verbal strata. Since film and television both have verbal and discursive components, theorization has generally focused on those aspects, even as they overlooked the array of minimal unintentional signs that make up the media viewer's engagement with image and sound.[50] From a semiotic point of view, television floods the viewer with nonverbal signs. Television leaks.

Poststructuralism has done much to rid theory of the totalizing claims of structural linguistics and the French semiological tradition. Its shift from the static structures of the text to the structuring of subjectivity allows for a more direct involvement with the mass culture that produces television. Yet, vestiges of the language bias persist even in newer theories of society as a construction of social discourses.[51] Implicit in the poststructuralist discursification of pleasure is the philosophical assumption that experience must be codified and categorized before it can be understood or engaged. Such a premise is clearly problematic if one looks at, or cares about, actual subjects (people) within, and across, cultures. One recent theoretical work on television, for example, turns what is only a vague presupposition in Foucault—that discourse dominates experience, that naming dominates pleasure—into its own sweeping and totalizing claim: "What passes for reality in any culture is the product of that culture's codes, so 'reality' is always already encoded, it is never 'raw.'"[52] Try greeting families of the *desaparacidos* in El Salvador or victims of the shelling and brutality in Bosnia-Herzegovina with this form of denial. According to such a view, experience and pleaure are never directly accessible. Such an account disregards, however, the large body of research on nonverbal communicational response that describe certain traits as physiologically hard-wired across cultures.[53] This television scholar, however, does not equivocate. The human subject is always mediated and encoded. The phenomenal world has been encrypted.[54] Experience has been caged.

The intellectual work it takes to abstract sex or sight or sado-masochism into discourse is cumbersome. The practice of sex, for example, suggests a more direct and less discursive kind of modeling, politics, and interpersonal semiosis.[55] The same theorists who celebrate heterogeneity are clearly shortsighted when they make global assumptions about the way that meaning is "always" made. The determinism of the extreme cult of encoding, for example, is as much a hindrance to political agency or worldly action as any preoccupation with style and appearance. Poststructuralism is not inevitably iconoclastic, only its recurrent and persistent "prisonhouse of language."[56]

Connectivity: Infrastructure over Experience

Financial interests in the electronic media also incarcerate the image with an aggressive but more benign-looking predisposition than those found in either the academy or the fourth estate. Future speak has so colonized industry agenda setters that architects of the information superhighway typically wield as fact what is actually little more than self-serving fiduciary speculation. What was seen in 1993 as an extension of cable television's leap toward a 500-channel environment was widely discussed by 1994 as something much more intrusive and pervasive: the information superhighway. This overhaul—from a fairly traditional programming paradigm that simply multiplies choice to a consensus celebration of infrastructure—elides history even as it neatly ignores human and social agency. Technological determinism, apparently, rules our national agenda for the future. The industrial obsession with connectivity and the infrastructure that comes with it stands as a fifth iconoclastic commitment, one that works vigorously against a number of new technological and stylistic alternatives, including HDTV.

Buzzwords and headlines notwithstanding, there is no consensus about what the information superhighway actually is. Two major institutional forces—telephony and cable television—have locked horns in a studied grip with ideological implications that far surpass the duel between competing delivery systems: fiber optics or coaxial cable. Given the simplifying glosses that make the highway a unified phenomenon in both popular and trade publications, it is worth examining more closely one of the intellectual models that fuels the highway rhetoric. Although high theory tends to presuppose cynicism of industry, superhighway and fiber-optic devotees are not bereft of their own intellectual and academic master theories. George Gilder's *Life After Television* presents itself both as a prophetic national call to arms and as a form of intellectual think-tank legitimation for architects of the superhighway.[57] The national superhighway summit as well as countless trade articles that followed the book's publication typically share an unabashed advocacy of television's nemesis and promised substitute, the telecomputer. According to this vision, the explosive key to both the technological revolution and to the free-market libertarian social revolution that will inevitably accompany it is none other than silica—the mundane material source of glass, the substance that makes possible both the ecstasies of fiber-optic interactivity and the mind-boggling geometrically expanding memory potential of the microchip.

Connectivity people like Gilder argue that user-friendly and decentralized microelectronics and fiber-optic technologies sound the death knell for the top-down corporate dinosaurs and propped-up monopolies that America has blindly come to cherish as network television. Utilizing the logic of the microcosm, proponents argue that efficiency—not exponentially growing and finally incapacitating complexity—grows as the number of interconnected switches grows.[58] A tripartite leap in logic then follows—from the behavior of subatomic particles to the computing power of the microchip to a vision of the world in which infinite individual needs are met through interactivity and technological responsiveness—and forms the basis for what might be called a cult of connectivity. Connectivity, and the public relations organs that promise it, now seem to fuel everything from developers of the internet to the federal government's recent Telecommunications and Information Infrastructure initiatives. A common liturgy now resounds among the wire people (the telephone, computer, and microchip industries): "Connect, and they will come."

The "experience" folks at Disney, Warners, Tri-Star, and NBC, on the other hand, are far less effusive about the on-line world. Program providers and producers can take heart. A national crusade to fund and lay the fiber-optic infrastucture ignores one of the most basic necessities of any technological revolution: a *trigger application* whose massive popularity can provide the economies of scale needed to justify widescale superhighway development. Most audiences have yet to clamor for the headaches of menus, interactive branching, and nonlinearity. They want Schwarzenegger and production value, and they want it now. Many audiences do not want Nintendo and e-mail either, they want narrative and character. The hardware and wire people, that is, have underestimated the centrality of software and content in the development and adoption process. This convenient gloss is no mystery, however, since AT&T, MCI, and PacBell currently have few stated financial interests in content. They simply plan to own the connections—the cable television industry be damned—and to charge the inevitable tolls. If they do, they win. No wonder the wire people and their ad agencies publicly place the ecstasies of the infobahn in the connections and network rather than in the content. Yet, a thorn in the flesh of the infrastuctural players persists. Movies on demand are much more likely to survive as a catalyst for superhighway development than the efficiency-inducing regimen of the networked microchip.

When gurus of connectivity, driven by nationalism, call for the dismantling of broadcast network television by castigating it as a morally vacuous pipeline for the brain dead they reflect an ignorance of the instrumental role that entertainment has played in the perpetuation of all mass cultural technologies in the twentieth century. When they argue for the abandonment of high-resolution HDTV in the United States—and the government-sponsored megacorporations that prop it up[59]—they ignore two even more basic factors behind the emergence of televisual programming: the proven popularity of *cinematic and narrative pleasures* among television audiences. The cult of connectivity then, everpresent in daily accounts of leveraged media buyouts and merger mania, is neatly blind to the fundamental reasons that an audience returns to the medium: to indulge—in the virtual world of stories, in the excesses of embellishment, and in the *ontology of the photographic image.* The lesson of the internet is instructive. While superhighway players ape the oversold, word-driven, on-line world the internet—"it's the pipeline, stupid"—the programming and production worlds return to a much more basic question. Who can possibly fill the twenty-four-hour world of the 500-channel universe with engaging programming? Higher and higher production values have sold well, and will continue to do so, even as the costs needed to create those values head south.

This fact was lost in the concerned shock that greeted the FCC in 1994, when the regulatory agency finally allowed Bell Atlantic to cross-over into the once-protected world of cable television and to offer customers what it called a "video dial-tone." The sensory meagerness of the dial-tone designation is instructive; the tactical corporate victory provisional. Infrastructure is both the cash cow and economic genius of connectivity. The home—long an El Dorado for TV programmers and admen alike—is now being securely locked-on within the electronic cross-hairs of multinational, hard-wired consumerism. Yet telephony—even in the era of fiber optics—has always lacked the vision people. On the superhighway, the aesthetic capital of Hollywood and the expertise of its emerging clones will be no less bankable—and leveraged—than in the past. Presentational distinction will continue to have a price tag on the infobahn. Bill Gates traveled to Hollywood to join the "dream team," not vice-versa.

John Chancellor, Neil Postman, Sigmund Freud, Stephen Heath, and George Gilder: less five horsemen of mass culture's apocalypse than masters of totalizing, iconoclastic bluster. Strange bedfellows indeed: the reporter, the humanist communications scholar, the psychopathologist, the neo-Marxian poststructuralist, and the libertarian techno-futurist. Yet each represents recurrent iconoclastic influences in media study. Consider their shared penchant for sweeping theoretical statements: "pictures steal our mind"; "the visual is nonexistent"; "language is the most important sign system"; "pictures short-circuit introspection."[60] Such similarities are uncanny given the theorists' extreme differences in political commitment. I am certainly not interested in disproving semiology, scopophilia, journalismo, or Gnosticism in communications study, only interested in showing that scholarship based on these intellectual commitments has been prone to over-statement and to exclusion. These excesses have typically worked to castigate im-age, style, and audience as secondary or unimportant phenomena—if not downright dangerous—in recent media studies. Semiology is not wrong because it deals with language-based semiosis, only wrong to claim that language is at the root of all communication. Language *is* important in film and television semio-sis, but no more so than sight. Scopophilia is not wrong to describe viewing as powerfully pleasurable, only wrong to deduce from this that viewing is patho-logical. Visual pleasures, after all, come in many forms, both regressive and pro-gressive. Poststructuralism is not wrong to claim that viewing is constrained by social discourses, only isolationist and ascetic if it suggests that no experience is possible unless it is first encoded. Television *is* heterogeneous as ideological criti-cism implies, even though poststructuralist applications frequently extrude expe-rience into a singular mold of verbal analysis and discursification. Human subjects are dominated by social discourses, but such a process cannot completely elimi-nate biological factors. *The presentational demeanor of television is persistently more complicated than these master paradigms suggest when invoked in critical practice.*

Theoretical Practice

Some of media theory's most persistent iconoclastic and scopophobic ghosts will simply not die. Their institutional resilience makes it essential to consider alter-native and more proactive ways by which critical theory and cultural studies can engage the televisual. Given the centrality of the visual regime in society and of stylistic exhibitionism in television, intellectual culture would do well to recon-sider the tools by which it addresses and accounts for mass culture. All of the televisual modalities examined earlier in the book come packaged with gaps, soft spots, and contradictions—ideological fissures that can be worked in both ana-lytical and practical ways. Consider briefly, for example, how televisuality impli-cates and challenges three areas of intellectual culture: historical analysis, aesthetics, and the reproduction and distribution of knowlege (pedagogy, forms of scholarship, and the relationship between media theory and production). Televisual culture deserves to be hailed differently.

Historiography: Style and/as History

First, television and mass communications studies have made *caricaturing texts or (historical) contexts* a favored intellectual and institutional ploy, an exclusion-ary ritual used to legitimize one sort of media research or another. As the previ-ous pages have perhaps indicated, this book treads uneasily on the contentious ground that lies between opposing camps in the text versus context debate. The field of television study seems hopelessly split between those who write off criti-cal study as merely qualitative (as impressionistic, subjectivist, and undisciplined) and those who write off social scientific study as naive and intellectually duplici-tous (as scientist posturing that imposes outmoded and innappropriate frame-works on aesthetic and human phenomenon).[61] Critics of social scientism are correct in pointing to the inability of the field to account for enunciative subtleties, narrative processes, and symbolic dimensions of program texts.[62] Social scientists tend to ignore the fact that even methodologically pure data are precoded and overdetermined inventions of researchers, and that all forms of data—surveys, statistical findings, and experimental results—can only be accessed and understood as texts. Audience questionnaires are no less symbolic and artificial than the programs they refer to. They can only acknowlege data that has first been delimited and allowed as input. Statistical information, then, is no less slippery than dramatic or televisual form; both demand interpretive operations that stand apart from the artifice/data in question. A little more self-consciousness and honesty about the centrality and inevitability of this hermeneutic burden, would help alleviate the worst caricaturing in television's pitched academic battle. *All contexts are also texts.*

On the other hand, the kind of textual analysis favored in critical theory *is* frequently speculative. Textual studies have been accused of erecting dramatic ideological conclusions from unrepresentative samples, of mining small and inconsequential texts in order to make extensive generalizations. Yet such criticisms betray an inherent cynicism about aesthetic substance and depth. If textual analysis reifies small texts at the expense of cultural applicability, then content analysis reifies numeric breadth at the expense of phenomenological significance. There is, in a philosophical sense, no more substance to one than the other. Critical theorists and communications researchers alike tend to find what their assumptions predetermine and allow them to find. The inability to extrapolate from a limited textual base is no less reductive than the inability to interpolate from wide—but semiotically shallow and delimited—data. The depth and significance of a television program, then, is one of degree and choice, although theorists seldom admit to this intellectual prefiguration. *All texts are also contexts.*

This book has been less an attempt to dance between these two theoretical commitments, than to examine a programming trend through several complementary perspectives not typically utilized within the same study. Six emblematic shows and subgenres have, in addition, been textually examined in-depth, since no historical survey by itself can suggest the richness, the subtlety, and the semiotic complexity provided by patient and close analysis. The systematic, inductive, and formal analysis of presentations should be an obligatory prerequisite for any kind of television study, for the need for in-depth analysis is not limited to interpretive tasks only. Such spadework also provides essential descriptive tools needed for any credible analysis, quantitative, historical, or otherwise. A repeated look back at the actual forms and practices of television by researchers would do much to keep academic accounts of the medium honest, to allow them to continually test propositions and conclusions against actual practice.

Even though this book has emphasized critical and cultural analysis, I hope that it also, in some small way, has suggested that television history might be approached differently. Early histories of television, like Erik Barnouw's, are deemed by critics to be detailed, impressionistic, and aesthetic.[63] The best recent histories, like William Boddy's *Fifties Television*, are both less anecdotal and less teleological and muster a comprehensive and exhaustive examination of primary industrial and regulatory documents. Ironically, Boddy's sophisticated analysis of aesthetic trends in the television industry depends entirely on written (rather than visual) and verbal (rather than aesthetic) sources for documentation. John E. O'Connor's *American History, American Television*, an anthology that self-consciously established television historiography as a discipline, also shows an overwhelming bias toward verbal and written evidence. Many of the good case studies about historical programs within the volume—on Murrow, on *The Selling of The Pentagon*, on *You Are There*, on the Watergate hearings, on *Roots*—seem uninterested in, if not downright oblivious to, the actual visual evidence available to researchers in those shows. Perhaps because most of these television scholars come from the tradition of American history, their chief strategies aim to re-create historical backstory through comprehensive contextual evidence. Historical study has traditionally used this deference to, and dependence upon, factual substantiation and backgrounding as a mark of credibility and accuracy.

What is left out in this approach, however, is the most obvious and most important historical document, the television program itself. Once again, context (hidden but true) rules over text (visible but illusory). Whereas such work originally fulfilled an institutional function by validating and defending television as a legitimate academic subject for historical study, the approach also sold historical analysis short. A methodological fixation on contextual substantiation not only ignores the fairly basic theoretical complications of ontology and narrative, but the attitude also cuts off consideration of the medium's more immediate material presence in culture. *The image is also an historical artifact. Television historians seldom treat it as such.*

The best defense, then, against those who would caricature textual analysis as a qualitative and speculative method directed at soft data is historiographic rather than aesthetic. From a methodological perspective the very television program itself—its visual and aural presence—has been written out of history. The past is not a singular thing to be referred to, but is rather a construction of disparate and competing discourses. Those who write off aesthetics as an inconsequential or tangential concern—a secondary reality, something that needs contextual propping up—blind themselves to one of the most concrete examples of television's historicity: its visual evidence. Grasping the significance of such evidence is a historical rather than merely stylistic imperative. Serious stylistic analysis helps provide history with more accurate terms and effective frameworks. As a critical work, this book was not intended to revise history, but to intervene into one of television history and theory's recurrent and favored blind spots: the televisual image. Style is very much historical.

Resuscitating Aesthetics (as Culture)

Despite the broadly negative political assessments of postmodern theory and the sometimes overwhelming scale of global communications, the new television is not without points of access. There is reason for optimism. I have used the term "televisuality" to describe an important historical moment in television's presentational manner, one defined by excessive stylization and visual exhibitionism. In no way does this terminology mean a kind of Kantian aesthetic essence or spirit of the age principle. Rather, televisuality has become an active and changing form of cultural representation, a mode of operating and a ritual of display that utilizes many different individual looks. Televisuality, as I have been using the term, is less a defining aesthetic than a kind of corporate behavior and succession of guises.

Although stylistic analysis may seem out of fashion in critical theory, it is important to recognize that televisual form is also clearly a cultural operation with marked political consequences. Can there be any politics without style? This work has sought to bring together two media areas that are typically seen as contradictory in the study of mass culture—the aesthetics and the politics of style. Resuscitating the aesthetic may seem curious given the fact that most European-derived critical theory has postured the aesthetic as a regressive class-based imposition that disallows social and psychological considerations. But poststructuralism typically bashes the concept only as a strawman. What theorists usually write off is an historically limited, post-Enlightenment concept that originates in German thought, both as a category and product of high culture. I take the term in its broader and more original sense as a reference to the sensate and material world. The visual image provides one of the most accessible and pervasive phenomenon for analysis. The workings of the televisual image are far from innocuous or inert. They help make up the very heart of television's engagement with viewers. Even the aesthetic economy described in the second section of this book thoroughly implicates social, political, and audience issues.

Surely in the current climate of multiculturalism we no longer need to drag out the concept of high art as a foil for our scholarly breakthroughs. All cultures—high and low, mainstream and marginal—are involved in systematic processes of aesthetic and visual representation. Television, interestingly, is fast becoming a central player in these cultural representations and rituals. Television aesthetics, then, is also very much an anthropological and cultural problem. This is especially true in urban areas like Los Angeles where immigrant, tribal, and exilic cultures continually undergo flux and hybridization.[64] To ignore the centrality of stylization in current American television then, is to ignore an important arena in which power is represented and cultural identities are formed. Avoidance, that is, ignores even the way that emerging social groups and subcultures survive.

By borrowing, fabricating, and hybridizing cultural icons and mannerisms as their own, many groups forge new, hybrid identities even as they negotiate what was once reductively thought of as *the* American mass culture. Televisual aesthetics, then, also implicates and indicts traditionally privileged communications studies notions of the mainstream and cultural homogenization. The aesthetic is both political and individualist; style becomes lifestyle in an endless cycle. Televisual mixing and matching is not unique to street-smart urban high-schoolers accustomed to district communiques in ten languages either. Culture organs show that hyperactive aesthetic-cultural mixage is as much a province of the affluent crowd out in Malibu and up on Mulholland when they announce, "Ethnic diversity isn't just a fact of life, it's a fashion theme for spring. These days, Eurocentric clothing pales in comparison to the lively wardrobes of the global village, cross-cultural looks–punk–Hare Krishna, mixed saris, Chinese calligraphy–print leggings, combat boots, and nose rings."[65] Ignore aesthetics? Disengage. Television—with its full-bore delivery of hip-hop, gangsta, grunge, and couture to young and wannabe-young style-hungry multiethnic bricoleurs—has become a crowded and central arena in the cultural process of symbolic survival.

Desegregating Theory and Practice

Signature television, industrial semiotics, and the ancillary afterlife are all cultural symptoms suggesting that media studies would do well to question business as usual in the academy. If scholars are in fact interested in intervening in mass culture, they need to rethink not just the ideological dangers of the media spectacle, but the self-imposed limitations of their own critical vantage points. In a major collection of work codifying cultural studies as a field, one scholar voiced an important fear, "that cultural studies will be just another listing in the college catalog under the letter 'C,' . . . it's going to be denied its political meaning. We need discussions about how we can intervene in the institutions in which we work."[66] If intervention is in fact cultural studies' raison d'étre, then the way that the academy reproduces media knowledge must also be overhauled. Floating interdisciplinary cultural scholarship across the academy does deny dominant culture the power to control the categories that frame research. Yet the tactic also tends to dehistorify television study, detach intellect from industry, and disembody progressive theory from viable forms of progressive practice.

One of the institutional strategies that most debilitated media study in this regard was the historic separation of doctoral media programs into specialized ghettos that had no contact with production programs. Although film studies as we know it would never have survived in the United States without its nurture in the womb of English and speech communications departments in the 1950s and 1960s, its emergent legitimacy as an autonomous academic field in the 1970s and 1980s left behind some moorings that would prove essential in the future. The idea of integrating theory and production in scholarly research may horrify some—especially university research professors bruised by the dumbing down process that typifies many basic television production courses. An alternative, however, lies not in perpetuating the same outmoded forms of production pedagogy, but in clearing away a beachhead for cross-over forms of production that have helped energize both theory and practice.

Some of the most important theoretical work in the last two decades has, arguably, been production work. Consider the influence and precedent of those that have aggressively plied the no-man's land between theory and practice: Peter Wollen and Laura Mulvey, Jean Rouch and Sol Worth, Chris Marker, Jean-Luc Godard, Martha Rosler, Laura Kipnis, Trinh T. Minh-ha, Deep-Dish and the Paper-Tiger Television collective. Even high-theory practitioners have converted: Thierry Kunzel from semiology to video art; Jean-Pierre Gorin from Marxist interrogation to documentary. From its activist niche, *POV* programs weekly doses of racial theory and sexual discourse from the margins, even as Marlon Riggs and Errol Morris made deconstruction a bankable docufiction genre. The Quixotesque white-trash exploits of *TV Nation* play nationally, while producer Michael Moore—a combination Gramsci and Studs Terkel for the *Beavis and Butt-head* generation—delivers left and labor critiques of capitalism for network NBC. While earlier auteurs like Paul Schrader and Peter Bogdanovich had shown that theory could be pulled into the American motion-picture mainstream, Peter Greenaway, Derek Jarman, and other theory-driven artists at Channel 4 showed the extent to which the televisual spectacle could be acutely stylized for English

living rooms. As this study has attempted to show, however, these signature figures are merely icing on a much bigger industrial cake. Many others have imported theory to television in utter anonymity: commercial directors who flaunt
semiotics and new wave; equipment manufacturers who hard wire art history;
lighting designers, art directors, and gaffers who parade designer couture for the
knowing viewer.

Some who integrate theory and production represent the moving image as
theory, while others make aesthetic theory a requisite or ancillary part of even
mainstream television viewing. In either case, the critical object facing television
study has changed. Of course the kind of desegregation that I am proposing challenges the very forms of scholarship that the academy is willing to sanction. Writing theory about the televisual is one thing. Visualizing theory about the televisual
is another. Manipulating images and sounds in the intellectual arena demands a
different kind of technical support for scholars: electronic postproduction, not just
word processors; image digitizers not just photocopiers; image banks and fiber
optic webs not just on-line catalogs. The same institution that protected and blessed
the VCR over a decade earlier, pornography, has helped insure the financial success and survival of the CD-Rom, an affordable visual platform for computer
interactivity. This kind of visual interactivity allows scholars new ways to illustrate their work, to do systematic research that can be repeated and verified by
scholars elsewhere, and to teach differently in the classroom. One of the reasons
visual research has essentially been absent in empirical and experimental mass
communications research is that scientific visual parameters have been impossible to maintain. Rigidly controlling constants and variables will be out of the
question as long as visual elements cannot be cleanly identified and isolated on a
frame-accurate and repeatable basis. Until visual artifacts from television can be
uniformly disseminated, the ability to generalize about the results of visual research and to apply those results in other settings is also severely shortchanged.

Several forces, however, clearly complicate changes in the way that media
knowledge is reproduced, and so resist desegregation. There are, for example, important legal ramifications. A great deal of current debate surrounds the long-
term status of the image as property. Television archives that once freely gave
access to scholars are now being strong-armed by network donors. Distraught when
Vanderbilt University publicized the availability of their television collection on
the internet in 1993, CBS played the heavy and threatened to cut back or end
network donations of footage in the future. The very same ancillary afterlife that
was fueling entertainment exhibitionism on primetime, then, was also transforming educational and informational news footage into property. Scholarship and
free access continue to be shunted in the commercial twilight of network television. As recession-era public universities shift in desperation to a commercial
model of support—with students as customers and clients and the private sector
as padrone—the notion of fair use will further deteriorate. In the shadow of the
500-channel environment, owners or producers of any image now regularly reserve the right to cash-in on that footage at some future date—in the next decade, in the next century. Some educational experts now see distance learning as
the answer both to declining support and to corporate competition that is entering higher education with a turn-key vengeance. In addition to breaking labor,

distance learning also promises to transform scholarship into a commodity—packaged property bought and sold over the very same cable or fiber-optic lines that wire the 500-channel environment.

The relationship, then, between theory and practice, academy and industry is not just a conceptual one. Both poles are caught up in a much broader transformation. The many other and sometimes benign practices that resist desegretating theory and practice (for example, ingrained fraternal orders of refereed scholarship, conference venues, and print-defined institutions of scholarly publication) will pale in the face of higher-education's corporate privatization. Unless proactive and aggressive initiatives are undertaken to protect and extend the federal doctrine of fair use for research, scholarship, and criticism, the snowballing logic of the new technologies and its owners will simply steamroll the faint pleas of artists and scholars for access. Self-inflicted intellectual iconoclasms, then, will be meaningless: televisual culture will vanquish scholars from the marketplace with disinterest, not vice versa. If this happens, media scholars may be left with perhaps the only truly revolutionary aesthetic of recent years—the scratch and digitizing mode of rap; an art of theft and commentary, based on the refabrication of existing properties. The FCC-mandated public-service definition of broadcasters as utilities in the early years of television—even if it was only lip service—will seem Edenic by comparison. A glorious, privatized marketplace of images awaits us on the electronic superhighway, one that allows no distinction between education and entertainment, critic and consumer.

In January 1994, the captains of electronic media converged on West Los Angeles, obsessed with the infrastructure that would predestine television's 500-channel future, and staked claim to the very analytical terms by which this future could be engaged. Even as the audience at home watched, C-Span showed audience members mulling over summit truisms and then calling in their stock orders by cellular phone. This was not, then, just a conversational summit. This was both a trial run for speculators and an exercise in virtual business. One had to be only mildly conscious to pick up the summit's recurrent themes: unabashed technological determinism, interactivity, and the ability of viewers to buy underwear while dressed only in underwear. Time and again, clashing moguls took the altruistic high ground by boasting about the emancipatory potential of interconnected communities. Yet one of the few common technical models that survived the partisan fusillades between warring coaxial, DBS, and fiber-optic zealots, was the predicted need for a bottomless software pit, for a video server–storehouse–video stream, for a repository that would package experience and make it available for video on-demand. Building the superhighway's infrastructure would be costly, we were warned. This risk was certainly less troubling, however, than bland acknowledgments by summiteer and CEO Notebaert that roadkill would be an inevitable part of the highway's on-line service to America. By admitting that carnage would be a part of the new high-speed communications environment, Notebaert was merely extending the slaughterhouse theme that the FCC chairman had earlier introduced when he christened summit deliberations with the symbolic blood of nineteenth-century Chicago meatpackers.

The carcass figure proved to be an apt, if unintended, metaphor for video in the post-television age. While industry apologists threw out the comforting prospect of live networked interactivity as candy to influential adoptors and potential critics, the summit simultaneously exposed gaps in the snowballing illusion of superhighway consensus. Brought together by the academy as part of a fabricated national agenda to converse about our new shared communal and networked future, several moguls betrayed a proprietary relationship with democracy through loaded slips of the tongue: we plan to do this with *my* network, or *my* cable–fiber-optic–direct broadcasting satellite service. The infrastructural players on the highway (the Malones and Dillers, the TCIs, and QVCs) are already in place, as are the softwarehousers of canned experience (the film and television industries, the Igers and Katzenbergs, the ABCs and Disneys), without which infrastructure is just so much costly real estate. The genius of the televised summit was that it allowed controlling interests, typically relegated invisibly offscreen to the board rooms of industry, to act as the highway's live and on-screen historiographers and aestheticians—and to do so long before America was either wired or on-line. Well in advance of the highway's building, then, its history was being written and prefigured. Its experiential carcasses, entertainment programs, were being drawn and quartered, not for the agora of the internet, but for cold storage in the video-server's packing houses. Liveness, immediacy, and connectivity have always been the most oversold parts of television. The summit was no exception. Stylization, on the other hand, has become the industry's preferred method of canning. Like Gustavus Swift's slaughterhouse-borne, cartel-busting carcasses, style packages experience for the cavernous vaults of tommorrow's digital video-server.

The televisual universe near millenia's end remains fairly callous to the speculations of high theory: its iconoclasms and scopophobia, its closed-loop deconstructions, and its totalizing cults of information and language. Even terminally hip critics who cashed in on 1980s visual culture, now act like angry iconoclastic prophets shouting in the wilderness, now rip persons who "learn with and communicate by pictures as barbarians."[67] Contrast these rote calls for separatism—voiced by those who have come to see the pure light of the word and the god of informational truth—to one major studio CEO's grand design for the 500-channel environment: "The hardware may be dazzling, but *it won't make any difference if our intellectual product isn't right.* If you are to be truly successful you'll be marketing, promoting, and publicizing programs with *inherent quality*" (italics mine).[68]

With or without the academy, then, an industry elite and its MBAs have placed wagers on two of the academy's most cherished distinctions—the intellect and the aesthetic—as basic corporate organizing principles. Neither ignorance nor naiveté is the issue. The televisual marketplace—the world of Eisner, Katzenberg, Tartikoff, Bochco, Brand and Falsey—now claims legitimacy as a site for both intellectual and aesthetic commerce. The self-imposed exile of media theory, however, need not be total. Desegregationist theorists like Jarman, Kunzel, Wollen, Rouch, Rosler, Trinh T. Minh-ha, and Michael Moore prove well that it need not be. Reincarnations of much older accusations, however, continue to mortar the wall around an intellectual culture that looks more and more ascetic and monastic—more like the alienated Gnostic purges of Theophilus than the electronic world of a 500-channel citizenry.

Notes

Whenever television programs or films are cited in the notes that follow with a "PVA-" designation followed by a two- to four-digit number, these titles can be viewed when reserved in advance at the UCLA Film and Television Archive, UCLA, 405 Hilgard Avenue, Los Angeles, CA 90028. Many of the other shows cited are current or continue in wide syndication in broadcast or cable.

Preface

1. *Northern Exposure,* "War and Peace," #77406, written by Robin Green and Henry Bromell, directed by Bill D'Elia, original airdate May 13, 1991. The synopsis was included as backstory for screenwriters of future episodes, in the production company's *Bible,* March 30, 1992.

2. I am of course borrowing Fredric Jameson's narrative term and applying it to the televisual. *The Political Unconscious: Narrative as a Socially Symbolic Act* (Ithaca: Cornell University Press, 1981).

3. I am referring here primarily to the primetime program production industry, not to local broadcast news operations. In the 1960s and 1970s industrial production was dominated by nontheatrical and non-network 16mm film production companies, who did at times sense the profit potential that came from cheaply aping cutting-edge looks. By the late 1970s and early 1980s, non-union industrial production began to be dominated by video—a technology that proliferated along with aesthetic pretense.

4. Hans Magnus Enzensberger, "Towards a Constituent Theory of the Media," *The Consciousness Industry* (New York: Seabury Press, 1974).

5. I consider this less a process of debilitating cooptation than a corporate exercise in radical self-annunciation. The reasons for the shift are complex and multiple: some cloned, some aped, and some parodied the avant-garde. While punk was being dulled and merchandized as new wave for the suburban mainstream, MTV was commissioning New York art world video and performance artists to produce regular spots for their programming. The discourse of many production people, of course, evoked a technological determinism and basis for experimentation, a subject discussed in chapter 3. For a fuller discussion of broader historical and ideological changes that emerged in culture at large—that is, *external factors* that helped create the conceptual appetite for change—see chapters 2 and 9.

6. There are precedents for the mass marketing of the avant-garde, as when directors Vincent Minnelli caricatured Van Gogh and Alfred Hitchcock celebrated Salvador Dali in Hollywood, or when Madison Avenue postured as Pop, Op, and Swiss graphic minimalism in the 1980s. Yet these earlier homages could not have been farther from the antagonistic spirit of their avant-garde subjects. Hollywood's Dali, like the artist's thousands of mass-produced "original" prints in shopping malls, fits well the easy world of the interior designer and the middle class, but contradicted the original, angry, and irrational aesthetic violence of surrealist leaders like André Breton. Pop was also an easy heist for advertisers in the 1960s, who completely missed the disruptive bite of Warhol's blank aesthetic.

7. In the heady air of deconstruction, scholars might counter that the deconstructive "making-ofs" on *Entertainment Tonight* merely reconstitute culture's dominant logics and institutions. Clearly, however, the intellectual personas that literary deconstruction institutionalize are no less regressive as ideological and class-based fabrications. The difference is merely one of focus. Deconstruction reifies intellectual machinery and academic surplus while television reifies production equipment and technological process. Both forms serve stardom.

8. Typical of the flight away from radical form (now that it is "owned" by primetime), avant-garde producers frequently reacted by rationalizing and justifying a more minimal and restrained style for their projects. In discussions following a public screening of his feature-length, large-screen, experimental video documentary entitled *Cage/Cunningham*

(on modernist patriarchs John Cage and Merce Cunningham), director Elliot Caplan noted defensively that if "I had known AT&T was going to later utilize the same kind of intercutting between black-and-white footage and color in their national television spots, I never would have used it in my work." The greatest insult to an avant-gardist, apparently—and the basis for a retreat back to a focus on content and formal restraint—is the recognition that radical style also has basic commercial value. From public comments by Caplan, Long Beach, California, November 22, 1993.

9. Although the avant-garde has *never* dominated mass culture, this period is important because mass culture at least sensed the value in masquerading as radical and oppositional.

10. How activist video did this in the 1980s and 1990s deserves an in-depth historical study in its own right.

11. "The Dream Sequence Always Rings Twice," *Moonlighting,* ABC Circle Films, October 15, 1985.

12. *Northern Exposure,* CBS, November 8, 1993.

Acknowledgments

1. "Non-Verbal Semiotics in Film and Video," International Summer Institute of Semiotic and Structural Studies (ISISSS), 1986, and *Televisuality: The Performance of Visual Style in American Television* (Evanston: Northwestern University, 1991).

Chapter 1. Excessive Style

1. Quoted in Laurence Jarvik and Nancy Strickland, "Cinema Very TV," *California* (July 1989), 198.

2. Quoted in Christina Bunish, "The Search for Realism: Directors David Steinberg, Ron Dexter and Bob Eggers Face the Challenges of Capturing Reality," *Film and Video* (September 1990), 66.

3. Jeff Kaye, "Sex, Mud and Rock and Roll," *Los Angeles Times*, November 9, 1989, F1.

4. Kevin Cosgrove, "Regis' Recipe for a Healthy Life," *TV Guide,* March 6, 1993, 8–11.

5. I choose the term "mass-market" television rather than the more traditional concept of "network" television, since broadcasting had clearly been overhauled and pluralized by 1990. "Mass-market" expresses a kind of programming and economic scale that is not limited to a few privileged broadcast corporations, but rather encompasses other institutions that work over national media markets with high-production value programming. This term, then, would include Fox (the fourth network), superstations, large-scale syndication companies, CNN, MTV, and other media corporations that produce and program on a national level.

6. Retheorization refers to how changes in practice and production discourse evidence shifts in working assumptions and orienting perspectives, not to the intentional and conscious formulation of theoretical premises and principles as ends in themselves.

7. Performative aspects of media have traditionally been associated with dramatic and theatrical elements, whereas style is typically postured as a static and fixed formal property owned by works of art. Here, with the concept "performance of style," I hope to indicate a shift away from an assumption of style as static property toward style as a hyperactive presentational process.

8. "Individuation" is a popular psychological term referring to a person's development of distinct ego boundaries and distinguishable personal behaviors. Stylistic individuation could, however, be as easily described in economic terms as a kind of product differentiation. Looking at the overt discourse of practitioners in the media industry suggests, however, the importance and degree of investment given to construing creative personas behind program looks and production accomplishments. The psychologizing of the media discourse

by practitioners should not be taken as a refutation of broader economic interests, however, for the two strata are probably intricately tied to each other.

9. Still one of the most engaging and detailed accounts of the golden age of live anthology drama in the 1950s is Eric Barnouw's *Tube of Plenty: The Evolution of American Television* (New York: Oxford University Press, 1975, 1992), a book that isolates skilled and serious artists at the center of broadcasting's live and dramatic showcases during that period.

10. The extent to which this type of production practice challenges privileged academic theories—like postmodernism and cultural studies—is discussed more fully at the end of this introduction, and in the chapters that follow.

11. John Dempsey, "More Mags Will Fly in the Fall: Too Much of a Good Thing?" *Variety*, April 12, 1989, reprinted in Marilyn Matelski and David Thomas, *Variety: Broadcast-Video Sourcebook I* (Boston: Focal Press, 1990), 27. *Entertainment Tonight* is produced by Paramount Television; *A Current Affair* is a production of *20th Century Fox Television*; and *Inside Edition* is produced and syndicated by King World.

12. Jim Van Messel, executive producer, *Entertainment Tonight* as quoted in Mike Freeman, "*Entertainment Tonight* Turns 3,000," *Broadcasting and Cable*, May 8, 1993, 30.

13. Said the critics: "What they've done is close to a miracle. They were dead and found the fountain of youth." TV Critic Tom Jicha, quoted in Harry A. Jessell, "New Wave Newscasts Anchor WSVN Makeover: Ex-Affiliate Finds a New Niche," *Broadcasting*, October 12, 1992, 24.

14. Joel Cheatwood, vice-president of news, and Bob Leider, executive vice-president, WSVN-TV, quoted in ibid.

15. Bruce Sandzimier, vice-president of editorial, Universal Television. Katherine Stalter, "Working in the New Post Environment," *Film and Video* (April 1993), 100.

16. "Help Wanted, Program Production and Others," *Broadcasting*, November 30, 1992, 57. In mundane station management activities like hiring, WRAL-TV5 in North Carolina, in personnel upgrades, now expected its graphic designers to have visual arts design degrees and to be proficient in Paintbox, animation, and still-store technologies. Even verbal- and text-oriented broadcast positions, like that of the news promotion producer at WRAL, were advertised and keyed to essential visual communications skills.

17. David Poltrack, CBS's senior vice-president of planning and research. Richard Zoglin, "The Big Boy's Blues," *Time*, October 17, 1988, 59.

18. This diversification trend mirrors, of course, developments over the last century in advertising.

19. These examples are described in Jeffrey Wells, "Is It the Reel Thing: Big Name Directors Try to Bring Film Magic to Coke Ads," *Los Angeles Times*, February 17, 1993. F1, F6.

20. Ibid., F7.

21. See discussion in chapter 10 of this book, "Televisual Economy," of the changes in business structure that CAA's appearance as a major agency player in 1993 caused.

22. *Variety*, September 30, 1987. *Variety*, September 13, 1989, reprinted in Matelski and Thomas, 73.

23. By 1994 the network's growth was predicted "to cost ABC at least $75 million per year in lost revenue." Tom Wolzien, as quoted in Geoffrey Foisie, "Fox Hounds ABC-TV, *Broadcasting and Cable*, June 14, 1993, 65.

24. "The real killer" of *Wonder Years* "was economics not prudishness." Coupled with escalating cast salaries, "the budget soared to $1.2 million per half-hour episode. Many hour-long dramas are shot for less." Steve Weinstein, "Reeling in the Bittersweet 'Wonder Years'," *Los Angeles Times*, May 12, 1993, F1, F6.

25. "Zapped: The Networks Under Attack," *Time*, October 17, 1988, 56–61.

26. *Broadcasting*, July 4, 1989. Cited also in J. Fred MacDonald, *One Nation Under Television* (New York: Pantheon, 1990), 253.

27. Two recent books underscore the recurrent academic view that the sitcom works to reinforce status quo values. Ella Taylor, *Primetime Families* (Berkeley: University of California Press, 1989), demonstrates the resilience of the myth of the nuclear family on network television up to its conservative reconstitution in 1980s sitcoms. Darrel Hamamoto,

Nervous Laughter: Television Situation Comedy and Liberal Democratic Ideology (New York: Praeger Publishers, 1990), makes a compelling case that the sitcom—even in its liberal manifestations—has systematically worked to elide racial and ethnic issues and threats to the mainstream, white, status quo.

28. *Full House* (ABC, 1987–1993).

29. Raymond Williams, *Television, Technology, and Cultural Form* (New York: Schocken, 1975).

30. David Marc, *Demographic Vistas: Television in American Culture* (Philadelphia: University of Pennsylvania Press, 1983), 5.

31. These exceptions include, of course, Jane Feuer et al. *MTM: Quality Television* (London: British Film Institute, 1985), and Horace Newcomb and Robert S. Alley, *The Producer's Medium* (New York: Oxford University Press, 1983).

32. The mercantile analogy of the "loss leader" is applied in the analysis of epic forms of televisuality, like the miniseries, in chapter 6.

33. This criticism holds true for fine art up to and including the period of high modernism. Although the intellectual-theoretical crutch that I speak of has always been more important to the art world than to Hollywood, various conceptual, video, and performance artists since 1968 have attempted to break through the art world's ideological props by engaging and critiquing the institution's support systems and economic industry.

34. If one were to view the research of this book within the tradition of film theory, the idea of a media-specific "language" certainly has ample precedent. Here, however—in a project that seeks to explicate the nature of *nonverbal* semiosis in television—the linguistic term creates problems of its own for analysis. Given my focus then, the analysis that follows will target: (1) visual modes of presentation distinctive to television; (2) aesthetic modes that are borrowed and redefined by television; and (3) cultural practices that impinge upon and inform these modes. The presentational modes that I refer to are generally and predominantly nonverbal and differ from orthodox film language. Media language models tend to focus on and privilege editing, narrative structure, and syntax. Televisuality, by contrast, privileges images—with *simultaneous* components typically displayed within a shared frame. In general, the televisual modes that I am theorizing represent a divergence from classical narrative cinema and television.

35. Having summarized my project in this way, a few additional words about terminology and definitions are in order. In television programming's shift toward visuality—both as a formal trait and as program content—stylistic signifiers are regularly stripped from their traditional signifieds and made open to continual redefinition and reuse. Given the centrality of what postmodernism terms disembodied signifiers, however, televisuality can also be profitably seen as an *industrial process* of assigning and bestowing value, not just as a look. But televisuality should be seen as more than just an industrial shift to, and preoccupation with, visual imagery. In a less macroscopic but no less important sense, televisuality also refers to that trait now common in television whereby programs intentionally engage the viewer with *multiple* and *simultaneous* layers of perceptual and discursive information, many times overwhelming him or her by combining visual, spatial, gestural, and iconic signals. Televisuality is, in this sense, a phenomenon of communicative and *semiotic overabundance*. Although isolated examples of the phenomenon existed in earlier periods in television history, and frequented feature film history, television popularized and cashed-in on this semiotic process and display of overabundance in the 1980s. Finally, televisuality implicates more than just industrial and aesthetic issues, and the final chapters will attempt to address the trend's historical and ideological significance.

36. I am utilizing "ideology" in the way that E. Ann Kaplan defines the Althusserian variant of the term. In *Rocking Around the Clock* (New York: Methuen, 1987), 188, she describes ideology as a "series of representations and images, reflecting conceptions of "reality" that any society assumes. Ideology thus no longer refers to beliefs people consciously hold but to myths that a society lives by, as if these myths referred to some natural, unproblematic reality." In this sense image and style-practice can be seen as part of any ideology and cultural mythology.

37. One key to understanding the ideology of style can be found in the contradictory and competing aspects at work within an emerging paradigm or myth. Structural anthropol-

ogy has taken the narrative process (by which contradictions are covered over and resolved) to be a key to a culture's mythology. See Claude Lévi-Strauss, "The Structural Study of Myth," *Structural Anthropology*, trans. Claire Jacobson and Brooke Grundfest Schoepf (New York: Basic Books, 1963), 206–231. Mimi White in her essay "Ideological Analysis and Television," *Channels of Discourse*, ed. Robert Allen (Chapel Hill: University of North Carolina Press, 1987), 134–171, has argued that it is precisely television's textual contradictions that expose the workings of ideology.

38. In describing the emergence of visuality as "uneven, partial, and irregular," I do not aim to lessen its importance as a distinct and identifiable phenomena. Rather, I hope to show its presence and power as an historical phenomenon—in the same way that Nick Browne's analysis of post–May 1968 French film theory demonstrated that "history assumes the aspect of an ensemble of unevenly developed, stratified and shifting relations enacted in a new social setting. Old connections are broken and displaced; new structures and commitments are in the process of emerging. The sense of uneven, fragmented movement of diverse but associated themes makes the ensemble of these texts an unfinished work site." Nick Browne, ed., *Cahiers du Cinéma: 1969–1972, The Politics of Representation* (Cambridge: Harvard University Press, 1990), 1. I will argue that the same scale of contentious shifting has occurred in recent television, but for very different reasons.

39. I hope when analyzing media to synthesize Kuhn's influential concept of epochal paradigm shifts, with the less cognitive perspectives of cultural-ideology studies and social mythology. Thomas Kuhn, *The Structure of Scientific Revolutions* (Chicago: University of Chicago Press, 1962).

40. The "writers of *thirtysomething*," *thirtysomething stories* (New York: Pocket Books, 1992).

41. Newcomb and Alley, *The Producer's Medium*.

42. Jack Kuney, *Television Directors on Directing* (New York: Praeger Publishers, 1990).

43. And surely with the proliferation of infomercials and talk shows, more of these low-resolution forms of TV are on the way.

44. John Fiske states that "the term 'culture,' used in the phrase 'culture studies,' is neither aesthetic and humanist in emphasis, but political. . . . Culture is not, then, the aesthetic product . . . but rather a way of living in an industrial society." "British Cultural Studies and Television," in *Channels of Discourse*, 254. A good collection of more recent cultural studies works is Tony Bennett, Susan Boyd-Bowman, Colin Mercer, and Janet Woollacott, eds., *Popular Television and Film: A Reader* (London: British Film Institute, 1981).

45. A number of these shows will be discussed in more detail in chapter 2. Pastiche and mocking parody were a requisite part of *Your Show of Shows* ("Ten From *Your Show of Shows*," n.d., (PVA-1906t). The textual fold-in of the military took place in *I Love Lucy*, "Lucy Gets Drafted," December 24, 1951 (PVA-81t), and the *Loretta Young Show*, "Dateline Korea," March 13, 1955, (PVA-8890t).

46. In current media theory, poetics has become a problematic concept. David Bordwell, in *Making Meaning* (Cambridge: Harvard University Press, 1989), 263ff., makes a case for poetics as a substitute for the excesses and shortcomings of most current interpretive-based film study. While the notion of poetics is caught up in this current debate, it is worth noting that its earlier usage by the Russian formalists assumed that the framework included a cultural and political dimension. More current theorists like Michael Renov, in a keynote address at the Thirteenth Annual Ohio University Film Conference on Documentary, called for a new project to develop a systematic "poetics of documentary" (Athens, Ohio, November 1990). Certainly the focus of poetics on stylistic formation, materiality, perceptability, and function are worth addressing and applying to contemporary television practice. Such a strategy is not incompatible with ideological analysis.

47. Colin MacCabe, ed., *High Theory/Low Culture: Analyzing Popular Television and Film* (New York: St. Martin's Press, 1986).

48. Given this semiotic density and the abundance of channels involved in perception, it is likely that if one has not accurately defined the stylistically complicated object of investigation, then one is not even accounting for or controlling the constants and variables involved in the process. In short, a lot of close textual and aesthetic work needs to be done even before the broadcasting scholar can do good science. Academic broadcasting studies

have tended to overlook the extreme complexity of the actual viewing situation and of the televisual text itself, in lieu of a dominant concern with master interpretive allegories, verbal content, or social meanings and effects. Much work remains to be done in terms of accurately describing the texts and intertexts that present and perform such contents and effects.

49. The history of film theory clearly demonstrates the limitations of formal taxonomies, which tend to reify description and overvalue aesthetic norms. It is worth noting in this regard that perhaps the two most influential media taxonomists, Christian Metz and Sergei Eisenstein, at least eventually changed course by correctly seeing and describing taxonomies within and as a part of an ideological and psychological dynamic. Eisenstein's writings clearly demonstrated a tension between syntactical taxonomies and constructivist inclinations on the one hand (tendencies he shared with other montage theorists like Lev Kuleshov and V. I. Pudovkin), and his overarching sense or obligation, on the other hand, to account for film's formal taxonomies as social and political weapons. *Film Form: Essays in Film Theory,* ed. and trans. by Jay Leyda (New York: Harcourt Brace, 1949) and *Film Sense,* ed. and trans. by Jay Leyda (New York: Harcourt Brace, 1942). Christian Metz first emerged in the 1960s as a theorist whose work promised scientific and endless noninterpretive cross-sections of film form and film structure in works like *Film Language,* trans. by Michael Taylor (New York: Oxford University Press, 1974). By 1976, Metz's *The Imaginary Signifier* (Bloomington: Indiana University Press, 1976) seemed to turn its back on the dead syntactical categorization of his earlier work by leaping to complicated conjectures about the spectator's social and psychological self. The scope of analysis is limited, once this impulse in theory has categorized, or promised to categorize, all of its formal options and taxonomies.

50. "It is TV that looks at the world; the TV viewer glances across TV as it looks. This delegation of the look to TV and *consequent loss of intensity in the viewer's own activity of viewing* has several consequences." John Ellis, *Visible Fictions: Cinema, Television, Video* (London: Routlege and Kegan Paul, 1982), 164.

51. Richard Adler continues: "The inevitable commercial interruptions virtually preclude prolonged absorption." "Introduction: A Context for Criticism," *Television as a Cultural Force,* ed. Richard Adler (New York: Praeger, 1976), 6.

52. Both Ellis and Williams sought to describe the fundamental components of the medium and experience of television. But glance theory also partakes of and elaborates an earlier academic schema, that is, the globalizing polar dichotomies that Marshall McLuhan described when comparing television to film. Raymond Williams, *Television, Technology and Cultural Form* (New York: Schocken, 1975).

53. "The mode of the TV image has nothing in common with film or photo, except that it offers also a nonverbal gestalt or posture of forms. With TV the viewer is the screen. . . . The TV image is visually low in data. . . . The film image offers many more millions of data per second, and the viewer does not have to make the same drastic reduction of items to form his impression. He tends instead to accept the full image as a package deal." Marshall McLuhan, *Understanding Media: The Extensions of Man* (New York: McGraw-Hill, 1964), 272.

54. McLuhan comments that "the viewer of the TV image, with technical control over the image, unconsciously reconfigures the dots into an abstract work of art, on the pattern of Seurat or Rouault. . . . The TV image is now a mosaic mesh of white and dark spots." *Understanding Media,* 273.

55. Ellis, in fact, makes the incredible generalization that: "The broadcast TV viewer is not engaged by TV representation to any great degree. broadcast TV has not so far produced a group of telephiles to match the cinephiles who have seen everything and know the least inconsequential detail about the most obscure actor and directors. Broadcast TV does not habitually offer any great incentives to start watching TV." (*Visible Fictions,* 162.) Williams argued that since television does not exist in discrete and isolated programs, but rather is constantly interrupted and linked to other programming in the evening, analysts should study the expanded programming sequence rather than individual units. Because of this basic understanding, later theorists would conclude that such a fragmented and cluttered aesthetic object would not logically entice viewers to the kind of intense engagement that cinephiles experience at the cinema.

56. In addition to the substantial popular publishing industry that focuses on television, television programming itself encourages style and detail consciousness on the part of viewers. Shows like *Entertainment Tonight* and *Arsenio*, and many other clones on cable, constantly spotlight the content and style of television.

57. In all fairness to Ellis, and others, glance theory may have accounted more credibly for television in the 1970s or early 1980s. Regardless of its origins, however, glance theory is made suspect by current television practice. The primetime television industry especially fashions programs and nonprogram materials with increasing style- and fashion-consciousness. Even when Ellis was writing, *MTV* and *Miami Vice* and miniseries like *Shogun* had established highly *visual* arenas for narrative, music, and drama. In the decade that followed a growing concern with stylishness evolved out of these forms. Some shows mimicked *MTV* and *Vice*. Others made their own claims for unique visual style (*Crime Story, Max Headroom, Hill Street Blues, L.A. Law*). By 1989, shows like CBS's *Beauty and the Beast* were mise-en-scène–strong and narrative-weak. Writers for such shows were faced with script assignments requiring many viewer pages and having long nonverbal scenes displaying auspicious and expressionist lighting effects. Directional lighting, colored gels, smoke, and synth music permeated primetime programming. The artistic reference in look was closer to Rembrandt than it was to the "ideology of inattentiveness" that glance theorists promoted. *TV Guide* promoted the special nature of such shows: these were not shows to be glanced at. Television by 1990 was in many cases self-consciously hip and excessively styled. There was no longer a zero-degree formal syntax and style at work here. The idea of a neutral and colorless writing, in short, a de-aestheticized style in the work of Alain Robbe-Grillet was promoted by Roland Barthes in *Writing Degree Zero* (New York: Hill and Wang, 1953), xvi, 76. The zero-degree style of TV in the 1950s and early 1960s was a dominant rather than radical or modernist tendency. Glance theory simply missed the opportunity to elucidate and explain the newer and important televisual forms—a refusal that helped reinforce television theory's denigration of the image.

58. In a critique of Ellis's book, "Television at a Glance," by Brian Winston (*Quarterly Review of Film Studies* 9, no. 3 [Summer 1984], 256–261), the British theoretical tradition from which Ellis's work comes is interrogated and rejected. Winston attacks almost all of Ellis's central assumptions: his ideas about television and enigma, the medium's essential temporal regularity, the importance of the nuclear family, the emphasis on the glance as a defining factor, and his underlying attitude and valuation of popular culture in general. Strangely enough, few have followed through on Winston's suggestions. Among my goals here are: (1) to show that glance theory and the related mythology of liveness predated the work of Ellis in American media studies by many years; (2) to demonstrate that glance theory continues to be widespread and popular in more contemporary critical work and cultural studies; and finally, (3) to suggest how the assumption and misperception of inattentiveness may actually be a key to television's underlying logic and appetite for embellishment, ornamentation, and stylishness.

59. In his explanation of the symbolic social use of television in the home Dennis Giles states: "But given the distractions of home viewing, given the fact that the TV image rarely dominates a room by its size alone and that it competes against other possible objects of vision in a lighted room, TV pictures are less forceful than theatrical movie images in holding the viewer's attention." Dennis Giles, "Television Reception," *Journal of Film and Video* 37, no. 3, (Summer 1985): 12–25.

60. Since Giles is concerned with the symbolic function of television set as furniture and icon, his remarks are suggestive. They tend, however, to devalue the aural-visual spectacle that television increasingly tries to heighten.

61. "[Television] emphasizes another invocatory drive: hearing; *sound dominates* [and] ensures continuity of attention," says Robert Deming, in "The Television Spectator-Subject," *Journal of Film and Video* 37, no. 3 (Summer 1985): 49.

62. A recent analysis by Valerie Walkerdine shows the extent of this TV-as-background assumption: "There is a specific way in which television is watched. This differs from the fascinated concentration of the spectator in the darkened cinema, and also from the way that television is often *used as a backdrop to domestic routines*." Valerie Walkerdine, "Video

Replay," in Manuel Alverado and John O. Thomson, eds., *The Media Reader* (London: British Film Institute, 1990), 349.

63. "The *totalizing, womblike effects of the film-viewing situation represent* [for Baudry], the activation of an unconscious desire to return to an earlier state of psychic development, one before the formation of the ego, in which the divisions between the self and other, internal and external, have not yet taken shape." Sandy Flitterman-Lewis, "Psychoanalysis, Film, and Television," in Robert Allen, ed., *Channels of Discourse,* 182.

64. This "womblike," "artificially psycho[tic]" state of the film viewer as a dreamer is for Flitterman-Lewis tied directly to the visual emphasis of cinema.What links this process to the cinema is the fact that it occurs in terms of visual images—what the child sees at this point (a unified image that is distanced and objectified) forms how he or she will interact with others at later stages in life. Ibid., 183.

65. Flitterman-Lewis contrasts television to this process in several important ways. Thus television substitutes liveness and directness for the dream state, immediacy and presentness for regression. It also modifies primary identification in ways that support its *more casual forms of looking.* The television viewer is a distracted viewer.

66. See especially the discussion of the live-remote mode in chapter 9, and the militarist guises of crisis coverage in chapter 11.

67. "It is the total involvement in all-inclusive *nowness* that occurs in young lives via TV's mosaic image." McLuhan, *Understanding Media,* 292. "Ours is a brand new world of *allatonceness.* 'Time' has ceased, 'space' has vanished. We now live in a global village . . . a *simultaneous* happening" (italics mine). Marshall McLuhan and Quentin Fiore, *The Medium Is the Message: An Inventory of Effects* (New York: Bantam Books, 1967), 63.

68. Peter H. Wood, "Television as Dream," in *Television as a Cultural Force,* Richard Adler, ed. (New York: Praeger, 1976), 23.

69. Horace Newcomb, *Television: The Most Popular Art* (New York: Anchor Books, 1974), 258.

70. "What is video then? Video is a process of expression that is *instantaneous*, electronic, and playable on one or more screens, through images and sound *transforming time into experience* and altering the habitual way the audience has of seeing. The soul of video is change, not permanence." Jonathan Price, *Video Visions: A Medium Discovers Itself* (New York: New American Library, 1972), 4.

71. What started in part as a countercultural and social movement to appropriate the tools of television production became within this aesthetic a way to alter personal consciousness. A focus on liveness, real time, and simultaneity could be wielded, in short, for both social and political ends.

72. Since video is a medium of *real time,* that is, because it transmits the temporal quality of the process being recorded, it alters our experience of our own memory, of history, and of daily life. Frank Gillette, "Masque in Real Time," in Ira Schneider and Beryl Korot, eds., *Video Art* (New York: Harcourt, Brace, Jovanovich, 1976), 219.

73. The industry wishes, or feels obligated, to maintain the illusion of *immediacy*, which it defines rather precisely as "the *feeling* that what one sees on the TV screen is living and actual reality, *at that very moment taking place*" (italics mine). David Antin, "Video: The Distinctive Features of the Medium," in *Video Art,* 177. It followed from this aesthetic that important video art (video art worthy to be curated, collected, and funded) was videowork that exploited the property of liveness and real time.

74. "These are the two features of the everyday use of medium that are suggestive for a discussion of video: the simultaneous reception and projection of an image, and the human psyche as a conduit." Rosalind Krauss, "Video: The Aesthetics of Narcissism," in *New Artists Video,* Gregory Battcock, ed. (New York: Dutton, 1978), 45.

75. The popularity among critics of real time and narcissistic video was due in part to the fact that such work overtly illustrated Lacan's "mirror stage"—a heuristic and psychoanalytic paradigm that became increasingly fashionable in intellectual circles during the decade. Yet, such work, legitimized only by prescriptive theory in the 1970s, ultimately had little impact on television production in general. Krauss, "Video: The Aesthetics of Narcissism," 55.

76. "Television viewers have come to expect a higher degree of *realism and authenticity*

in every aspect of television, from news and documentaries to entertainment and sports programming. Shooting on location is one way to enhance a production"(italics mine). Alan Wurtzel, *Television Production* (New York: McGraw-Hill, 1979), 510.

77. "You should think of television performing *its most distinctive function, the live transmission of events*. . . . Contrary to film, the basic unit of television, the television frame, consists of an ever-changing picture mosaic. . . . Each television frame is in a continual state of becoming. . . . As such, the sequence of the actual event, cannot be reversed when shown on television. . . . *The now of the television event is equal to the now of the actual event* in terms of objective time, that is, the instantaneous perception by the observer of the actual event and by the television viewer" (italics mine). Herbert Zettl, *Sight, Sound, Motion: Applied Media Aesthetics* (Belmont, Calif.: Wadsworth, 1973), 263.

78. "In terms of mode of address, I have argued that notions of 'liveness' lend a sense of flow which overcomes extreme fragmentation of space." Jane Feuer, "The Concept of Live Television: Ontology as Ideology," in *Regarding Television*, E. Ann Kaplan, ed. (Los Angeles: The American Film Institute, 1983), 19.

79. Feuer's explanation is so good because it ties the liveness ideology to one of the most influential concepts in television critical theory, the flow. Yet, assertions about the centrality of liveness made from the analysis of one talk show raise other problems. Williams, *Television*, 86–118.

80. "Television's self-referential discourse plays upon the connotative richness of the term 'live,' confounding its simple or technical denotation with a wealth of allusiveness. Even the simplest meaning of 'live'—that the time of the event corresponds to the transmission and viewing times—reverberates with suggestions of 'being there' . . . 'bringing it to you as it really is.' The contradictory television coinage 'live on tape' captures the slippage involved." Feuer, "The Concept of Live Television," 14. This view is important in correcting earlier glosses and essentialisms of liveness *theory*, but wrong, as I hope to show later, if it implies that liveness is a dominant myth in television *practice*. Other important myths are also at work; myths that suggest neither simultaneity, presentness, or "being there." With the emergence of pictorialism and the preoccupation with individuated program looks as common objectives in recent television, there is no reason to position liveness as *the* determining ideology. Rather liveness has become one stylistic item on the larger menu of visuality. This view, then, is an inversion of Feuer's. Whereas Feuer argues that stylistic codes produce realism and liveness, I am suggesting that liveness is a visual code and component of a broader stylistic operation. It is a look that can be marshaled at will, feigned and knowingly exchanged with ontologically aware viewers.

81. Robert Vianello critiques the conflation of the live and the real by showing that liveness is a complicated and political construct in "The Power Politics of 'Live' Television" (*Journal of Film and Video* 37, 3 [Summer 1985], 39), yet overvalues liveness by claiming that its promise "pervades every moment" of broadcast: "It is on these instantaneous and spontaneous transmissions that television truly establishes itself as a social institution of the real. . . . Television becomes the perpetual possibility of making contact with the real; it is this possibility which pervades every moment of broadcast."

82. Todd Gitlin, "Introduction: Looking Through The Screen," in Todd Gitlin, ed., *Watching Television* (New York: Pantheon, 1986), 6.

83. Gitlin, "Car Commercials and Miami Vice: 'We Build Excitement,'" in *Watching*, 136–161.

84. Marc, *Demographic Vistas*, 21.

85. Margaret Morse, "The Ontology of Everyday Distraction," in *Logics of Television: Essays in Cultural Criticism*, Patricia Mellencamp, ed. (Bloomington: Indiana University Press, 1990), 193–221.

86. "Successive, simultaneous time, measured by regular, on-the-half-hour programming . . . indefinitely multiplied by cable and satellite transmission, hypostasized by familiar formats and aging stars in reruns and remakes, trivialized by scandal and gossip, is disrupted by the discontinuity of catastrophe coverage. So-called heterogeneity or diversity ceases as do commercials and TV continuity time as we focus on a single event. . . . TV time of regularity and repetition, continuity and 'normalcy,' contains the potential of interruption, the thrill of live coverage of death events." Patricia Mellencamp, "TV Time and

Catastrophe: Or Beyond the Pleasure Principle of Television," *Logics*, 243–244. Mellencamp has reversed the logic of the liveness myth, while at the same time acknowledging and presuming its centrality. Unlike earlier liveness theorists, for her "simultaneity" does not stand for liveness but for a massively and artificiality constructed temporality. That is, television is no longer seen as simultaneous with *live events*, but as simultaneous with itself and with *other programs* that happen at the same time. In Mellencamp's reversal, "liveness" stands not for the the dominant norm in television, but as the potentially disruptive agent that can attack and expose conventional programming pleasures.

87. Ibid., 261–262.

88. This view of catastrophic temporality seems to devalue the fact that even catastrophes are immediately stylized and constrained as pictures and endless loops, with encrusted graphics, in a process of almost immediate representation that can make even the Kennedy assassination or fires caused by arson in Malibu in some sense pleasurable. Certainly stylized catastrophe loops are more pleasurable to most than coverage of the same events on the radio.

89. The full quote reads: "The more rapid internationalization of television via the *immediacy* of satellites on a *global* allocation of an electromagnetic spectrum never imagined as nationally determined, replicates the *emphasis on transmission*." Patricia Mellencamp, "Prologue," in *Logics of Television*, 3. (italics mine).

90. Probably the most forceful indication of the continuing centrality of liveness in contemporary media theory is Daniel Dayan and Elihu Katz's *Media Events* (Cambridge: Harvard University Press, 1992), published after this chapter was written. A very good critique of Dayan and Katz, and a reconsideration of ontological aspects of liveness, is James Friedman, "Live Television: Ceremony, (Re)presentation, Unstructured and Unscripted Events," presented at the Screen Studies Conference, Glasgow, Scotland, June 1993.

91. Sharon D. Moshavi, "Niche Cable Networks Attract Advertisers of Same Genre," *Broadcasting and Cable* (March 8, 1993). 47.

92. Louis Sahagun and Michael Kennedy, "FBI Puts Blame on Koresh for Cultists' Death," *Los Angeles Times*, April 21, 1993, A13.

Chapter 2. Unwanted Houseguests and Altered States

1. *Beauty and the Beast* (CBS, 1987–1990). Premiere episode, September 25, 1987, PVA-3218t.

2. While this chapter examines televisuality within the history of broadcast programming, chapter 3 will focus on stylistic exhibitionism in light of more contemporary changes in the production of commercials, advertising, and production technology.

3. Air dates, network information, and archival inventory numbers from the UCLA Film and Television Archives collection for these videotapes are cited in the endnotes. When available, these archival references are typically designated by four- or five-digit numbers preceded by PVA or T.

4. *Route 66*, "Journey to Ninevah," September 28, 1962 (CBS, 1960–1964), PVA-10074.

5. *The Continental*, March 20, 1952 (CBS, 1952), PVA-5759t.

6. *Ernie Kovacs Show* (NBC, 1951–1952, 1955–1956; CBS, 1953–1954).

7. *Ernie Kovacs: TV's Original Genius*, n.d., PVA-722t.

8. *Ted Mack's Original Amateur Hour*, May 1949 (Dumont, 1948–1949), PVA-2109t.

9. *Photographic Horizons*, November 10, 1948 (Dumont, 1948), PVA 8230t.

10. See Lynn Spigel *Make Room for Television* (Chicago: University of Chicago Press, 1992), and George Lipsitz, *Time Passages: Collective Memory and American Popular Culture* (Minneapolis: University of Minnesota Press, 1990).

11. I am referring here of course to Walter Benjamin's essay "Art in the Age of Mechanical Reproduction," one of the most influential and reprinted theoretical works on the radical possibilities and Marxian potential of photography, film, and, by implication, television. *Illuminations* (New York: Schocken, 1969), 217–252.

12. During the period of high modernism, the "Family of Man" photographic exhibition (New York: Museum of Modern Art, 1955), attempted to connect the white-walled sanctum

of the high-culture museum with the vast human spirit it construed at the core all cultures. Photographic critics have stated that this show "was probably the most publicized and widely seen photographic exhibition of all time." Barbara London Upton, *Photography* (Boston: Little, Brown, 1985), 418.

13. *Window on the World*, March 25, 1949 (Dumont, 1949), PVA-9728t.

14. *Voice of Firestone* featured regular guests and performing artists from high-brow culture, including the Metropolitan Opera Company (NBC, 1949–1954; ABC, 1954–1963).

15. *The Honeymooners*, "TV or Not TV," October 1, 1955, PVA-6035t. "Better Living Through TV," November 12, 1955 (CBS, 1952–1957), PVA-6035t.

16. Herbert Gans, *Popular Culture and High Culture: An Analysis and Evaluation of Taste* (New York: Basic Books, 1974).

17. See Lipsitz on debt, *Time Passages,* 48–68.

18. *The Burns and Allen Show*, "Gracie Wins a TV Set," February 8, 1954 (CBS, 1950–1958), T-29491.

19. William Boddy, *Fifties Television: The Industry and Its Critics* (Urbana: University of Illinois Press, 1990).

20. Eric Barnouw, *Tube of Plenty: The Evolution of American Television* (New York: Oxford University Press, 1975, 1992).

21. *Texaco Star Theater*, "Starring Milton Berle," January 18, 1949 (NBC, 1948–1953), PVA-6033t.

22. Denise Mann, "The Spectacularization of Everyday Life: Recycling Hollywood Stars and Fans in Early Television," in Lynn Spigel and Denise Mann, eds., *Private Screenings: Television and the Female Consumer* (Minneapolis: University of Minnesota Press, 1992).

23. *The Martha Raye Show*, January 3, 1956 (NBC, 1954–1956), T-676.

24. *The Jack Benny Program,* "Jack Casts a TV Show," October 29, 1963 (CBS, 1950–1964), PVA-5015t.

25. *The Donna Reed Show*, "The Career Woman," April 28, 1960 (ABC, 1958–1966), PVA-5814t.

26. Ibid., "April Fool," April 1, 1959, PVA-5814t.

27. *Leave it to Beaver*, "Box Office Attraction," February 28, 1963 (CBS, 1957–1958; ABC, 1958–1963), PVA-8660t.

28. Ibid., "Beaver on TV," February 21, 1963 (CBS, 1957–1958; ABC, 1958–1963), PVA-8660t.

29. *The Defenders* (CBS, 1961–1965). *Dr. Kildare* (NBC, 1961–1966).

30. J. Fred MacDonald, *One Nation Under Television: The Rise and Decline of Network Television* (New York: Pantheon Books, 1990), 172.

31. *Ben Casey,* "A Certain Time, A Certain Darkness," December 11, 1961 (ABC, 1961–1966), PVA-2477t.

32. It is significant that the producers of the Ben Casey series chose Klee, a moderate and unthreatening maker of small colorful pictures, to represent the dangers of drug addiction/modern art, rather than, say Breton, Duchamp, or Pollack. "Dangerous" art from Hollywood's point of view, actually turns out to be palatable within middle-class interior home decoration.

33. *Run for Your Life*, "Hang Your Head Down and Cry," n.d. (NBC, 1965–1968), PVA-8919t.

34. *Easy Rider*, directed by Dennis Hopper, Raybert Productions-Pando Co., 1969, PVA-1947m.

35. *Stagecoach*, directed by John Ford, United Artists, 1939, PVA-3648.

36. *All in the Family,* "Lionel Moves into the Neighborhood," March 2, 1971 (CBS, 1971–1983), PVA-1692t.

37. *Mary Hartman, Mary Hartman*, episodes one, and two (syndicated, 1976–1977), PVA-2096t.

38. Whereas *Roseanne,* with episodes on issues of teen sex, homosexuality, and unemployment, comes closest to Lear's politics, the show's narrative equivocations are shared with the Fox family sitcoms, a tendency that strips these later shows of both a Lear-styled partisanship and Lear's implicit posture of moral superiority. *Roseanne*, "Toto: We're Not in Kansas," March 28, 1989 (ABC), PVA-7043t. *Married with Children*, "Requiem for a

Barber," February 12, 1989 (Fox), PVA-6872t. *The Simpsons*, "Call of the Simpsons," February 18, 1990 (Fox), PVA-7397t.

39. Lynn Spigel has given a particularly good account of how television was theorized by the industry and by experts within the aesthetic framework of theater. *Make Room for Television* (Chicago: University of Chicago Press, 1992).

40. John Fiske has shown how cultural and aesthetic forms can be discussed in economic terms in "The Cultural Economy of Fandom," in *The Adoring Audience*, Lisa Lewis, ed. (New York: Routlege, 1992), 30–49.

41. The logic of this textual search for contradiction and rupture is articulated in Mimi White, "Ideological Analysis and Television," in *Channels of Discourse*, Robert Allen, ed. (Chapel Hill: University of North Carolina, 1987).

42. *Window on the World*, March 25, 1949 (Dumont, 1949), PVA-9728t.

43. *Help Thy Neighbor*, September 20, 1952 (unidentified station, Los Angeles), PVA-1600t.

44. *Ted Mack's Original Amateur Hour*, May 1949 (Dumont, 1948–1949), PVA-2109t.

45. *Texaco Star Theater: The Milton Berle Show*, January 18, 1949 (NBC, 1948–1953), PVA-6033t. Dave Garroway and Steve Allen celebrated these limitations in early television as well, and naturalistic theater has always forced the audience to confront and accommodate a severely limited technical apparatus. In early television, recognizing these technical limitations was part of the fun, too. Pointing out how the apparatus worked or failed to work, even as one watched the performance, was probably never seen by viewers as a kind of deconstruction, and the practice predated the consciousness of postmodernism by many years.

46. *Colgate Comedy Hour, With Dean Martin and Jerry Lewis*, November 4, 1951 (NBC, 1950–1955), PVA-7051t.

47. David Marc, *Democratic Vistas: Television in American Culture* (Philadelphia: University of Pennsylvania Press, 1984), 21.

48. *Captain Video*, "Chauncey Everett," 1952 (Dumont, 1949–1955), PVA-122t.

49. *The Honeymooners*, "TV or Not TV," October 1, 1955, PVA-6035t.

50. I use the term "sloppy direction" not to judge this type of production as inferior, only to more accurately describe the textual logic of such shows as inconsistent. Shows with breaks in continuity and programs that threatened to come apart at the seams were about as far from the seamlessness of classical Hollywood form as one could imagine.

51. *Beulah*, n.d., 1950 (ABC, 1950–1953), PVA-5981.

52. Denise Mann has shown how aging stars played up the second-rate status of television as a form of consensus building with audiences. By constantly playing with TV's technical limitations and artifice, such stars worked to secure a different kind of female audience.

53. Boddy, in *Fifties Television*, describes how limited sets and artifice provided an important degree of challenge for directors, writers, and performers, who needed to move through such low-tech worlds in real time, for sixty or ninety continuous minutes.

54. *Goodyear Playhouse*, "Marty," May 24, 1953 (NBC, 1951–1957), PVA-1450t.

55. *Playhouse 90*, "Requiem for a Heavyweight," November 11, 1956 (CBS, 1956–1960), PVA-1358t.

56. See Boddy, *Fifties Television*, 103–104.

57. In programming strategy, flow-through conceives of the whole evening on a network as an integrated programming unit, not an individual show or episode. "Hammocking" is the tactic by which new and unproven shows are launched in an hour or half-hour timeslot between two other successful or highly rated shows. These twin peaks promise to develop a successful audience for the new show. "Counter-programming," Weaver's claim to fame, does not compete with strongly rated shows in the same time slot, but targets a completely different genre and demographic. In the cutthroat world of programming, even a successful program can be beaten, since a counter-programmed special, for example, gathers together a dispersed audience uninterested in the previous time-slot winner.

58. *General Electric Star Theater* (CBS, 1953–1962).

59. *General Electric Star Theater*, 'Atomic Love,' November 22, 1953 (CBS, 1953–1962), PVA-9734t.

60. Barnouw, in *Tube of Plenty,* provides a good account of the acquiescence of Hollywood film producers to the inevitability of television.

61. *Disneyland,* "The Disneyland Story," October 27, 1954, and July 13, 1955 (ABC, 1954–1958), t-390437.

62. RKO, for example, not only acquiesced when it made its film library available for television broadcast, but the studio eventually died.

63. *Father Knows Best*, "Hero Father," n.d. (NBC, CBS, 1954–1962), PVA-1713t.

64. *Father Knows Best*, "Formula for Happiness," n.d. (NBC, CBS, 1954–1962), PVA-135t.

65. *77 Sunset Strip,* "Downbeat," May 8, 1959 (ABC, 1958–1964), PVA-60t.

66. Review of *77 Sunset Strip, TV Guide*, November 22, 1958, 30.

67. Review of *77 Sunset Strip, Variety*, October 15, 1958, 34.

68. *Wanted: Dead or Alive*, September 6, 1958 (CBS, 1958–1961).

69. This insider account of the story behind *Lucy*, featuring daughter Lucie Arnaz and a compilation of home movies showing the happy side of the Arnaz family, may have been a forceful revision of, and reaction to, an earlier dramatization aired by CBS on February 10, 1991, *Lucy and Desi: Before the Laughter*—a made-for-TV movie that was publicly criticized for it inaccuracies by both Lucie and Desi Arnaz, Jr.

70. *I Love Lucy*, "Drafted," December 24, 1951, PVA-81t. "Lucy Does a Commercial," May 5, 1952, PVA-416t. (CBS, 1951–1957).

71. *Metropolis* was a landmark accomplishment of German expressionism and UFA, directed by Fritz Lang and produced in 1926.

72. It must be pointed out that the generic look of the telefilm was not just a product of more efficient production economics. The stylistic control and safe style that came with it was also safe from the conspiracy, threat, and volatility of live shows: game shows, dramas. A proficient but nondescript form worked well during an era in which Hollywood and television were keeping their political heads low.

73. *Now, Voyager*, directed by Irving Rapper, Warner Bros. (1942), PVA-5935m.

74. *Meet John Doe*, directed by Frank Capra (1941), PVD-451m.

75. *The Goldbergs*, "Acceptance at the Dance," May 4, 1954 (CBS, 1949–1951; NBC, 1952–1953; Dumont, 1954), PVA-5235t.

76. *2001*, directed by Stanley Kubrick (1968).

77. A promotional designation that assigned the woman to a raceless netherworld outside of blackness and whiteness altogether.

78. *Mod Squad*, "Keep the Faith Baby," March 25, 1969 (ABC, 1968–1973), PVA-2520t.

79. Jonas Mekas (*Diaries, Notes, and Sketches,* 16mm, 1964–1969) and Stan Brakhage (*Dog Star Man*, 16mm, 1965) were both influential figures in the development of the American underground cinema in the 1960s, and proponents of the kind of stylistic and romantic ecstasies suggested here in *Mod Squad.*

80. *Julia*, "Mama's Man," September 17, 1968 (NBC, 1968–1971), PVA-6398.

81. Doctrinaire editing and directing wisdom have it that viewers seldom notice cuts between successive shots that are more than 30 or 45 degrees different, while transitions between shots from very similar angles jump on the screen in ways that draw attention to the cutting rather than the action.

82. This whitening effect of the studio style is important to note, both because Julia was the first starring role for an African-American actress since *Beulah* in the late 1940s and early 1950s and because the show proved to be successful with a broader-based multiracial audience. See J. Fred MacDonald, *Blacks on White TV: African Americans in Television Since 1948* (Chicago: Nelson Hall, 1992).

83. *Kung Fu*, series Pilot/made-for-TV movie, February 22, 1972 (ABC, 1972–1975), PVA-3345t.

84. *Laugh-in*, September 16, 1968 (NBC, 1968–1973), PVA-2024t.

85. Dwight Whitney, "Look Who's in the 'In' Crowd," *TV Guide,* September 21, 1968.

86. Richard Nixon's guest appearance included the kind of hip adage, "Sock it to me," that few associated with the retrenched Republican party and its silent majority during this period.

87. *Laugh-in* plays up excessive videographic style, excessive within constraints of 1960s 2-inch quad videotape and the cumbersome splicing of edits that came with the format.

88. "Zaniest Team on TV," *Reader's Digest*, January 1969.

89. *The Beverly Hillbillies*, rebroadcast July 6, 1993 (CBS, 1962–1971; TNT, 1993–1994). Jethro is garbed in film auteur fashion, with striped leisure shoot unbuttoned to the waist, gold chains covering his open chest, and dark psychedelic glasses. He wonders about his ability to get a woman back in the hills, now that he is sophisticated and hip.

90. *Maude*, "Walter's Heart Attack," September 23, 1974 (CBS, 1972–1978), T-14855. *The Jeffersons* (CBS, 1975–1985).

91. *Goodtimes* (CBS, 1974–1979).

92. The Emmys that were awarded to *All in the Family* included: best series (1971, 1972, 1973), best continuing performance by an actress (1971, 1972), best continuing performance by an actor (1972), and best writing (1972).

93. This analogy to the "golden age" is repeated in Newcomb and Alley, *The Producer's Medium*, 197.

94. *Beverly Hillbillies* (CBS, 1962–1971), *Petticoat Junction* (CBS, 1963–1970), *Green Acres* (CBS, 1965–1971).

95. "The term "zero-degree" comes from Alain Robbe-Grillet's influential treatise on modernist literature, *Writing Degree Zero* (New York: Hill and Wang). The term has been applied since that work to characterize the blank and seamless style of classical Hollywood cinema. I use it as an apt description for the reemergence for an empty and meager studio style on television in the 1970s.

96. *Rhoda*, "Rhoda's Wedding," October 28, 1974 (CBS, 1974–1978), PV-1920t.

97. *Mary Tyler Moore*, "Mary's Three Husbands," February 26, 1977 (CBS, 1970–1977), PVA-938t.

98. *Happy Days* (ABC, 1974–1984), *Mork and Mindy* (ABC, 1978–1982), *Laverne and Shirley* (ABC, 1976–1983).

99. Gary Marshall reemphasizes the importance of writers as the defining factor in his series, in *The Producer's Medium*, 238–253.

100. Horace Newcomb and Todd Gitlin share this view. See *The Producer's Medium*, 197.

101. Ibid., 210, 203.

102. Ibid., 202.

103. *Columbo* (NBC, 1971–1978), *Quincy* (NBC, 1976–1983), *Delvecchio* (CBS, 1976–1977), *The Incredible Hulk* (CBS, 1978–1982), *The Six-Million Dollar Man* (ABC, 1973–1978), *The Bionic Woman* (ABC, 1976–1977; NBC, 1977–1978), *Knight Rider* (NBC, 1982–1986).

104. *CHiPs* (NBC, 1977–1983).

105. The "magic hour" has become a cinematographer's ideal: the half-hour or so immediately following sunset when the suns rays are refracted into a painterly and diffused spectrum of pastel colors. Terrence Mallick's *Days of Heaven* is held up as an important example of magic hour cinematography.

106. Alex MacNeill, *Total Television: A Comprehensive Guide to Programming, 1948 to Present* (New York: Penguin, 1991), 143. *Charlie's Angels* (ABC, 1976–1981).

107. *Dukes of Hazard*, Warner Bros. (CBS, 1979–1985).

108. This "jiggle" scene, involving the highway patrol officers and a weekend carwashing event, was part of *CHiPs* aired in syndication on TNT, July 6, 1993.

109. André Bazin, *Questions of Cinema* (Berkeley: University of California Press, 1967).

110. Both quotes are from Leslie Fishbein, "Roots: Docudrama and the Interpretation of History," in *American History, American Television*, ed. John E. O'Connor (New York: Ungar, 1983), 280, 297.

111. *Full House* (ABC, 1987–), *Major Dad* (CBS, 1989–), *Coach* (ABC, 1989–).

112. *Roseanne* (ABC, 1988–).

113. Marc assigns the stand-up comedian to the presentational form of comedy found in the comedy-variety genre. By the 1980s stand-ups had gained a dominant role in the sitcom as well. (*The Cosby Show*, NBC, 1984–1992; *Roseanne; Seinfeld*, NBC, 1990–). No longer

participants in the subtle world linked to the sitcom by Marc, 1990s sitcoms also "wear their badge of artifice." Marc, *Demographic Vistas,* 21.

114. *Whose the Boss*, December 1988 (ABC, 1984–1992).

115. *Seinfeld* (NBC, 1990–).

116. *Roots* (ABC, 1977), *Roots: The Next Generation* (ABC, 1979).

117. *Monty Python's Flying Circus* (syndicated, U.K., 1970), PVA-221. *Second City T.V.* (syndicated, 1977–1978). *SCTV*, July 11, 1981, PVA-2209t. *Saturday Night Live,* "With Howard Cosell," November 1, 1975 (NBC, 1975–), PVA-5082t.

118. *Hotel* (ABC, 1983–1988), *St. Elsewhere* (NBC, 1982–1988).

119. *M.A.S.H.* (CBS, 1973–1983).

120. *M.A.S.H.,* "The Interview," February 24, 1976, PVA-780t.

121. Alex MacNeill, *Total Television,* 457.

122. *High Noon*, directed by Fred Zinneman, United Artists, 1952.

123. *M.A.S.H.,* "The Interview," February 24, 1976, PVA-780t. "Goodbye, Farewell, Amen," February 28, 1983, PVA-2953t.

124. The primetime soap *Peyton Place* (ABC, 1964–1969) in the 1960s and the miniseries *Rich Man, Poor Man* (ABC, 1976–1977) were precedents that demonstrated the narrational power of the extended family, *Roots* and *Roots: The Next Generation* showed that there was popular interest in regional and international geography.

125. *Dallas*, "Who Dun It?" November 21, 1980 (CBS, 1978–1991), PVA-8659t.

126. This finale of *Dallas*, structured around J.R.'s astral travel was aired May 3, 1991.

127. Gitlin, "Hill Street Blues: 'Make It Messy'," *Inside Primetime* (New York: Pantheon, 1985).

128. *Police Tapes*, directed by Susan Raymond and Alan Raymond, 1976, and August 3, 1978 (PBS, 1978), PVA-6646t.

129. *Hill Street Blues*, "Hill Street Station," premiere, January 15, 1981 (NBC, 1981–1987), PVA-1981t.

130. One of the early " 'rules' was that there would be no brick, no reds, and no browns." MacNeill, *Total Television,* 496. By the 1986–1987 season the popular press echoed the producer's PR that the show's look was being "shifted" and designed around darker colors, blues and grays.

131. The guest appearances of original music-video sequences designed around popular rock tracks by Glenn Frey, Eric Clapton, Mick Jagger, and others became a regular part of *Miami Vice*'s dis-integrated presentational form. See *Miami Vice*, the premiere episode, September 16, 1984 (NBC, 1984–1989), PVA-2222t.

132. L. S. Kim has written about the "racial economy" in shows like *Beulah* in "The Erasure of Difference and the Denial of Ethnicity: The Ethnic Domestic in Television," unpublished ms., UCLA, December 1992, later presented at the Screen Theory Conference (Glasgow, Scotland), June 1993.

133. *Beulah*, no date, 1952 (ABC, 1950–1953), PVA-5981t.

134. MacDonald, *Blacks and White on TV.*

135. Mildred Lewis, "Paradigms of Race: Televisuality and Otherness, an Examination of *Roc*," unpublished ms., UCLA, March, 1993.

136. *The Goldbergs.*

137. A Jamaican expert on reggae, as quoted in J. Hoberman, "Sex, Drugs, and Dreadlocks," *Premiere* (August 1993), 50.

138. *Dodge* automobile ad, "Commercials: International Broadcasting Awards" (Hollywood Advertising Club, 1959), PVA-932t.

139. *Route 66.*

140. *Bachelor Father*, "Boys Will Be Boys," June 5, 1962 (CBS, 1957–1959; NBC, 1959–1961; ABC, 1961–1962), PVA-9933t.

141. *The Leave It to Beaver*, "Box Office Attraction," February 28, 1963 (ABC, 1958–1963), PVA-8660t.

142. *77 Sunset Strip*, "Downbeat," May 8, 1959, PVA-60t.

143. *The Dick Van Dyke Show*, "Laura's Little Lie," October 9, 1963 (CBS, 1961–1966), T-16426. High-culture fashion as a moral issue to be resisted in the name of domesticity and

familial peace. "There's more to life than fashion, careerism and city values." Ella Taylor points out a shift in this suspicion of fashion has been altered so that by the early 1960s, in *The Dick Van Dyke Show*, women are allowed "contemporary" fashion, but only at home. The show allows women in the mannish workplace but only if they wear "suits." Ella Taylor, *Prime-Time Families* (Berkeley: University of California Press, 1991).

Chapter 3. Modes of Production

1. C. F. Jenkins, "Radio Photographs, Radio Movies, and Radio Vision," *Journal of Society of Motion Picture Engineers* (May 1923), 81.

2. R. R. Beal, "RCA Developments in Television," *Journal of Society of Motion Picture Engineers* (August 1937), 143.

3. A good and accessible source for this early technical literature is Jeffrey Friedman, ed., *Milestones in Motion Picture and Television Technology: The SMPTE 75th Anniversary Collection* (White Plains, NY: Society of Motion Pictures and Technical Engineers, 1991), especially 97–233. The word "television" was later added to the professional organization's prewar name, SMPE.

4. Raymond Williams, *Television, Technology and Cultural Form* (New York: Schocken Books, 1974).

5. William Boddy, *Fifties Television* (Urbana: University of Illinois Press, 1990), especially chapter 1.

6. *American Cinematographer* (June/July 1993).

7. Ben Gradus, *Directing the Television Commercial* (Los Angeles: Directors Guild of America, 1981), 4, 5, 34.

8. "Close-ups: Michael Oblowitz," *Millimeter* (February 1989), 196.

9. The most notable critique of intentionalist theory is the work of Monroe Beardsley and William K. Wimsatt, "The Intentionalist Fallacy," in *Problems in Aesthetics*, ed. Morris Weitz (New York: Macmillan, 1987), 347–360. The irony of raising this issue is that while the industry assumes it is the guarantor of the meanings of its productions, academic theory ignores even provisional discourses by the industry in favor of the spectator's supposed meanings and reception. My study focuses on the terrain between these two polarities. I aim to reconsider the theorization of the industry, but also to take that theorization as a mythology and framing practice that legitimates, naturalizes, and explains its operations.

10. From a review in *American Cinematographer* (January 1987), 94.

11. From *Millimeter* (April 1988), 148.

12. The most extensive explication of the classical Hollywood style is found in David Bordwell, Janet Staiger, and Kristin Thompson, *The Classical Hollywood Style: Film Style and the Mode of Production to 1960* (New York: Columbia University Press, 1985). This relatively neutral-looking style still pervades much of the serious dramatic programs in primetime television.

13. Katherine Stalter, "Working in the New Post Environment," *Film and Video* (April 1993), 100. A couple of universal shows—*Murder She Wrote,* and *Columbo* MOW's—are still cutting film. "They do that on *Murder She Wrote* because the producer who was associated with that show from the beginning was a film person . . . we're still cutting film because that's how its been done. Other studio shows, like *Law and Order* is cut on the CMX-6000. Dick Wolf and the other producers like that system; it works well for them. But they are also conscious of trying to reduce expenses," claims Bruce Sandzimier, vice president of editorial, Universal Television.

14. Bruce Sandzimier, VP of editorial, Universal Television continues: "A case in point would be *South Beach,* a new one-camera dramatic series we just put into production. The executive producer . . . also produces *Law and Order* and *Crime and Punishment* for Universal. His shows have been editing on the CMX-6000, which you get through the Post Group. So he—and we, the studio—developed a relationship with the Post Group." Stalter, *Film and Video* (April 1993), 100.

15. Arthur Wooster, second unit director, *Covington Cross,* as quoted in Josephine Ober,

"Cover Story: Team Work on *Covington Cross*," *In Camera*, Eastman Kodak (Spring 1992–1993), 4.

16. From comments by Ray Peschke, quoted in *Film and Video* (November 1991), 74.

17. "Episodic Television has always had to narrow its scope because of lack of money, but digital technology can open up all kinds of possibilities," states Hal Harrison, vice president of postproduction, Viacom. Stalter, "Working in the New Post Environment," 100.

18. WRAL-TV5, Raleigh, North Carolina, advertisement, "Situations Wanted Technical," *Broadcasting* (November 30, 1992).

19. SMPTE time-code editing refers to an electronic timing and identification scheme standardized by the Society for Motion Picture and Television Engineers. This time-code system laid down a stream of digital audio blips onto an existing audio or address track on the recorded videotape stock. Once done, each video frame had now been assigned an identifying address that any standardized editing controller could find and cut on automatically. Frame accuracy in editing was but one of the advantages of this system.

20. Sony, direct-mail advertising insert, included as a promotional insert in *Video Systems* (June 1993), 16–17.

21. Stalter, "Working in the New Post Environment," 100.

22. Iain Blair, "*Needful Things*: Producer Jack Cummins, Director Fraser Heston and Cinematographer Tony Westman Bring Their production of the Stephen King Novel to the Pacific Northwest," *Film and Video* (April 1993), 99.

23. Quote from "Small Screen Shooters: Four Distinguished Cinematographers Discuss the Craft of Shooting Film for Episodic Television," *Millimeter* (April 1988), 143.

24. Ibid., 142.

25. I have organized my analysis of the status of the image in the industry around two poles that correspond to the current film versus video look debate. Whereas in chapter 5 I will address one major area implicating visual style as central in the practice of television (that is, the industrial discourse centered around electronic post and videographic effects), I will in this chapter limit my analysis to the discourse surrounding a more conventional production arena (that is, the ways that producers and directors talk about shooting television programs for primetime and advertising).

26. See especially Harry Matthias and Richard Patterson, *Electronic Cinematograhy: Achieving Photographic Control Over the Video Image* (Belmont, Calif.: Wadsworth, 1985), and Anton Wilson, *Anton Wilson's Cinema Workshop* (Los Angeles: American Society of Cinematographers, 1983), 243–297. Other books that examine both technical and aesthetic issues in the film versus video image discourse include David Viera, *Lighting for Film and Electronic Cinematography* (Belmont, Calif.: Wadsworth, 1993), and the introductory text, Larry Ward, *Electronic Moviemaking* (Belmont, Calif.: Wadsworth, 1990).

27. Quote of Eastman Kodak executive from "Electronic Imagery," *American Cinematographer* (June 1987), 89. Through most of the current season, some 80 percent of prime evening time schedules for the three major networks have been made up of programs originated on film. In addition, we are now seeing more fourth network and pay channel movies and other specialized programming originated on film. This might be one of our best years ever in terms of original negative used for TV production.

28. Ibid., 89.

29. I am, of course, comparing only the quality of *color* film stocks between the two periods. Some of the 35mm black-and-white negative stocks used in the 1950s, while much slower in light sensitivity than many modern color stocks, could achieve the kind of rich tonality impossible for any color stock to render, if lit properly. But this comparison is a bit like comparing apples and oranges.

30. Stuart Allen, "Lighting for Television: Faster Filmstocks Are Changing the Ways that Cinematographers Approach Their Work," *Film and Video* (June 1990), 47.

31. Alan Hume, DP, quoted by Ober, "Cover Story: Team Work on *Covington Cross*," 4.

32. Ad for Tiffen filters, *Film and Video* (November 1991), 71.

33. From Richard Schafer, "Choice of Transfers: Film to Tape," *American Cinematographer* (September 1986), 97.

34. Ibid., 97.

35. Shafer, "Choice of Transfers: Film to Tape," 99.
36. In practical terms CCDs replaced vacuum tubes in cameras with rectangular chips that were comprised of grids of light-sensitive microscopic materials. This field of points corresponded roughly to the pixels that make up the grid on a computer screen. The more points or pixels in a grid, the higher the visual resolution.
37. It is also likely that this transformation—this intensive reinvestment in primetime production—may be related to the slowly atrophying number of feature films that were being produced each year during the early 1980s. The availability of both ideational and labor surpluses in Hollywood during this period might partly explain the renewed focus on primetime production practice in the 1980s. The complexities of this relationship between feature film and primetime are, of course, beyond the scope of this book.
38. For an in-depth analysis of the ideology of MTV's visual style, see E. Ann Kaplan, *Rocking Around the Clock*; Todd Gitlin gives an excellent feel for the visual look and demeanor of *Miami Vice* in *Watching Television*.
39. Brian Winston, *Misunderstanding Media* (London: Routlege, Kegan Paul, 1986), discussed and applied to video in Roy Armes, *On Video* (New York: Routlege, Chapman and Hall), 206–208.
40. *Film and Video* (May 1993), 112–114.
41. Allen, "Lighting for Film," 14–15.
42. Fresnels are production lights faced with concentric-faceted lenses that focus the rays on a subject from a distance, without being intrusive. Despite its headline potential in the production trades in the 1980s, the kind of "ecstatic showcase for optical people" described here was also reminiscent of certain showcase feature directors in Hollywood, such as Alfred Hitchcock's choreographed optical events in *Rope* and *Under Capricorn*.
43. As discussed in chapter 2, gone are the days when one might describe TV's soundstage, high-key look as a zero-degree television style. The term is from Roland Barthes, *Writing Degree Zero* (New York: Hill and Wang, 1953).
44. Quote from Ed Plante, in Bob Fisher, "Cagney and Lacey: The New York Look in L.A.," *American Cinematographer* (January 1987), 88.
45. Ibid., 88.
46. Alan Hume, DP, *Covington Cross*, quoted in Ober, "Cover Story: Teamwork on Covington Cross," 3.
47. From Stevan Larner, ASC, in "Beauty and the Beast: God Bless the Child," *American Cinematographer* (April 1989), 71. The producer goes on: "It's a very creative show and any input is greatly appreciated."
48. Ibid., 71.
49. Eric Estrin and Michael Berlin in conversations with the author, California State University, Long Beach, California, March 1990.
50. Ibid.
51. Gerald Perry Finnerman, "*Moonlighting*: Here's Looking at You Kid," *American Cinematographer* (April 1989), 70–71.
52. This claim is curious since *Casablanca* was shot on panchromatic rather than orthochromatic stock as the cinematographers claim. The contrasty look we associate with the film probably results from its widespread dissemination on television. Nevertheless, the important factor is the highly conscious degree to which these television cinematographers manipulate and flaunt an awareness of cinematic codes.
53. Retrostyling is a concept popularized by Fredric Jameson in "Postmodernism and Consumer Society," in Hal Foster, ed., *The Anti-Aesthetic* (Port Townsend, Wash.: Bay Press, 1983), 111–125. The notion that retrostyling is a way to evoke and simulate another look, another time, or another style in shows like *Moonlighting* supports Baudrillard's view that simulation is a fundamental and defining factor in the culture of postmodernism.
54. The simple but sensitively lit studio interview style of documentary was associated with the acclaimed partisan oral history films of James Klein and Julia Reichert, like *Seeing Red* (1984), and with Buckner, Dore, and Sills's film *The Good Fight* (1984). Although Warren Beatty's *Reds* shows that even big-budget feature films can masquerade with the style.
55. I am using the term "taste culture" from Herbert Gans's work on the sociology of

popular art and culture, *Popular Culture and High Culture: An Analysis and Evaluation of Taste* (New York: Basic Books, 1974).

56. While many of the newer craftspeople and cinematographers have now actually had aesthetic and film historical training in university film schools, even middle-aged and older DPs who immigrate to primetime from feature filmmaking bring with them a tradition that values interdisciplinary research and aesthetic sensitivity to cultural image making. For one of the best single sources that betrays the DP's not uncommon interest in art, art history, still photography, design, and architecture, see Dennis Shaefer and Larry Salvato, *Masters of Light* (Berkeley: University of California Press, 1986). Of course, even if a contemporary television DP had never been to film school or mastered the art historical sensitivity of his ASC brethren (a consciousness seldom limited to an awareness of Rembrandt lighting), he or she would have to have been amnesiatic for the past decade to be ignorant of the design and art historical consciousness that has pervaded commericals and print advertising.

57. Horace Newcomb uses this tension between the static formula and the need for some generic change as a partial basis for his proposal that continuity is one of the chief aspects of a television aesthetic in Newcomb, *Television: The Most Popular Art* (New York: Anchor, 1974). Thomas Schatz does a similar thing in *Hollywood Genres: Formulas, Filmmaking and the Studio System* (New York: Random House, 1981).

58. From David Heuring, "The Street: Shooting Video with an Eye to Film," *American Cinematographer* (June 1988), 73.

59. André Bazin, "An Ontology of the Photographic Image," *What Is Cinema?*, trans. Hugh Gray (Berkeley: University of California Press, 1967), 9–16. What has changed, obviously, is the conventionality and cultural form of realism. That is, audiences and makers alike now can read degraded electronic realism, because they understand, to some degree, that the look results from technologies very different from those that produce photographic realism.

60. Robert Hilliard quotes S. J. Paul of *Television/Radio Age* as describing the temporal pressure in which "a mood is created . . . and a sales point is made," that is, a singular objective for both mood and point. Robert L. Hilliard, *Writing for Television and Radio* (Belmont, Calif.: Wadsworth, 1981), 41.

61. This information is from an article on the spots by Brooke Sheffield Comer, "Music Video That Looks like Film," *American Cinematographer* (September 1986).

62. Ibid., 95.

63. From Bruce Stockler, "Seducing Reality: Documentaries Mix Truth and Fashion," *Millimeter* (May 1988), 48.

64. I am taking the word "povera," or poverty, to describe this genre of televisuality, from the continental European tradition that described conceptual and environmental art of the 1960s and early 1970s as "art povera." In short, this low-tech, hand-made anti-art was seen during the period as an extreme form of aesthetic and cultural radicality. The radical intent of art povera is ironic given the product oriented aims of the Nike Corporation.

65. Stockler, 49.

66. Ads have always sold the sizzle rather than the steak, the sensation rather than the product. Within this tradition, however, Levi's anti-ads are worth noting for the *degree* to which they avoid both descriptions of the product and also representations of the product. Anti-ads are far removed aesthetically from the product shot aesthetic that glamorized goods in print and broadcast during the preceding two decades. Levi's anti-ads left one with crude apparitions and fragments of activities on the street, not sensations of the product. In advertising's ongoing tactic of sensory surrogacy, the sensory connection to the product in the anti-ad became more tenuous and open than ever.

67. Stockler, 48.

68. Ibid., 47.

69. Ibid., 47 (italics mine).

70. Ralph L. Holsinger, *Media Law* (New York: Random House, 1987), 388.

71. KCOP-Channel 13 versus hunger strikers, June 6, 1993.

72. KCBS versus "Fiesta Broadway," May 31, 1993.

73. CBS network's *48 Hours* treatment of public access producers on "Talk, Talk, Talk," May 1993.

Chapter 4. Boutique

1. After having characterized American television this way, Horace Newcomb and Robert S. Alley, in their book, *The Producer's Medium* (New York: Oxford University Press, 1983), state that their aim is to "shatter" the centrality of television's "anonymity," xi–xii.

2. Mary Murphy, "Tsk, Tsk, Tori," *TV Guide* (May 8, 1993), 19.

3. *Beverly Hills 90210*, Fox, April 27, 1993.

4. Richard Adler, *Television as a Cultural Force* (New York: Praeger, 1976), claims that "the performers . . . are the key creative figures of the medium. . . . In theater, the successful realization of a play is primarily a collaboration between playwright and director, while film . . . is widely recognized as a director's medium. *But in television, both writer and director must serve the performer,*" 8.

5. NBC promotional tapes (1988), PVA-6604t. *Midnight Caller* (NBC, 1988–1990).

6. Ad, *Broadcasting and Cable* (June 7, 1993).

7. Mike Freeman, "MCA Taps Big-Screen Producers for TV," *Broadcasting and Cable* (March 8, 1993), 23–24. If these major event ratings are good, one or more could be spun off into a weekly series block, even if no prior cable window was planned. Shelly Schwab, president of MCA-TV, argued that the key impetus behind this kind of project was Universal's historic involvement in Operation Prime Time (OPT, a coop of major suppliers in the 1970s), and Nalle's "ability to line up several marquee movie producers, who are often contracted under the Universal Pictures banner." Budgets were to range from $3 million to $5 million per episode. Directors lined up by May included John Landis (*Blues Brothers, Animal House, Trading Places*), Sam Raimi (*Dark Man, Army of Darkness*), stuntman turned director Hal Needham (*Smokey and the Bandit*), Rob Cohen (*Running Man, Bird on a Wire*).

8. Michael's Waspish, socially committed wife—a Princeton-educated woman named Hope Murdoch Steadman—became a weekly foil to the blandishments and lucre of the ad business. This fact meant that the signature center of the show oozed even more so with self-awareness about the series' complex significance.

9. A good example of a collection that combines various accounts critical of Gulf War coverage, balance, and objectivity is "Screening the War: Filmmakers and Critics on the Images that Made History," *International Documentary* (Spring 1991), 20–25. The analyses make no reference to the war's ideological effects on programming other than the coverage itself.

10. As spoken by news anchor Warren Olney during newsbreak on KCOP-Channel 13, on Monday, April 30, 1990.

11. From KTLA promotional materials as quoted by Howard Rosenberg, *Los Angeles Times*, May 1, 1990, F7. Second hostage Frank Reed was released the last week of April and arrived on U.S. soil on Friday, May 4. KTLA-Channel 5 in Los Angeles broadcast *Voyage of Terror* during primetime on two nights during this period (May 1–2, 1990).

12. The "Long before Noriega" campaign was used by NBC for *Drug Wars*, both in broadcast and print formats. See for example ads in *TV Guide* and *Television Times*, during the first week of January 1990.

13. I am drawing especially on the work of Mimi White, who in "Television: A Narrative, a History," *Cultural Studies* (1990), 282–300, describes the important process by which television both "produces" and "disperses" the idea of history.

14. Although it is likely, due to the lead time needed to schedule features on television, that *Salvador* was programmed in response to the December 1989 guerrilla uprising in Salvador (an event that, only a few weeks before, shocked Western experts by its ferocity), the actual week of the broadcast clearly placed the fiction within the context of the American invasion of Panama.

15. *Camarena* followed weeks of promos that preceded it by proclaiming "Long before Noriega." NBC, by the time it promoted the miniseries, apparently saw Noriega as the master paradigm for their docufiction. Or at least the Camerena story became interchangeable with the Noriega story. To underline the importance and acceptance of this conflation of fact with fiction, the Mexican government responded to an NBC news special on corruption with adamant protests. To them, both Tom Brokaw's special and the Camarena miniseries—

to which it was overtly linked in programming—were racist and unfairly indicted and linked the Mexican government to Latin-American drug corruption.

16. I am unable to discuss here other important hybridizing operations, such as censorship-related post-dubbing and automatic dialogue replacement (ADR), a group of hybridizing processes in televisual adaptation that originate in standards and practices departments and involve the replacement of dialogue in video postproduction. Such operations are usually associated with censorship—a phenomenon that has long interested media scholars. I am less interested in censorship per se, than in the formal processes by which censorship is wielded. The standard array of manifest behaviors that stimulate censorship include explicit sex, violence, and profanity. All three areas motivate censorship and deletion in the televised *Salvador*, but not in a consistent manner. Explicit violence and profanity do remain in the televised version, although excessive examples of such traits typically elicited formal covering-over tactics. This covering over usually takes the form of post-dubbing or ADR, whereby narration is recorded later and inserted into the original audio mix in order to replace offensive language. The resulting quality varies, and differences in ambient noise usually signals to an astute viewer that dialogue has been replaced. This type of substitution is widespread and is usually preceded by warnings that televised material is of "an adult nature" or that it "is not for sensitive viewers." Since censorship-based ADR is acknowleged by the industry as an obligatory practice, it lacks the apparent arbitrariness of many of the other, less obvious, hybridizing televisual operations that I will focus on.

17. Although not discussed in detail here, additional examples of pretextual forms in *Salvador*, included numerous clips and "actuality images" normally associated with factual, documentary, and reportage forms. These included maps, titles, newsreels, and video footage—all variants of televisual language utilized within the film production itself.

18. While exemplified in the work of photographers like Robert Capa and Eugene Smith, it was in the writings and work of Henri Cartier-Bresson that the "decisive moment" aesthetic was most completely theorized.

19. I am referring here to Nick Browne's concept of the super-text as described in "The Political Economy of the Television (Super)Text," *Quarterly Review of Film Studies* 9, 3 (1984), and to Raymond Williams's influential notion of the broadcast flow described in *Television, Technology, and Cultural Form*" (New York: Schocken, 1974).

20. See especially the display of overt visual production effects in recent shows like *Hard Copy, Eye on L.A., Unsolved Mysteries,* and others.

21. Lev Kuleshov was, of course, one of the most influential teachers and theorists of early Soviet film. His classic experiments showed that wildly divergent narrative worlds could be created simply by altering the sequential order of the very same visual shots. The semiological concern with the syntagmatic dimension of narrative film starting in the 1960s also focused on the *sequential order of visual images* as the basis for narrative intelligibility. Televisuality, then, does not just involve the invention of new visual configurations, it also hybridizes by reordering existing elements into new arrangements that appear simultaneously.

22. This widespread view of narrative structure is found in many popular and how-to works on screenwriting and film, including Syd Field, *The Art of Screenwriting* (New York: Dell, 1979).

23. Robert Scholes's view of "narrativitous counter-tendencies"—a force that actively transforms textual material into narrative, rather than a passive adjective that merely identifies some component as part of a narrative—is described in "Narration and Narrativity in Film," *Film Theory and Criticism*, eds. Gerald Mast and Marshall Cohen (New York: Oxford University Press, 1985), 390–403.

24. The rapid-fire preview collages: (1) "General Manuel Noriega faces the bench of justice in Miami."; (2) "Lunch at a restaurant in the Crenshaw district, ends in a hail of bullets"; "And then . . . ," (3) "Prostitutes and the spread of AIDS."

25. The line referring to his "many women" was used both by the mass media in reference to Noriega's lifestyle and as a condemnation by Maria of Boyle's lifestyle in the *film Salvador*.

26. The Ron Kovic–Oliver Stone–*Born on the Fourth of July* story was covered on television in everything from *Sneak Previews* to *Entertainment Tonight* during the period December 1989–January 1990.

27. "Viet Vet Kovic May Challenge Dornan," *Los Angeles Times*, January 5, 1990, A3.

28. Film preview, *Born on the Fourth of July*, broadcast during television presentation of *Salvador*, January 4, 1990.

29. Boyle's dialogue spoken by James Wood in the film *Salvador*, 1982.

30. From Panamanian video footage, prominently featured in the news broadcast immediately following *Salvador*, January 4, 1990.

31. See Raymond Williams, *Television, Technology, and Cultural Form*, and Nick Browne, "The Political Economy of the Television (Super)Text," 174–182.

32. While very much aware of the temporal effects of breaks on dramatic form, most practitioners in the primetime production industry tend to be focused on the world within the fictional brackets. Academic theorists typically disregard this intratextual constraint. See Tania Modleski, *Loving With a Vengeance* (New York: Methuen, 1982).

33. A basic summary of the contents of the break is as follows: there are eleven commercial advertisements; six bits of station-related material (one news break, three show identifications, two previews); and one studio-produced movie preview.

34. See especially Sandy Flitterman, "The Real Soap Operas," *Regarding Television* (Los Angeles: AFI, 1983), 84–96.

35. Reporter Bill Lagattuda exchanged the names of the two leaders on the evening news of KNBC-Channel 4 in Los Angeles, as reported by KABC radio on the morning of January 13, 1990.

36. As reported in the *Los Angeles Times*, April 27, 1990.

37. I would like to thank the Scholarly and Creative Activities Committee, the Office of Research and the California State University, Long Beach, for providing the time necessary to complete the research for an early draft of this chapter.

Chapter 5. Franchiser

1. "Thomson Digital Image/TDI print ad, *Broadcasting and Cable* (June 1993), 70. The full ad for "Metaballs" reads: "Metaballs Blob Modeler: for TDI Explore, TDImage 3-D animation, visualization packages; Boolean operations generate smooth holes, cavities with soft edges; optimizes number of polygons to create smooth curves; real time model checks with polygon meshing; blending of Blobs and conversion to polygons." The ad continues: "Metaballs Blob Animator: attach blobs or blending groups to skeletons for animation of muscle and joint movement; combine blending groups to create melting objects, moving molecules. . . . " Such are the graphic ecstasies in the new world of metaballs.

2. Rob Wyatt, New York design house, Telezign, as quoted by Sean Scully, in "Low-Cost High Tech Is Graphic Equalizer," *Broadcasting and Cable* (June 14, 1993), 58. Practitioners still suffer from apolitical, technological determinism, as the following explanation from this essay attests: "For example, in the mid 1980s computer animators created the first 'flying logos.' Suddenly everyone was doing them and the public began to expect it. The same thing happened with morphing, That's a prime example of how technology leads technique. . . . What happened to the good old-fashioned dissolve?"

3. John Ellis, *Visible Fictions: Cinema, Television, Video* (London: Routlege, and Kegan Paul, 1982), 162.

4. Valerie Walkerdine, "Video Replay: Family, Films, and Fantasy," Victor Burgin, et al., eds., *Formations of Fantasy* (London: Methuen, 1986), reprinted in Manuel Alvarado and John Thompson, eds., *The Media Reader* (London: British Film Institute, 1990), 349.

5. Richard Adler, "Introduction: A Context for Criticism," *Television as a Critical Force*, Richard Adler, ed. (New York: Praeger, 1976), 6.

6. Print advertisement for Boss Film Studios, in *Millimeter* (October 1989), 127.

7. Diane Struzzi, "Scandals, Rumors, and Rock 'N' Roll: On the Edge with Margeotes, Fertitta, and Weiss," *Millimeter* (October 1989), 126.

8. Helen Davis, "Designer's Role Within the Graphic Environment," *Computer Pictures* (August/September 1989), S20.

9. Classical film theory was greatly concerned with the aesthetic models utilized in both film analysis and production. Theorist Dudley Andrew demonstrated how central im-

ported paradigms from noncinematic fields were in early speculations about film. See *The Major Film Theories* (New York: Oxford University Press, 1976). *Figuration* for Eisenstein, for example, meant viewing the art of film like a living organism or an engineered machine. Central figures for Bazin, on the other hand, likened cinema to the death mask, footprint, or mathematical asymptote. *What Is Cinema*, vols. I and II, trans. Hugh Gray (Berkeley: University of California Press, 1971). Such figures had hermeneutic rather than just analogical importance, for the way one constructed the object of film analysis from these paradigms greatly influenced the types of results the theorist could arrive at.

10. Margaret Morse, "The Ontology of Everyday Distraction: The Freeway, the Mall and Television," in Patricia Mellencamp, ed., *Logics of Television* (Bloomington: Indiana University Press, 1990), 193–221.

11. George Comstock, *The Evolution of American Television* (Newbury Park, Calif.: Sage Publications, 1989), 13–40.

12. By "low theory" I refer to the frequently complex framing and interpretive discourses used and performed by practiners, producers, and equipment manufacturers in the television industry. I choose the term "low" not to judge or relegate this practice to a low rung in an academic hierarchy of value, but to draw attention to the fact that, with a few important exceptions, intellectuals have ignored this area almost entirely.

13. Lynn Spigel, *Make Room for TV: Television and the Family Ideal in Postwar America* (Chicago: University of Chicago Press, 1992).

14. William Boddy, *Fifties Television: The Industry and Its Critics* (Urbana: University of Illinois Press, 1990).

15. A classic work on linguistic method by Kenneth Pike, *Language in Relation to a Unified Theory of the Structure of Human Behavior*, part I (Glendale, Calif.: Summer Institute of Linguistics, 1954), 8–27, describes two fundamentally different approaches to describing and understanding alien languages. The first, an *-etic* approach (as in phonetics), scientifically analyzes structures, patterns, and codes from outside the speaking group. The analyst deciphers a new language by classifying its parts according to a universally established linguistic system, logic, and method. This kind of work into the deep structure of language can be done from recordings analyzed far from their source of origin. The second approach, an *-emic* approach (as in phonemics), analyzes linguistic structures and systems first from within the speaking group. It seeks to "discover and describe the patterns" of a particular language and to relate those dialectical patterns to the organization of that specific culture. An -emic approach depends to a great degree upon the use of indigenous informants for explanation and interpretation. The formalist approach of the -etic method, as one might guess, is not therefore as popular with linguists or anthropologists who use field research. Their dependence upon informants in the field during the process of linguistic analysis values the intentional apects of language as a cultural practice. By contrast, the -etic approach assumes that languages work through more universal laws and logics. Although this distinction originates with linguistics, its earliest focus with Pike was upon behavior and the social construction of meaning. For this reason, the -emic approach is also well suited for the analysis of culturally symbolic systems, imagery, music and, I will argue, television.

16. I utilize "iconoclastic" here according to its original definition in medieval Byzantium: as a religious purging and theoretical denigration of the visual imagery. For a fuller discussion of these tendencies, see the postscript at the end of this book.

17. The -etic bias has dominated media critical theory, at least after the early days of *Cahiers du Cinéma*, when theorists frequently thought and wrote from a filmmaker's perspective. American broadcasting studies, on the other hand, have at least paid lip service to the industry by appropriating its prized marketing and research categories in quantitative and audience studies. Empirical broadcasting studies of this sort have, however, hardly ever given the same kind of unqualified consideration to the noncorporate complexities that make up the production world of television's image makers and the aesthetic perspectives of its practitioners.

18. Pike, *Language in Relation*, 27.

19. From "Hard News With Soft Edges," *Post* (October 19, 1989), 44.

20. A stated commonplace in the production world is that style without substance or

context or narrative motivation is an empty gesture ("button pushing," "eye candy"), yet the disembodied signifier is precisely one fundamental characteristic of televisuality. Hence, there appears to be a frantic effort in production and technical discourses to give an interpretation or a definition for each effect. This is what I mean by an attempt "to own signification"—button pushing and stylistic flourishes desperately need a motivation, a cause, a logic for effective marketing and adaptation in the field. The extent to which industry public relations and marketing people can succeed at this control of the interpretive discourse grants them proprietary privilege over both the form of signification and the apparatus that produces that signification.

21. It is shortsighted to limit an examination of the challenges to the television apparatus to those myths centered around electronic broadcast origination only (liveness, immediacy), since film origination, and film-style aesthetics still dominate a great percentage of primetime network programming.

22. From promotional flyer on Paintbox, "Paintbox in the DPC," Quantel (Fall 1989).

23. From ad for 8200C postproduction switcher, Crosspoint Latch Corporation, Union, N.J., 1989.

24. From direct marketing brochure, "DaVinci," 1989.

25. From direct marketing brochure, "Encore," 1989.

26. Ken McGorry, "Animate Objects," *Post* (September 1989), 118–120.

27. The spot created by directors Gabor Csupo and Gary Schwartz was a video opening for a show entitled "Radio Vision International," produced by Klasky/Csupo, Los Angeles, 1989.

28. To appreciate the ambiguity of the video screen's surface, try focusing a single lens reflex (SLR) 35mm still camera at close range on a screen displaying an image. Inevitably, moire patterns will result given the acute problems that result in this situation from shallow depth of field, and the *various distances inside of the glass* at which one can choose to focus.

29. From "All Bent Out of Shape," *Post* (September 18, 1989), 42.

30. From the ad "Now it costs less to get more in 3D video effects: Including page turns for under $20,000," *Videography,* vol. 14, no. 10 (October 1989).

31. Roland Barthes, *Mythologies* (New York: Hill and Wang, 1979).

32. From promotional brochure for the "Eclipse CVE 200," DSC Corporation, Gainesville, Florida, 1989.

33. From promotional brochure for A-53D Digital Special Effects device, Abekas, Redwood City, Calif., 1988.

34. From the print ad "If you're into page turns, don't turn this page," *Video Systems* (September 1987).

35. From Microtime brochure, n.d.

36. Lars Tragardh and Paul Meyers, "Graphics: Software Update/ An Industry in Flux," *Video Systems* (December 1987), 30.

37. From direct marketing brochure, "Abekas: Tools for Creative Expression," Abekas, Redwood City, Calif., 1989.

38. From advertisement for Dubner DPS-1, in *Video Systems* (February 1987), 77.

39. From advertisement for AT&T Topas, in *Video Systems* (January 1989), 63.

40. From direct marketing brochure, by "Pastiche," 1989.

41. From print ad "Six Levels of Video in a Single Pass" by Crosspoint Latch Corporation, *Video Systems* (October 1989), 136.

42. This approach (of generalizing depth through layered planes and atmospheric changes in contrast) differed from Northern European painters, who relied on stricter graphic and linear structures to model space and depth.

43. Quotes and stills of Coca-Cola spot produced by Polycom are from Quantel Corporation's direct marketing portfolio, "Every Picture Tells a Story," 1989.

44. For a comparative analysis of the fiction effect in both cinema and the unconscious see, Christian Metz, "The Fiction Film and Its Spectator," *The Imaginary Signifier* (Bloomington: Indiana University Press, 1982), 99–147.

45. From Delta 1 print ad, *Millimeter* (October 1989), 76.

46. From direct advertising brochure on the "Pastiche," from Electronic Graphics corporation, 1989.

47. From direct marketing ad by Colorgraphics Systems corporation for its real-time color corrector, 1989.

48. Jill Kirschenbaum, "Dynamic Duos: Director-Editor Teams Look for Acceptance," *Millimeter* (October 1989), 147.

49. Ellis, *Visible Fictions*, 127–171. Ellis's influential notion that television viewing is fundamentally different from cinema viewing, and that television assumes a relationship with the viewer based on the glance rather than the gaze, has pervaded many other accounts in television critical theory in the past decade. The notion that television viewers are by definition "distracted" and therefore "innattentive" has been argued as a fundamental aspect of the televisual apparatus. *Recent applications* of Ellis's glance theory include Sandy Flitterman-Lewis, "All's Well That Doesn't End," in Lynn Spigel, ed., *Private Screenings* (Minneapolis: University of Minnesota Press, 1992), 217.

50. Rick Altman, "Television/Sound," in Tania Modleski, ed., *Studies in Entertainment: Critical Approaches to Mass Culture* (Bloomington: Indiana University Press, 1986), 39–54.

51. Ibid., 44–50.

52. Flow is discussed in Raymond Williams, *Television, Technology, and Cultural Form* (New York: Schocken Books, 1975).

53. Some of the best of this work is found in Lisa Lewis, ed., *The Adoring Audience* (New York: Routlege, Chapman, and Hall, 1991); Ellen Seiter et al., eds., *Remote Control: Television Audiences and Cultural Power* (New York: Routlege, Chapman, and Hall, 1989); and Henry Jenkins, *Textual Poachers: Television Fans and Participatory Culture* (New York: Routlege, Chapman and Hall, 1992).

54. Jean-Louis Comolli, "Technique and Ideology: Camera, Perspective, Depth of Field," and Jean-Louis Baudry,"Ideological Effects of the Basic Cinematographic Apparatus," both reprinted in Phillip Rosen, ed., *Narrative Apparatus, Ideology* (New York: Columbia University Press, 1986), 421–443 and 286–298.

55. Pierre Bourdieu, *Distinction: A Social Critique of the Judgement of Taste* (Cambridge: Harvard University Press, 1984).

56. Brian Winston, *Misunderstanding Media* (Cambridge: Harvard University Press, 1986).

Chapter 6. Loss Leader

1. "Fourteen in the Fast Lane: Profiling the Chief Programmers for the leading cable Networks," *Broadcasting and Cable* (June 7, 1993), 62.

2. *Shogun* (NBC, September 1980). *The North and The South*, a twelve-hour, $25 million production, was the top-rated miniseries of the season (ABC, November 1985). *Peter the Great* was an eight-hour, $26 million miniseries (NBC, February 1986). *Amerika*, a fourteen-hour miniseries, was set in 1997, after America had fallen to Soviet domination (ABC, February 1987). *The Winds of War,* an eighteen-hour, $40 million miniseries, was shot in six countries from a 962-page script that included 1,785 scenes (ABC, February 1983). *War and Remembrance* promised to out-do them all. According to Alex McNeill, *Total Television* (817), the show was shot in ten different countries, with a (mind-boggling) total of 757 different sets and a budget of $110 million. Analysts estimated that the miniseries lost $20 million when telecast (ABC, November 13, 1988–May 14, 1989).

3. *The Day After*, (ABC, 1983), *In the Line of Duty: Ambush at Waco* (NBC, May 23, 1993), *Amy Fisher: My Story* (NBC, December 28, 1992).

4. This risk is no more apparent than in the yearly rituals of baseball's seven-game world series and the NBA's seven-game championship final. Bids by broadcasters for these kinds of expensive high-technology productions presuppose or gamble on such series going the distance. A four-game series, for example, can spell disaster for a network, since a large percentage of its ad revenues abruptly vanish with the lack of competition.

5. John Lippman, "All Webs Spinning Record Coin," *Variety* (November 8, 1989), 47.

6. Executive Perry Simon, as quoted in Beth Kleid, "Escape Routes: NBC Lathers up Prime Time with Two Soapy Mini-Series," *Television Times* (August 22, 1993), 15.

7. Mike Freeman, "Co-op Advertising: Trick or Treat," *Broadcasting and Cable* (June 14, 1993), 56.

8. Stuart Kaminsky, *American Television Genres* (Chicago: Nelson-Hall, 1985). John Cawelti, *Adventure, Mystery and Romance* (Chicago: University of Chicago Press, 1976).

9. David Marc and Robert J. Thompson, *Prime Time, Prime Movers* (Boston: Little, Brown, 1992). Histrionic claims of "originality" by programmers run head-on into the pervasive view that television—regardless of its content—is a mundane art. Apart from high-culture humanistic critics who see the structure of broadcasting as banal, even an influential leftist theorist like Walter Benjamin also cast television as mundane (without "aura"), with his progressive variant of technological determinism. "Art in the Age of Mechanical Reproduction," *Illuminations*.

10. Raymond Williams, *Television, Technology, and Cultural Form* (New York: Schocken Books, 1975).

11. *Bradymania* (ABC, May 19, 1993), *Legend of the Beverly Hillbillies* (CBS, May 24, 1993).

12. George Lipsitz, *Time Passages: Collective Memory and American Popular Culture* (Minneapolis: University of Minnesota Press, 1990).

13. Mimi White introduces and discusses a process called the "overproduction of history" in her essay, "Television—A Narrative, a History," *Cultural Studies* (1990), 282–300. For a particularly good account of the relationship between history and liveness see the chapter "Television and Its Historical Pastiche," in James Schwoch et al., eds., *Media Knowledge: Readings in Popular Culture, Pedagogy, and Critical Citizenship* (Albany: State University of New York Press, 1992), 3–4. What I describe as "historical exhibitionism" (a stylistic and mass communitarian ritual that reconstructs identity), White describes as the "overproduction of history."

14. Lynn Elbers, "Jewish Culture and History Get a Dramatic Workout on Network Series," *TV Times* (May 16–22, 1993).

15. Ibid.

16. See especially Mimi White, *Tele-Advising: Therapeutic Discourse in American Television* (Chapel Hill: University of North Carolina Press, 1992).

17. Michel Foucault, *The History of Sexuality,* vol.1., trans. Robert Hurley (New York: Pantheon, 1976), and Michel Foucault, *Discipline and Punish* trans. Alan Sheridan (New York: Random House, 1977).

18. Print ad for "History TV," *Broadcasting and Cable* (June 7, 1993), 48–49.

19. Voice-over in the opening of *War and Remembrance.*

20. Intertitle graphic broadcast with miniseries.

21. Voice-over narration in opening of miniseries.

22. Kristin Thompson in *Ivan the Terrible: A Neoformalist Analysis* (Princeton: Princeton University Press, 1981) has done much to elaborate on the function of formal excess in film narration. Portions reprinted in Phillip Rosen, ed., *Narrative Apparatus Ideology* (New York: Columbia University Press), 130–142. Yet in systematic surveys of narrative theory like David Bordwell's, the role of style in narrative is treated much less directly than are paradigmatic and rhetorical aspects of realist film form. See David Bordwell, *Narration and the Fiction Film* (Madison: University of Wisconsin Press, 1985), 49–53, which devotes only a few pages to the role of visual style as a "principle(s) of narration." Other comprehensive narrative theorists like Seymour Chatman do not seem concerned at all with the role of visual style. See Seymour Chatman, *Story and Discourse: Narration in Fiction and Film* (Ithaca: Cornell University Press, 1978).

23. Bordwell, 53.

24. As summarized by Bordwell (53) from Thompson, *Ivan the Terrible.*

25. Christine Gledhill has summarized Geoffrey Nowell-Smith's work ("Minnelli and Melodrama," *Screen* 18, 2 [Summer 1977]) on melodrama by describing instances of formal excess as radical "breakdowns" in the dominant system. Such breakdowns "represent

the 'ideological failure' of melodrama as a form, and so its *progressive potential*" (italics mine). In Pam Cook, ed., *The Cinema Book* (New York: Pantheon, 1985), 76.

26. Jane Feuer, "Melodrama, Serial Form, and Television Today, *Screen*, 25, 1 (January–February 1984. Reprinted in Manual Alvarado and John O. Thompson, eds., *The Media Reader* (London: BFI Press, 1990), 253–264.

27. Feuer, 258.

28. Two things are worth noting before turning to a more in-depth analysis of *War and Remembrance*. First, Feuer was basically defining excess in television within the context of *film style mise-en-scène*. While the miniseries form does have generic roots in film melodrama, its connection is much looser than that of, say, the link between primetime soaps and family melodrama films of the 1950s. The evolution of the soaps into one of television's premier continuing forms suggests, however, that a different kind of logic—the continuing form specific to television—works to subjugate its formal excesses under a *dominant narrative*. Yet, the first episode of *War and Remembrance* shows a very different, unmelodramatic, and uncinematic kind of formal excess. Excess here results as much from videographic flux as from filmic mise-en-scène. That is, when compared to the primetime soap's dramatic film-style excess, which Feuer describes, stylistic excess in the miniseries (its visual flow and its infiltrations of image and text) is less predetermined or locked down by narrative in the first place. The extent of subversion in the excessive text on television, then, remains an open question. Excessive televisuality in *War and Remembrance*, in the final analysis, is not defined by cinematic style—this is ABC Circle Films, not Douglas Sirk—so it is difficult to utilize Feuer's explanation to fully account for the miniseries.

29. Although sound is beyond the scope of this study, the heavy and *repetitious* orchestral score also sets-up (that is, "inputs" and preinterprets for viewers) the excessive motifs that will recur *endlessly* throughout the miniseries.

30. Robert Burgoyne, "The Cinematic Narrator," *Journal of Film and Video* 42, 1 (Spring 1990), 3–16.

31. The reappraisal of work by early twentieth-century Russian Formalists lead neoformalists like Bordwell and Thompson to reject several main tenets of enunciation theory. The idea of an idealized and artificially constructed authorship/narrating source, for example (a concept popularized by enunciation theorists), was rejected.

32. In addition, Bordwell rejects the kind of partisan ideological implications that the enunciation theorists linked to their methods. "Suturing," or the way that continuity editing inscribed the viewer within a scene, was to the French enunciation theorists evidence of political subjugation. Spectator positioning was seen as a textual process of political emasculation and sexual repression. These conclusions are described as premature and naive speculations by theorist Bordwell. This rejection of the enunciation theorists as naive, is found in Bordwell, *Narration in the Fiction Film,* 110–111.

33. While this somber quote from the novel could have been used to bring the viewer up to speed by starting the story in medias res, the statement has no time-place information or backstory detail. As a kind of speculation on the meaning of life, keyed over immortal blackness to millions of viewers who have never read the novel, the device fulfills anything but a pedestrian narrative function. It deifies authorial wisdom and virtue.

34. John Fiske and John Hartley, *Reading Television* (London: Methuen, 1978), 83.

35. Raymond Williams, "Interview with Raymond Williams," in Tania Modleski, ed., *Studies in Entertainment: Critical Approaches to Mass Culture* (Bloomington: Indiana University Press, 1986), 15.

36. See Robert Scholes, "Narration and Narrativity in Film," Gerald Mast and Marshall Cohen, eds., *Film Theory and Criticism* (New York: Oxford University Press, 1985), 390–404.

37. Ibid., 396.

38. Christian Metz states that the bracketed syntagma "is frequently . . . different successive evocations [that] are strung together through optical effects (dissolves, wipes, pan shots, and less commonly, fades)." It is a reference that evokes the minimally narrative ordering of shots in the endless opening of *War and Remembrance. Film Language* (New York: Oxford University Press, 1974), 126–127.

39. These televisual forms do not fit easily into Metz's model of the grand syntagmatic. The extreme diversity of shots in the overall segment bear at times a sense of alternation, since certain characters reappear and are intercut with many others, but the sequence as a whole is not clearly linear or chronological. Of any of his categories, by minute twenty of the sequence this long flow of autonomous fragments can be viewed retroactively and alternately as episodic and ordinary sequences. See Metz, *Film Language,* 108–148.

40. The problem I describe here, the difficulty of controlling large scale narratives, may be symptomatic of television in general. The excessiveness and discursive density operative in this genre, then, may only be the miniseries' solution to the problem of handling massive narratives. Other genres address the task differently.

41. Discursive aspects of television frequented the work of Modleski, Morse, Kaplan, Altman, and others. See especially E. Ann Kaplan, ed., *Regarding Television: Critical Approaches, An Anthology* (Frederick, Md.: University Publications of America/American Film Institute, 1983); Tania Modleski, ed., *Studies in Entertainment: Critical Approaches to Mass Culture* (Bloomington: Indiana University Press, 1986); Rick Altman, "Television/Sound," in Modleski, 39–54; and a special issue of the *Journal of Film and Video* (Summer 1985), for a collection of articles utilizing discursive analysis.

42. Margaret Morse, "Sport on Television: Replay and Display," in Kaplan, *Regarding Television,* 44–66. The quote is from Margaret Morse, "The Television New Personality and Credibility," in Modleski, *Studies in Entertainment,* 56.

43. Pierre Bourdieu, *Distinction: A Social Critique of the Judgement of Taste* (Cambridge, Mass.: Harvard University Press, 1984).

44. Dan Harries, "Fringe Benefits/The Subculture of Film Cult(ure)," paper presented at the Society for Cinema Studies Conference, Los Angeles, May 26, 1991.

Chapter 7. Trash TV

1. As quoted in Katherine Stalter, "Animation: New Techniques Continue to Push Animated Projects into Spotlight," *Film and Video,* November 1992, 46.

2. *Remote Control* (MTV, 1987–1990). *Fraggle Rock* (HBO, 1983–1988). *The Jim Henson Hour* (NBC, 1989). *Pee-Wee's Playhouse* (CBS, 1986–1991).

3. *Wheel of Fortune* (NBC, 1975–1989, 1991–).

4. Peter Greenbaum, "Making a Good Impression: Television Show Opens Must Hook the Viewer While Conveying a Sense of Style," *Film and Video* (May 1993), 112–114.

5. *Tiny Toon Adventures,* Channel 11, Los Angeles, April 28, 1993 (Fox, 1992–1993).

6. Thomas Fields-Meyer and Richard L. Meyer, "Bedrock Values," *New York Times,* March 27, 1993, A19.

7. "Morning Report: Television," *Los Angeles Times,* Friday August 27, 1993, F1.

8. Charles Solomon, " 'Rangers,' 'Bonkers!' Not Kiddie Fare," *Los Angeles Times,* September 6, 1993, F11.

9. In *Camera Obscura* 17 (May 1988). Reprinted in Constance Penley, *The Future of Illusion: Film, Feminism and Psychoanalysis* (Minneapolis: University of Minnesota Press, 1989), 162.

10. Ian Balfour cites several of the same gags (the male-female kissing spoof, the phallic pencil sketch, etc.) to back up his argument that the show is about the "rhetoric of sexual difference." In Balfour, "The Playhouse of the Signifier," *Camera Obscura* 17 (May 1988): 155–168.

11. The clearly divergent agendas of the industry and high theory are evident by comparing the technological focus of "Special F/X," *Millimeter* (September 1986), 107–110, to the theoretical preoccupation with sexuality in "Pee-Wee Herman: The Homosexual Subtext," *Cine-Action* 9 (Summer 1987), 3–6.

12. Tania Modleski, "The Incredible Shrinking He(r)man: Male Regression, the Male Body, and Film," *Feminism Without Women: Culture and Criticism in the "Post-Feminist" Age* (New York: Routlege, Chapman, and Hall, 1991), 90–111, also criticizes the *Camera Obscura* approach, but in a very different way. She attacks the shortsighted logic of the journal's critics, for embracing a form that is no less misogynistic than other sexist forms

that are concerned with "bigness." To her, Pee-Wee's obsession with "smallness" and sexual regression is simply a response to *male insecurities*. Ultimately the show's overemphasis on the phallus and on male anxiety "undermines the feminist project." I am less interested in arguing about the usefulness or correctness of interpretations of the show, than in suggesting that there is an alternative and more provisional way to account for the pleasures of its reception—one that pays greater and more immediate attention to the program's complicated play of signs.

13. Peter Wollen, "Godard and Counter Cinema: Vent d'Est," *Readings and Writings: Semiotic Counter-Strategies* (London: Verso, 1982), 79–91.

14. Wollen's work is really a synthesis and crystalization of the modernist aesthetic tradition espoused by Berthold Brecht. See especially *Brecht on Theater*, trans. John Willett (New York: Hill and Wang, 1986), 37–38, 69–74. Brecht's list of radical elements in epic theater strongly resembles Wollen's later list of polarities. Godard was strongly identified, even by his contemporaries in the popular press, by his use of disruptive technique, discontinuity, and plays with image and text. For a dominant cultural indictment of Godard's counter-cinematic forms, and as a contrast to Wollen's apologetic and homage, see Tom Milne, ed., *Godard on Godard* (New York: De Capo, 1972), 196–200.

15. Sandy Flitterman, "The Real Soap Operas: TV Commercials," E. Ann Kaplan, ed., *Regarding Television* (Los Angeles: The American Film Institute, 1983), 84–96.

16. This "secret word" convulsion was also a homage to Groucho Marx's show *You Bet Your Life* (NBC, Syndicated, 1950–1961), in which everyday words and biting verbal salvos from Groucho regularly made part of the pleasure of watching antipleasurable.

17. Pee-Wee's perpetual and off-handed adult asides are merely an extreme form of a split audience technique, that was used more covertly in *Rocky* (ABC, 1959–1961) and *Bullwinkle* (NBC/ABC, 1961–1973) cartoons and *Soupy Sales* (Syndicated, 1965–1967). The earlier, more restrained shows tended to camouflage their multiple audience appeals.

18. For an influential account of the dissolution of the object and subject in postmodern and postindustrial culture, see Jean Baudrillard, *The Ecstasy of Communication* (New York: Semiotext(e), 1987), 11–96.

19. Susan Stewart, *On Longing: Narratives of the Miniature, the Gigantic, the Souvenir, the Collection* (Baltimore: Johns Hopkins University Press, 1984), 152.

20. Stewart is describing an essay written by E. L. Magoon in 1852 called "Scenery and the Mind," which appeared in *The Home Book of the Picturesque* (New York: G. F. Putnam, 1852), 1–48. Comment is from Stewart's book, *On Longing,* 75.

21. If accumulation teaches desire and lack, then bricolage (a lifestyle activity) implies that there is a legitimate reason for all of the stuff in the first place. Bricolage, then, does not just teach creativity. In the face of apparently endless resources, it really is good management training for the next generation of MBAs.

22. I am borrowing this term from Mimi White, *Tele-Advising: Therapeutic Discourse in American Television* (Chapel Hill: University of North Carolina Press, 1993). As I hope to make clear in the analysis that follows, however, by viewing this program as a problem-solving operation, I am less interested in the concept of therapy in a classical sense, than I am in televisuality as a form of social therapy—as an activity and play of signs that rationalizes and constructs a natural place for the individual within culture.

23. Lasch's book gives a historical account that explains well some of the cultural impulses behind the performance and ornament of *Pee-Wee's Playhouse:* it describes the ideology of privatism in American culture and the resulting narcissistic preoccupation with self. Christopher Lasch, *The Culture of Narcissism: American Life in an Age of Diminishing Expectations* (New York: W. W. Norton, 1978).

Chapter 8. Tabloid TV

1. Howard Rosenberg, "Whatever Happened to Privacy?" *Los Angeles Times*, November 25, 1992, F1.

2. John Dempsey, "More Mags Will Fly in the Fall; Too Much of a Bad Thing?, *Variety* (April 12, 1989). Reprinted in Marilyn Mattelski and David O. Thomas, eds., *Variety: Broadcast-Video Sourcebook I* (Boston: Focal Press, 1990), 27–28.

3. James Friedman has suggested that television's contradictory designation "live-on-tape" is analogous to "frozen fresh" in retail grocery discourse.

4. Roger Ailes, "The Truth About Tabloid TV," *TV Guide* (May 7, 1993), 5.

5. From Margy Rochlin, "The Unblinking Eye: David Lynch and Mark Frost Chronicle America," *International Documentary* (Fall 1990), 11.

6. Frost and Lynch *act* not the least bit aware of the tradition of poetic documentary that goes back to the work of Walter Ruttman and Joris Ivens in the 1920s. Their definition of the concept as a "lyrical, visual style" that tells the story fits as well the documentaries *The Plow That Broke the Plains* (Pare Lorentz, 1936)—the type of film that John Grierson would call "aestheticky"—and *Song of Ceylon* (Basil Wright, 1934), films that have been characterized as highly personal, lyrical, visually exquisite, and complex. Jack C. Ellis, *The Documentary Idea: A Critical History of English Language Documentary Film and Video* (Englewood Cliffs, N.J.: Prentice-Hall, 1989), 67–68, 90, 102.

7. While Wright and the French Impressionist filmmakers provide apt but ignored precedents for Lynch and Frost's poetic aspirations, Robert Flaherty's films predate Lynch and Frost's alienated (and romantic) pseudo-anthropological airs by at least the same number of decades.

8. Now classic works on the ideology of the technological and cinematic apparatus include Jean-Louis Baudry, "Ideological Effects of the Basic Cinematographic Apparatus," *Film Quarterly* 28, 2 (Winter 1974–1975), 39–47; Baudry, "The Apparatus: Metapsychological Approaches to the Impression of Reality in the Cinema," *Camera Obscura* 1 (Fall 1976), 104–128; Jean-Louis Comolli, "Technique and Ideology: Camera, Perspective, Depth of Field," parts 3 and 4, *Narrative, Apparatus, Ideology*, ed. Phillip Rosen, trans. British Film Institute (New York: Columbia University Press, 1986), 421–443. The single most influential text articulating the gaze theory is Laura Mulvey, "Visual Pleasure in Narrative Cinema," *Screen* 16, 3 (Autumn 1975), reprinted in Rosen, *Narrative, Apparatus, Ideology*, 198–209.

9. Brian Winston, "Great Artist or Fly on the Wall: The Griersonian Accommodation and Its Destruction," *Visual Explorations of the World: Selected Papers from the International Conference on Visual Communication*, Jay Ruby and Martin Taureg, eds. (Aachen, GDR: Edition Herodot im Rader Verlag, 1987), 190–204.

10. The best utilization and interrogation of the apparatus theory from the French and psychoanalytic tradition when applied to documentary is William Guynn, "The Ambivalent Spectator: Je Sais Bien, Mais Quand Meme," paper presented at the Ohio University Film Conference on Documentary, November 9, 1990, and *A Cinema of Nonfiction* (Cranbury, N.J.: Farleigh Dickinson University Press, 1990).

11. Alan Rosenthal, ed., *New Challenges for Documentary* (Berkeley: University of California Press, 1988), 345–424.

12. Bill Nichols, *Ideology and the Image: Social Representation in Cinema and the Other Arts* (Bloomington: Indiana University Press, 1981), is certainly an exception to this tendency.

13. I take this perspective from Erving Goffman, *Frame Analysis: An Essay on the Organization of Experience* (Boston: Northeastern University Press, 1974).

14. The Fox network's importation of David Lynch for documentary purposes (and for economic gain) evokes strongly a production strategy that the network's predecessors undertook in the 1920s. When William Fox imported German star director F. W. Murnau in 1926, it was done more for show and prestige than for the actual box office that Murnau could or would deliver. Showcasing Murnau and the German artistic look offered the B-movie mill of Fox the chance to rise to a new level of cultural distinction. This process is discussed in economic terms, as product differentiation, in Robert Allen and Douglas Gomery, *Film History: Theory and Practice* (New York: Alfred A. Knopf, 1985), 86–108.

15. Laurie Thomas and Barry Litman describe well the process by which the Fox Network used distinctive programming tactics to counteract their inferior economic and regulatory position in the late 1980s, in "Fox Broadcasting Company: Why Now?: An Economic Study of the Rise of the Fourth Broadcast 'Network,' *Journal of Broadcasting and Electronic Media* 35, 2 (Spring 1991), 139–157.

16. This combination of references makes one wonder if this is a documentary or science fiction posing as a documentary.

17. *POV* is comprised of recent and existing documentaries produced by independent filmmakers and is funded by a consortium of Public Broadcasting stations and foundations. Requests for submissions to its fifth consecutive season were made for the 1991–1992 programming year. This went on in spite of a growing backlash to a perceived liberalism and penchant for radical topics. Marlon Riggs's poetic and sensitive look at part of the African-American gay subculture, entitled *Tongues (Un)Tied*, and the documentary *Stop The Church*, loudly critical of Catholic social policies, were pulled from broadcast in the summer of 1991 by many affiliate PBS stations, and threats from the right wing continued against *POV.*

18. *Tribes* on *American Playhouse* represented somewhat of a breakthrough for the stylized documentary organized around personal vision. The program contrasts starkly, after all, with the standard documentary venues on PBS that are thoroughly topical in nature (*Nova, Smithsonian, Nature*) or that bear the journalistic and ontological burden of the sociopolitical other (*Frontline, Bill Moyers Specials*).

19. If one were to view Mahurin's work through the lens of John Grierson's bipartite definition of documentary as "the creative treatment of actuality," *Tribes* intensifies and performs the "creative treatment" to a degree and in a way that overshadows the collected fragments of "actuality."

20. The dark, psychological demeanor of Mahurin contrasts with what Jack Ellis describes as the aesthetic tradition of impressionism that influenced Walter Ruttman in *Berlin: Symphony of a Great City* (1927), and Joris Ivens in *Rain* (1927). See Ellis, *The Documentary Idea*, 49–50. Impressionism, after all, was really just an outgrowth of scientific naturalism that was dominant through the nineteenth century. It was a tradition that placed great value on visual observation.

21. Metz describes how central the actions of "disavowal" and "delegation" are in the cinema of fiction. Both terms are based on a fetishistic retreat and a denial of the artifice of cinema. This is the very opposite of the process at work in the televisual documentary, for it involves a process that flaunts artifice and the physicality of form. Christian Metz, *The Imaginary Signifier* (Bloomington: Indiana University Press, 1976), 71–73, 99–148.

22. Guynn, "The Ambivalent Spectator," November 1990.

23. For a more detailed discussion of the new pictorial language of television, see chapter 5, "Franchiser."

24. Of course, shopping (unlike voyeurism) is not pathological, but popular culture has toyed with the idea that obsessive shopping is. Consider the bumper sticker that states, "When the going gets tough, the tough go shopping," as a metaphor for the cliché that middle class suburbanites with disposable income use shopping and the mall—not just as an escape—but as a therapeutic behavior and cathartic act. It is easy to see, however, why the collection-effect has more to do with a political problematic, with a dynamic of power and status, than it does with gender. The televisual documentary flow promises abundance. Its embellishment and excess are traits that can be appreciated by those with enough cultural capital get the inside references. Those who do not get the stylistic references, know they should. Like consumers who do not or cannot keep up with the latest fashions, such viewers know by implication that they are on the outside of something that they ought to understand. As a form of semiosis that makes constant demands on the viewer to discriminate, the genre also encourages the viewer to consume more mass culture. For through increased cultural consumption the viewer becomes a better player at decoding the semiotic style-driven consumer game. The televisual genre has also come to stand for the discriminating buyer-viewer, for the well-trained semiotic-shopper. Although connoisseurship of this sort is neither male nor female, it certainly is a political and class issue, for consumers vary widely in their ability to play mass culture. Even those who are not good players are made aware of their lack.

Chapter 9. Televisual Audience

1. Critic Jonathan Miller of *The New Yorker*, is quoted in David Littlejohn, "Thoughts on Television Criticism," in Richard Adler, ed., *Television as a Cultural Force* (New York: Praeger, 1976), 159.

2. Rick Marin, "Reading Wild Palms," *TV Guide,* May 15, 1993, 16.

3. Enunciation theory, based on the premise that film and television texts positioned, interpolated, or hailed the viewer continued to develop under the influence of psychoanalysis, feminism, and postmodernism in the 1980s. See especially, Beverle Houston, "Viewing Television: The Metapsychology of Endless Consumption," *Quarterly Review of Film Studies* 9, 3 (Summer 1984), who shows how the program flow constructs desire in viewer, analogous to the oral stage in psychoanalysis, where TV is positioned as a "maternal" source of pleasure. Sandy Flitterman-Lewis, extends and adapts the methods developed by enunciation theorists like Raymond Bellour, in her close shot-by-shot analysis of television sequences in "All's Well That Doesn't End—Soap Opera and the Marriage Motif," in Lynn Spigel and Denise Mann, eds., *Private Screenings: Television and the Female Consumer* (Minneapolis: University of Minnesota Press, 1992). To Flitterman-Lewis, the viewer is textually defined as distracted and glancing, given the frustrating presence of narrative irresolution, symptomatic of the soaps. John Wagner extrapolates from recent television, and the work of Lyotard and Baudrillard, more general conclusions about the postmodernist negation of the subject, in "Absolute TV," *Quarterly Review of Film and Video* 14, 1–2 (1992).

4. Traditional mass communications approaches frequently favored a cognitive persuasion and effects model which implied that mass viewers were somehow injected by propaganda, stimuli, and information. H. D. Lasswell, *Propaganda Techniques in the World War* (New York: Alfred A. Knopf, 1927). The incursion of the Frankfurt School—Adorno, Horkheimer, and others—into American social science in the 1930s reinforced the notion of the anonymous and passive mass. Criticized as a "magic-bullet theory" by subsequent theorists, research like that of Paul Lazarsfeld, et al., *The People's Choice* (New York: Duell, Sloan and Pearce, 1944), and Paul Lazarsfeld and R. K. Merton, "Mass Communication, Popular Taste, and Organized Social Action," in L. Bryson, ed., *Communication of Ideas* (New York: Harper and Row, 1948), 95–118, began to question the singularity of the false consciousness model in effects study. Uses and gratifications research would have absolutely no part of the passive audience presupposed by broadcasting social science. Using a functionalist rather than effects approach, this movement in the late 1950s and 1960s frequently invoked psychological models like Maslow's hierarchy, from A. H. Maslow, *Motivation and Personality* (New York: Harper and Row, 1954), to show how television, far from being dangerous, actually fulfilled many basic human and emotional needs. Elihu Katz argued, for example, that far from lacking control, viewers "selectively fashion what they see and hear," and that viewers cannot be influenced by a message unless they have some use for it in the first place. "Mass Communications Research and Popular Culture," *Studies in Public Communication* 2. In developing the field of content analysis, George Gerbner and his colleagues demonstrated just how difficult it is to leap from the manifest content of mass-mediated messages to general conclusions about the audience. George Gerbner et al., "The Demonstration of Power: Violence Profile No. 10," *Journal of Communication* 29, 3 (1979) 177–196. A good example of how the mainstreaming implications from this tradition can be used to expose hegemonic operations against minorities is Larry Gross, "Out of the Mainstream: Sexual Minorities and the Mass Media," in Ellen Seiter et al., eds., *Remote Control: Television, Audiences and Cultural Power* (New York: Routlege, 1989), 130–149. Attacking the passive-active dichotomy at the very crux of American mass-communications social science and effects research is Robert Kubey and Mihalyi Csikszentmihalyi, *Television and the Quality of Life: How Viewing Shapes Everyday Experience* (Hillsdale, N.J.: Lawrence Erlbaum Associates, 1990). Using neurophysiological methodology, the research not only finds passivity at the root of much television viewing, but justifies it as a therapeutic state that helps the human subject with its organizing and coping functions.

5. The ideological, class- and race-consciousness audience research of Stuart Hall, centered around the Birmingham group in the 1970s, was of course a formative influence in British Cultural Studies. Stuart Hall "Encoding/Decoding in Television Discourse" (1973), reprinted in Stuart Hall et al., eds., *Culture, Media, Language* (London: Hutchinson, 1981), and Stuart Hall and T. Jefferson, *Resistance Through Rituals* (London: Hutchinson, 1978), suggest both the centrality of the audience and the political importance of the viewer's activities in this tradition. Later works, like Dick Hebdige's *Subculture: The Meaning of*

Style (London: Methuen, 1979), influenced a number of subsequent studies that also fled from the text in search of the socioeconomic and class-defined groups that actually consumed and used culture. One of the best (but partisan) surveys of the lasting influence of David Morley's "nationwide" study, *The "Nationwide" Audience* (London: The British Film Institute, 1980) and the debates that followed its publication is David Morley, *Television Audiences and Cultural Studies* (New York: Routlege, 1993).

6. Television theorists who have popularized the ethnographic model in critical and cultural studies include, Ian Ang, *Watching Dallas: Soap Opera and the Melodramatic Imagination* (New York: Methuen, 1982), a study that surveyed cognitive viewer reports to account for television's multiple viewing pleasures and audience relationships. Although they make absolutely no mention of Ang's work on the very same subject ten years earlier, Elihu Katz et al., "On Commuting Between Television Fiction and Real Life," *Quarterly Review of Film and Video* 14, 1–2 (1992), also analyze the cross-cultural construction of meanings in *Dallas*, by examining viewer's "self-reports," and focus group discussions. As with Ang, viewers are allowed to consciously interpret their meanings and motives for watching—a radical notion, at least for traditional critical theory. Dick Hebdige's *Subculture,* an account that characterized subcultures as a kind of sociological avant-garde and style as a political phenomenon that impacted both oppositional practice and cultural assimilation, was also influential for critical and cultural study, for it demonstrated how ethnographic research did not have to leave political partisanship behind. Lisa Lewis has provided an anthology of affirmative fan studies that speak to the interpersonal concerns, questions of individual worth, and the tension between community and alienation that drives fan subcultures in Lisa Lewis, ed., *Adoring Audience: Fan Culture and Popular Media* (New York: Routlege, 1992). Henry Jenkins has provided the best recent television ethnography, by examining fans as more than just "counter-readers," but as producers and artists, who "filk" mainstream programming fare for their own ends, frequently as part of weekend fan subcultures and communities. Henry Jenkins III, *Textual Poaching: Television Fans and Participatory Culture* (New York: Routlege, 1992).

7. Kim Mitchell, "Fortunately, They're Not in the Demographic," *Multichannel News,* March 8, 1993, 6.

8. Joel Cheatwood, news vice president, WSVN—Miami, as quoted by Harry A. Jessell, "New Wave Newscasts Anchor WSVN Makeover: Ex-Affiliate Finds a New Niche," *Broadcasting,* October 12, 1992, 24.

9. "Cicely" is both the name of the town in which the ensemble of characters in *Northern Exposure* resides and the name of the turn-of-the-century woman who arrived to bestow upon the townspeople a sense of civility and purpose.

10. The hyperactive, broad-but-shallow *Cliff Notes* intellectualism that I posit at the center of excessive shows like *Northern Exposure* is very similar to the education and class-based recognition-pleasures that work in a Woody Allen film or a Cole Porter song.

11. "Morning Report: Where They're Watching," *Los Angeles Times,* March 11, 1993, F2.

12. Frazier Moore, "From Schools to Truck Stops, 'Place-Based' Media Flourish," *TV Times,* March 1993, 7.

13. Sydney Blumenthal, "The Syndicated Presidency," *The New Yorker,* April 5, 1993, 42–43.

14. Blumenthal argued that it was Clinton's technical endrun around the media that most infuriated reporters: "Now he is accessing the very technology that the press considered their monopoly: the electronic filigree, space satellites and computer networks."

15. Eugene Jackson, World African Network, Los Angeles, in "Monday Memo," *Broadcasting and Cable,* June 14, 1993, 98.

16. Advertisement by USA Network in *Broadcasting and Cable,* June 7, 1993, 36–37.

17. CBS president Jeff Sagansky as quoted by Rick DuBrow, "No.1 CBS Has Its Eye on Middle-Age Viewers," *Los Angeles Times,* May 21, 1993, F21.

18. NBC promo from *NBA Eastern Semifinal Series* between the New York Knicks and the Chicago Bulls, May 1993.

19. Mike Freeman, "Co-op Advertising : Trick or Treat," *Broadcasting and Cable,* June 14, 1983, 56.

20. "Do You Have Sony Style," ad, *Life*, August 1993, 48.

21. Scott D. Yost, "Viewpoint: Triumphs of the Tube," *Emmy*, August 1993, 19.

22. "Editorials: Getting the Word Out," *Broadcasting*, November 16, 1992, 90.

23. Sean Scully, "For Some, Interactive Future is Now," *Broadcasting and Cable*, June 14, 1993.

24. Leon E. Wynter, "Young People Are Hip to Minority TV Tastes," *Wall Street Journal*, May 3, 1993, B1. Nielsen ratings show that current young white audiences are skewed in their tastes more toward black and Hispanic shows than to other top-rated white shows.

25. Steve McClellan, "Grabbing the Grazers in a Crowded Field," *Broadcasting*, February 1, 1993, 14.

26. Andrea Ford, "Suspected Gang Members Held in Slayings of 2 Men," *Los Angeles, Times*, January 6, 1993, B1, B4.

27. Since videotapes had been used in the arrests of participants in the L.A. riots, camcorders on the street were no longer seen as benign.

28. See for example, *Four More Years*, produced by the TVTV Collective, 1972, distributed by Electronic Arts Intermix.

29. Dierdre Boyle, after chronicling the move of TVTV-types Michael Shamberg, Bill Murray, and Harold Ramis into the mainstream entertainment industry notes: "Few have come along to take up the challenge of guerrilla television's more radical and innovative past," in "Guerrilla Television," *Transmissions*, Peter D'Agostino, ed. (New York: Tanam Press, 1985), 213.

30. Brian Winston, *Misunderstanding Media* (London: Routledge and Kegan Paul, 1986).

31. This estimation of Paik as the technological visionary is widespread, but tends to overlook the fundamental role that engineer Shuye Abe played in developing the Paik-Abe synthesizer.

32. When independent tapes offered private looks at sensational or inaccessible figures the networks *would* find a way to air the tapes. A half-inch tape on radical Abbie Hoffman, for example, was aired by CBS news.

33. The development of the first generation of time-base correctors allowed engineers to correct the skew and timing errors caused by imperfections in small format half-inch videotape recorders. These TBCs essentially squared-up the blanked ends of each scanning line in the frame in order for taped footage to meet FCC requirements.

34. David Bordwell et al., *Classical Hollywood Cinema* (New York: Columbia University Press, 1985), 251-260.

35. Bruce Owen and Steven Wildman, *Video Economics* (Cambridge, Mass.: Harvard University Press, 1992), 266–275.

36. Owen, *Video Economics,* 276–277.

37. An earlier cassette tape format, EVR, had previously failed to achieve widespread use.

38. These now acceptable recordings were frequently produced on portable three-quarter–inch decks like the SONY V0-3800, a low-band predecessor to the BVU-100 and 200.

39. *America's Funniest Home Videos* (ABC, 1990–), *Rescue 911* (CBS, 1989–).

40. The impact of the industry-wide economic recession on television in the years following 1989 and 1990 is discussed more fully in the next chapter of this book.

41. Michel Foucault, *Discipline and Punish* (New York: Vintage, 1979).

42. Owen, *Video Economics*, 297–313.

43. Bordwell, *Classical Hollywood Cinema*. "Trended change" is discussed on pages 247 and 248.

44. Pierre Bourdieu, *Distinction: A Social Critque of the Judgement of Taste* (Cambridge: Harvard University Press, 1984).

45. KROQ, a Los Angeles–area radio station, ran this campaign for fan participation in U2's national tour, fall 1992.

46. *JFK*, directed by Oliver Stone, 1991.

47. The special on Kevin Costner and *Dances With Wolves* was played repeatedly during the film's release, in fall 1991, on MTV.

48. *The Making of a Live Television Show*, Los Angeles, 1971, PVA-9387.

49. *Zoo TV*/U2/Channel 13/Syndicated Special (1992).

50. Phillips, Clear Sound ad (1992).

Chapter 10. Televisual Economy

1. Bill Burns, Timothy Carlson, and Jerry Lazar, "Failed Shows, Few Replacements Fueled Shakeup at CBS," *TV Guide*, December 9, 1989, 51.

2. Amy Adelson, senior vice president, ABC Productions, quoted in Katharine Stalter, "Developing Quality Television: Executives Work with Producers and the Networks to Improve Programming," *Film and Video* (January 1993), 48.

3. "Grapevine: Ratings Rx," *TV Guide*, November 4, 1989, 40.

4. These production elements are all included in *Cop Rock* (ABC, 1990).

5. ABC made $150 to $200 million in profits in 1989 (compared to losses of $70 million and $4 million in 1986 and 1988 respectively); CBS reached almost $100 million in 1989 versus $45 million in 1988. John Lippman, "All Webs Spinning Record Coin," *Variety*, November 8, 1989. 1, 47.

6. Daniel Cerone, "Pioneer ABC Lost Its Way in the End," *Los Angeles Times*, July 24, 1991, F9.

7. Bill Carter, "Fall TV Schedules Go Back to Basics," *New York Times*, May 25, 1991, 48.

8. CNN rather opportunistically reported this fact in its special on the economics of network television, when Terry Keenan interviewed industry experts on *Moneyline*, August 26, 1991.

9. ABC's financial figures and Robert Iger's response are reported in Cerone, "Pioneer ABC," F9. Iger, entertainment president of the network, suggested solving the problem by giving over that portion of total programming that was losing money to affiliate stations, who could produce cheaper programming.

10. CBS increased the license fees for *Murder She Wrote* from $25 million to $40 million in 1990. NBC paid a record $48 million per half-hour episode for *The Cosby Show* in the same year. Both retreated from these commitments, as the networks were forced to deal with a decrease in revenues and viewership. John Lippman, "Tangled TV Webs Begin to Unravel," *Los Angeles Times*, August 8, 1991, F17.

11. S.C., "Twin Peaks Caps ABC Season," *Broadcasting* 118, 17 (April 23, 1990), 35.

12. "Rad TV," *Village Voice*, April 10, 1990, 32–42.

13. Lisa Kennedy, "Risque Business: Fox Stoops to Conquer," *Village Voice*, April 10, 1990, 37.

14. Although few awards resulted from the twenty-six nominations, critical recognition of Fox helped make this a watershed year for the young fourth network. The industry ceased to speculate on its demise or survival and began to see it for what it was: a long term competitor with a distinctive programming attitude.

15. As quoted in Rick Du Brow, "Emmy's 'Peaks' Season," *Los Angeles Times*, August 3, 1990, F1.

16. *Broadcasting*, April 23, 1990, 49.

17. Cerone "Pioneer ABC" F9.

18. Mirroring this extended scheduling misdirection and quandary, many viewers began to feel that narrative of *Twin Peaks* had also lost its way. Without its original dramatic center, all that remained by the time of its cancelation was its shell; its attitude and look.

19. Daniel Cerone, "Game Shows: The Price Isn't Right," *Los Angeles Times*, August 28, 1991, F1, 5–6.

20. Carter "Fall TV Schedules Go Back to Basics." 48.

21. These viewing and financial statistics are from A.C. Nielsen Co., NBC, and Economists, Inc., as reported in Lippman, "Tangled Webs Begin to Unravel," A1, 16.

22. One hour of *America's Funniest Home Videos* programming (two episodes) costs between $400,000 and $600,000 to make. The average price of a half-hour network show is $400,000, while the average price of a one-hour show is approximately $900,000. John Lippman, "Networks Push for Cheaper Shows," *Los Angeles Times*, February 19, 1991, D1, D10.

23. Even the Rodney King footage found itself in a seemingly infinite number of visual guises. Hardly any broadcasters showed the footage by itself without some sort of station mediation. At first, merely the original camcorder's time-date information appeared in the

frame. Quickly, however, each station that bought or stole the footage off the satellite feed from CNN added and keyed their station logo in some part of the image. Other stations played the footage in graphic boxes, thereby further distancing the viewer from the reality, while polite and attractive announcers and newsreaders chatted away about the event's significance. *Primetime Live* digitized the footage and made it a part of their sweeping graphic opening, even though they used it to lead into a previously made expose' on Chief Daryl Gates. Public affairs programming used video stills from the beating as a catchy background for their arguments and interviews. Video images were extracted and printed in newspapers and on posters. The maker of the image quickly lost control as the footage traveled instantaneously around the country and across various media. The final chapter discusses this process in more detail.

24. John Berger makes a similar argument about the oil painting as a form of possession in *Ways of Seeing* (London: British Broadcasting Corp. and Penguin Books, 1972). See especially chapter 5. Although painting and framing people of color expressed colonial ownership, the rich quality of oil paint itself provided the owner with an aura of quality and status, in addition to property.

25. Robert Iger rationalizes, "I think the remote control device actually has hurt the drama form more than anything else. Comedy requires less of a commitment on behalf of the viewer to watch. And I think it's sort of less susceptible to flipping around." Cerone, "Pioneer ABC," F9.

26. John Lippman, "Harbert Named ABC's Chief Programmer," *Los Angeles Times*, December 15, 1992, D2.

27. Stalter, "Developing Quality Television," 48.

28. Rick Du Brow, "Fox Comes Out Fighting with Its Fall Lineup," *Los Angeles Times*, May 26, 1993, F1, F12.

29. J. Fred MacDonald, *One Nation Under Television* (New York: Pantheon, 1990), 267, noted that as a result of this relaxation, each network would "regain the right to fill as much as 100 percent of its twenty-two primetime hours per week with its own shows." NBC, CBS, and ABC have all taken advantage of this deregulation, much to the horror of many primetime producers and independent studios, who still must seek approval from the big three—who now act as both *their competitors and clients*—in order to air entertainment series on the networks. "Fin-syn" is short for The Financial Interest and Syndication Rules (FISR), enacted by the FCC in 1971 to severely limit the networks' ability to monopolize syndication and primetime program production.

30. As the network business gets more restrictive and less financially rewarding, program ownership will be more important. "The in-house companies (like ABC Productions) are the future, because program ownership is the only way network television can be even remotely profitable." Amy Adelson, senior vice president, ABC Productions, as reported in Stalter, "Developing Quality Television," 48.

31. John Lippman, "Networks Tap into Television Production," *Los Angeles Times*, June 4, 1993, D1, D5.

32. Sean Scully, "Low-Cost High Tech Is Graphic Equalizer," *Broadcasting and Cable*, June 14, 1993, 58.

33. Ibid.

34. Sharon D. Moshavi, "Niche Cable Networks Attract Advertisers of Same Genre," *Broadcasting and Cable*, March 8, 1993, 47.

35. *Film and Video*, May 1993, 72–75.

36. Michael Drexler, president of BJK&E Media Group/N.Y., quoted in *Film and Video*, May 1993, 72–75.

37. Scott Williams, "Doing a Little Fall Arithmetic," *TV Times*, July 11, 1993, 7.

38. Tartikoff's explanation of the stylistic demands of the 500-channel environment was delivered to programming executives at the NATPE International Convention in January 1993, as quoted by Steve McClellan, "Grabbing the Grazers in a Crowded Field," *Broadcasting*, February 1, 1993, 14.

39. For example, Amy Adelson of ABC Productions explains that her decision to develop a series "is contingent upon that person's talent as a writer. Most producers are also writers and series are writer driven. My job is to nurture talent. I want to create an environ-

ment where writers are able to come up with ideas for shows. I help them organize their thoughts." Quoted in Stalter, "Developing Quality Television," 48. Scott Siegler, vice president of Columbia Pictures Television echoes these sentiments, but recognizes that there are other, less writer-centered, ways that programs are built: "What's always been important to me is the written word and the writer . . . trying to find concepts that are unique, scripts that are saleable, and using the written word as the foundation for building a series. That's not the only way of doing it. I worked for a guy at Warners who came out of casting. He was very interested in finding a performer and building a show from that foundation." Quoted in Stalter, 45.

40. Stalter, "Developing Quality Television," 48.

Chapter 11. Televisual Politics

1. Mary Rourke, "Fashion," *Los Angeles Times Magazine,* January 2, 1993, 25.

2. Numerous articles and op ed pages in newspapers and news magazines made and perpetuated this mythology of the deadening effect of overanalysis. See for example, Charles Hagen, "The Power of a Video Image Depends upon Its Caption," *New York Times,* May 10, 1992, 32, and "King: Video Blurs the Line in Beating Trial, Experts Say," *Los Angeles Times,* February 14, 1993, A34. This easy write off of visual analysis probably results more from journalism's traditional distrust of the deceptions and lies of images, than from anything else.

3. Such a miscarriage was all too troubling to many academics, who saw in the trial's mode their cherished method of intellectual deconstruction, an analytical approach that had dominated critical theory in recent years. Simi Valley seemed to indicate that deconstruction— without reconstituting a human center, for proof or for justice—was in fact an empty and politically impotent exercise.

4. See for example, Andreas Huyssen's discussion of mass culture as an antithesis to the masculinist impulses of high-modernist culture, "Mass Culture as Woman: Modernism's Other," in *Studies in Entertainment: Critical Approaches to Mass Culture,* Tania Modleski, ed. (Bloomington: Indiana University Press, 1986), 188–208.

5. Modleski's work was important for it showed not only that the institutional and narrative logic of daytime television is organized around the desires of women, but that such desires are not finally nor adequately satisfied by daytime soaps—a factor that leaves open the possibility of resistance. Tania Modleski, *Loving With a Vengeance: Mass Produced Fantasies For Women* (New York: Methuen, 1982).

6. Beverle Houston, "Viewing Television: The Metapsychology of Endless Consumption," *Quarterly Review of Film Studies* 9, 3 (Summer 1984), was to become one of many articles that displaced the phallocentric assumptions of psychoanalytic-based criticism that was then being imported to the new medium of television by academics. Houston's account of television stood in stark contrast to Laura Mulvey's influential theory of the male gaze that had driven classical Hollywood cinema and dominated high film theory.

7. Lynne Joyrich's discussion of "hypermasculinity" as a cultural and programming reaction to the threat of feminism is found in "Critical and Textual Hypermasculinity," *Logics of Television: Essays in Cultural Criticism,* Patricia Mellencamp, ed. (Bloomington: Indiana University Press, 1990), 156–172.

8. Sandy Flitterman-Lewis is especially persuasive in defining television's spectatorial modes and viewing pleasures in stark, polar opposition to the masculinist modes of film, in her chapter "Psychoanalysis, Film and Television," in *Channels of Discourse: Reassembled,* Robert Allen, ed. (Chapel Hill: University of North Carolina Press, 1992), 203–246.

9. It is important to reiterate that these masculinist modes evident in television (word-based journalism, fetishized production technology, etc.) *coexist* simultaneously or alternately with guises, genres, and narrative forms that have been characterized as feminine. That is, the history of programming shows that these two tendencies are neither mutually exclusive nor singular.

10. Sergeant Stacey Koons's description of Rodney King as "Mandigo" was from the manuscript of his book on the beating incident and was widely reported by both the print

and television media before and during the first trial. Koons described a confrontation be-tween King and a white, female California Highway Patrol Officer as a "Mandigo sexual encounter." Jim Newton, "U.S. Loses Bid to Question Koon on Manuscript," *Los Angeles Times*, March 26, 1993, A1, 18.

11. Donald Bogle has demonstrated the power and importance of racist stereotyping in his book *Toms, Coons, Mulattoes, Mammies, and Bucks: An Interpretive History of Blacks in American Film* (New York: Viking, 1973). One of the lasting contributions of books like this one and J. Fred MacDonald's *Blacks in White TV: African Americans in Television Since 1948* (Chicago: Nelson-Hall, 1992), is that they describe media in concrete terms as politi-cal—and they do so without equivocating or transforming race into intellectual abstrac-tions. Bogle stands against those who disguise or abstract the racist foundations of Hollywood, and MacDonald addresses head-on the important but frequently overlooked connections between television, politics, and civil rights. Yet studies of stereotyping can also become a more static kind of content analysis, an approach that tends to overlook the more dynamic process of racial othering as a cultural ritual described in works like Hamid Naficy, "Mediawork's Representation of the Other: The Case of Iran," in *Questions of Third Cin-ema*, ed. Paul Willemen (London: British Film Institute, 1990), 227–239. I hope in this study to examine some of the televisual rituals that "other" the person of color during crisis.

12. Paul Moyer's play-by-play commentary on KABC included the following: "There's some creep trying to break a window. . . . I'm sorry for using the word 'creep.' " Thursday, April 30, 1992.

13. Live and on-camera, field reporter Linda Moore repeated the sociological explana-tion that described the hoodlums as otherwise bored and idle perpetrators: "People just need an excuse to do this." KABC, Thursday, April 30, 1992.

14. On the second day of the uprising reporter Linda Moore recounted—with an air of professional self-satisfaction—her earlier plight at being trapped in a gun battle between rioters and Korean store owners. "I never thought that I would cover battle—and here I am in *my own backyard*—and I'm covering a war." Yet her actual presence on camera suggested more the air of tourism. Her claim that this was her backyard was undercut by the economic, racial, and professional signals that her performance gave off. In what possible ways, for example, could glamorous Anglo Moore claim that these armed Korean merchants, Salva-dorans, Chicanos, and African-Americans had always been her neighbors? Like other jour-nalists thrown into the fracas, she came across more like a picture-happy, career-building alien than a neighbor.

15. Command presence, the notion that some individuals are able to dominate social space by their physical appearance and authoritative demeanor and so encourage deference from others in the same social space, is a commonly appreciated ideal in the training of military officers, law enforcement officials, football coaches, and (shall we say it) aca-demic lecturers—all roles that institutionalize authority in the social system. As we shall see, this linkage between media's command presence and law and order is far from gratu-itous.

16. In art history and Catholic iconography, stigmata are bodily marks indicating or replicating the wounds or pains of the crucified Christ; marks that are frequently associated with religious ecstasy.

17. Mike Davis, *City of Quartz* (New York: Vintage, 1990), 244, 258, 268.

18. Neil Postman argues that pictures—a synonym for disinformation—shortcircuit in-trospection, analysis, and context, in *Amusing Ourselves to Death* (New York: Penguin, 1985), 101–108.

19. One of Michel Foucault's most enduring accomplishments was to demonstrate the fundamental importance and privilege given polar dichotomies by the institutions of power and oppression in Western culture. Television's race paradigm—white-black, us-them, in-side-outside—is a perfect example of how an impossibly complicated social phenomenon is reduced to polar terms that seal power on the side of the discursive speaker, in this case television.

20. Ella Taylor, *Primetime Families* (Berkeley: University of California Press, 1991), shows how the domestic, nuclear family was updated and situated in the workplace in the late 1960s and 1970s. The Caucasian *L.A. Law* family during the riots—even in its work-

place home—appears as paranoid about the racial and violent threats rumbling outside of its walls as any of the suburban families in the 1950s sitcoms were about what they defined as external threats.

21. There were of course many women reporters involved in rebellion coverage as well, yet their numbers and approach did not displace the authoritative masculinist command presence that held down the center of coverage, especially around the management functions associated with the in-studio anchor's desk.

22. I am especially thankful to Lynn Spigel for making this point, following a presentation of a draft of this chapter at the "Console-ing Passions" conference on feminism and television at the University of Southern California, Los Angeles, in April 1993. By connecting these binarist representations of race and masculinity to the generic distinctions between information and entertainment, she suggests that there is an institutional logic behind these awkward attempts to maintain racial distinctions.

23. Anna Everett, "The Emancipatory Use of Video in the Classroom," unpublished ms., UCLA, December 11, 1992. Beretta Smith, "Arming Youth: Video for the Revolution," unpublished ms., UCLA, December 1, 1992.

Postscript

1. Fredric Jameson, *Signatures of the Visible* (London: Routlege, 1990), 1.

2. Stephen Heath, *Questions of Cinema* (Bloomington: Indiana University Press, 1981), 203.

3. Although the field of critical theory is typically contentious, no lives have been lost over the issue of television style—*yet*.

4. Aiden Nichols, "The Vindication of the Icons," *The Art of God Incarnate* (New York: Paulist Press, 1980), 77.

5. Jean Baudrillard, *The Ecstasy of Communication*, trans. Bernard Schutze and Caroline Schutze (New York: Semiotext(e), 1988), 63. The material in this chapter on the denigration of the image in high theory, on scopophobia and iconoclasm was published in 1991 as part of my earlier work on the subject, *Televisuality. The Emergence and Performance of Visual Style in American Television*, Ph.D. dissertation (Evanston: Northwestern University, and Ann Arbor: University Microfilms Inc.), especially pages 1–100. The critique of French semiology and film theory included there was adapted from my "Nonverbal Semiotics in Film and Video," a paper for the International Summer Institute of Semiotic and Structural Studies, 1986. Since the final manuscript for the book at hand, *Televisuality*, was submitted in December 1993, I did not have the benefit of reference to Martin Jay's subsequent publication *Downcast Eyes: The Denigration of Vision in Twentieth Century French Theory* (Berkeley: University of California Press, 1993). Jay's important book demonstrates that the suspicion of visuality that I limit to and criticize in contemporary media theory extends back and develops over centuries within a much broader French intellectual tradition, one that includes Descartes, Bergson, and Sartre, and even Breton. While my critique of the scopophobic framework was initially a reaction to the imposition of French film theory into the new field of television studies in the early 1980s, the wider net cast here shows that the impulse to denigrate the image pervades several *American* traditions of media scholarship in fundamental and problematic ways as well.

6. These particular quotes—symptomatic of a broader intellectual climate—are from Hamid Mowlana et al., eds., *Triumph of the Image: The Media's War in the Persian Gulf—A Global Perspective* (Boulder: Westview Press, 1992), xi, 266. An important collection of essays entitled *Triumph of the Image*, for example, gathered non-Western accounts of the Gulf War conflict, but refracted them through a philosophical prism of Western design. The irony that this packaging was done in a book whose purpose was to critique the management of information from the Developing World is worth considering.

7. George Gerbner, "Persian Gulf War, The Movie," in *Triumph of the Image*, 243–248.

8. I am less concerned with recounting the political accomplishments of this work, which I endorse, than I am in reflecting on the theoretical and practical stakes that inevitably accompany this Manichean paradigm. In fact, as I hope this book shows, I am absolutely in

league with the authors in seeing plurality and decentralization as keys to attacking the new communications order. The tactical gains of *Triumph* are certainly impressive: it musters important evidence that serves as a persuasive alternative and powerful antidote to official histories of the Gulf War. The demonstration that coverage was orchestrated and managed for political interests and expediency is an important accomplishment of the book. So too is the tactic of allowing the journalistic voices of the unheard and the banished to be heard. Developed World scholarship should consider this a crucial and urgent task. Manicheanism was a syncretistic religious dualism originating in Persia in the third century A.D., which taught the release of the spirit from matter through asceticism.

9. *Triumph*, xii–xiii.

10. The "Persian Gulf War" essay makes the questionable claim that "Images of actuality . . . do not need logic to build their case," *Triumph*, 246.

11. "Persian Gulf War: The Movie," 244. The same essay reveals this determinism when it cites McLuhan and others and says of the technological means of communication, that "when the means change . . . the telling of stories, including history, also changes," *Triumph*, 244.

12. A synthesis of ancient Greek thought and early Christian theology, gnosticism held, for various cults in the first centuries A.D., that matter is evil and that emancipation could only come through an immediate awareness of spirituality.

13. Neil Postman, *Amusing Ourselves to Death: Public Discourse in the Age of Show Business* (New York: Penguin, 1985). Jerry Mander, *Four Arguments for the Elimination of Television* (New York: Morrow Quill, 1978).

14. Daniel Boorstin, *The Image: A Guide to Pseudo-Events in America* (New York: Atheneum, 1962).

15. While the relative impact of print versus radio was a component in Paul Lazarsfeld et al., *The People's Choice* (New York: Columbia University Press, 1948), the implications of Lazarsfeld's "magic bullet" theory presupposed that the audience was susceptible to the influence of mass media. By the time the United States entered the war, propaganda was no longer the province of the fascists, and the war department ran to Hollywood, and to the highly visual form of film, to turn America's boys into killers overnight. Frank Capra, *The Name Above the Title: An Autobiography* (New York: MacMillan, 1971). Carl Hovland et al., *Experiments in Mass Communication* (Princeton: Princeton University Press, 1949).

16. Phillip Tichenor, "The Logic of Social and Behavioral Science," in Guido H. Stempel and Bruce Westley, eds., *Research Methods in Mass Communication* (Englewood Cliffs, N.J.: Prentice-Hall, 1981), 19.

17. Patt Morrison, "Image Isn't Everything," *Los Angeles Times Magazine*, February 21, 1993, 14.

18. For a more complete interrogation of the Rodney King videotaped beating and subsequent Los Angeles rebellion, please see the analysis in chapter 11.

19. These unequivocal statements and journalistic principles are all from Morrison, "Image Isn't Everything," 14. Morrison elaborates her iconoclasm: "I can remember the milestones of TV families more clearly than some of my own. That is a powerful thing to hand over to image-makers, as every political adviser in America knows." Surely this statement challenges Morrison's own thesis. Clearly the danger she refers to could result as much from the abandonment of picture-making as from the refusal to think. This former journalism professor at USC invokes the facts that: John DeLorean was acquitted in a videotaped cocaine sting; Marion Barry, was acquitted of smoking crack-cocaine on videotape; and the NFL getting rid of instant replay "which solved nothing," to bolster her indictment of the image.

20. The references to and defenses of the libertarian press, to the fourth estate, and to rationalist definitions of information in journalism are too many to mention, although certain works helped codify the enlightenment vision of a free and unadulterated press. See especially Fred S. Siebert, Theodore Peterson, and Wilbur Schramm, *Four Theories of the Press* (Urbana: University of Illinois Press, 1963), a widely cited work, and obligatory reading for most journalism students in America.

21. Rich Brown, "Chancellor: 'Too Much Vox Populi,' " *Broadcasting*, February 1, 1993, 8.

22. Reuven Frank, "Television News: Chasing Scripts, Not Stories," *Broadcasting and Cable*, March 8, 1993, 12. "Videotape makes everything infinitely reducible, makes possible news libraries, and the frequent practice of simply illustrating written stories with a library image matched to each word and phrase by researchers. File footage is no longer even flagged."

23. One of the duo's shared accomplishments, in very basic terms, was to retheorize the world as "not what it appears to be." Although the Greek tragedians made a similar point several millenia earlier, they did not have to contend with the kind of resolute empiricism that dominated the nineteenth-century worlds of Marx and Freud. With the new negative hermeneutic, manifest social practices and personal behaviors were described as governed and guided by more fundamental but *hidden or repressed factors;* economics, sexuality, and the unconscious. Without the Marxian or psychoanalytic revolutions—both of which modeled experience in spatial terms (above and below; visible and invisible)—critical theory would probably look little like it does in today. Media theorists might revert to older critical guises, as cultural archivists, celebrants, pontiffs, or antiquarians—that is to roles that *accept what culture gives them at face value.* It became clear from the revolution and spread of interpretive suspicion, however, that the fundamental realities of the world were by nature hidden and that only rigorous intellectual analysis could bring those hidden forces and determining factors out into the open. The present study, of course, partakes in some of the same general suspicions, but with some specific reservations. I am taking and modifying the term "negative hermeneutic" from Fredric Jameson, *The Political Unconscious: Narrative as a Socially Symbolic Act* (Ithaca: Cornell University Press, 1981), who uses it to justify Marxism over other contemporary critical methods that are seen to lapse into assumptions of false immanence and classical ideals of unity. Consider the following statement by Jameson: "Such a view dictates an enlarged perspective for any Marxist analysis of culture, which can no longer be content with its demystifying vocation to unmask and demonstrate the ways in which a cultural artifact fulfills a specific ideological mission . . . and in generating specific forms of false conscious. . . . It must not cease to practice this essentially negative hermeneutic function (which Marxism is virtually the only critical method to assume today)." Jameson adds a balancing call for a utopian element in cultural study, yet he remains a resolute defender of the classical method that critiques false consciousness (291). Even with Jameson's poststructuralist reworking of Marxism, cultural phenomenon are still basically conspiring and duplicitous agents for other ideological impulses and are not to be trusted.

24. "Skoptophilia," here translated scopophilia, is described as a kind of obsessive and neurotic gazing-impulse, in Sigmund Freud, *A General Introduction to Psychoanalysis* (New York: Washington Square Press, 1962), 318, 379.

25. Freud, *A General Introduction*, 318. "Of the many types of symptom characteristic of the obsessional neurosis the most important are found to be brought about by the undue strength of one group of sexual tendencies with a perverted aim, *i.e. the sadistic group.* . . . Other forms of this neurosis are seen in excessive worry and brooding; these are the expressions of an exaggerated sexualization of acts which are normally only preparatory to sexual satisfaction: *the desire to see,* to touch, and to investigate" (italics mine).

26. Jean-Louis Baudry, "Ideological Effects of the Basic Cinematographic Apparatus," *Narrative, Apparatus, Ideology: A Film Theory Reader* (New York: Columbia University Press, 1986), 289–295. The dominant cinema, according to this view, exploited the viewer's scopic and specular drives in order to perpetuate the dominant political system.

27. Daniel Dayan, "The Tutor Code of Classical Cinema," in *Movies and Methods*, Bill Nichols, ed. (Berkeley: University of California, 1976), 442. Zealous to show that imagery in dominant cinema is regressive, Dayan allegorizes sight as a political example of Lacan's infantile mirror stage. The eye's relation to the image is a symptom of political subservience.

28. Christian Metz, "The Passion for Perceiving," *The Imaginary Signifier* (Bloomington: Indiana University Press, 1982), 58–61. Metz utilized Lacan's detailed description of the desire to see and described it as a fundamentally important sexual perversion that is manifest in voyeurism.

29. It is no coincidence that important theoretical works in the 1980s shifted from

image-related issues to the audio realm. See especially John Belton and Elizabeth Weis, eds., *Film Sound: Theory and Practice* (New York: Columbia University Press, 1985); the special issue "Cinema/Sound," *Yale French Studies* 60 (1980); Claudia Gorbman, *Unheard Melodies: Narrative Film Music* (Bloomington: Indiana University Press, 1987); and Kaja Silverman, *The Acoustic Mirror* (Bloomington: Indiana University Press, 1988).

30. Stuart Ewen, *All-Consuming Images: The Politics of Style in Contemporary Television* (New York: Basic Books, 1988.)

31. Stephen Heath, *Questions of Cinema* (Bloomington: Indiana University Press, 1981), 203.

32. The complete quote reads: "All this . . . is to emphasize the need now to understand the structured process of film in cinema, to pose the terms for that understanding—in which *language remains a constant and crucial point of reflection, a junction problem* for thinking cinema today. Heath, *Questions*, 217.

33. Film theory has done this, mostly by expanding the notion and guise of language, many times in stimulating but arcane ways. Since the field of general semiotics no longer assumes that language is the preeminent model for all sign systems, the discourse that film semiology has constructed (over its questionable linguistic base) is surely suspect. The point that I am making is *not* that film and television lack a linguistic capacity, nor that media semiotics is doomed because it was based on language. It is merely that by adopting a master-paradigm of language, media semiology forced itself into an untenable position—it was unable to adequately account for nonverbal semiosis in film and video. Once the linguistic commitment was made—to language as the model for all sign systems—even attempts to shift semiology by invoking the names of Freud, Benveniste, Lacan, or others avoided dealing with the efficacy of the semiological model itself. For this study, I am defining a sign not according to any linguistic concept, but simply as a process by *which one thing is made to stand for another.* Because television and film are so fundamentally referential in obvious ways (for example, images and sounds standing for other things, in an endless chain of reference), semiotics remains a vital reference through which to analyze film and television. The method by which one applies semiotics, however, needs to be reconsidered. This definition of the sign, "a sign . . . is something which stands to somebody for something in some respect or capacity," is from Charles Sanders Pierce, *Collected Papers*, vols. 1–8 (Cambridge: Harvard University, 1931–1958), vol. 2, 228. In many cases, the new theories simply psychologized linguistic discourse. Even with poststructural revisions, the semiological model remains insecure in lieu of actual stylistic practice and in the face of television's semiotic abundance.

34. Interestingly, one of the earliest forays into film semiotics introduced the possiblity of a nonverbal and specifically Peircian, rather than Saussurian, analysis of film. Peter Wollen, *Signs and Meanings in Cinema* (Bloomington: Indiana University, 1969). Wollen's 1969 auteurist method included the use of Peirce's tripartite model that categorized signs as either icons, indexes, or symbols. By doing so, his work provided a useful framework that could account for various imagistic and pictorial qualities in film without reference to language. Unfortunately, as Silverman's text on semiotics points out, the Peirce-Wollen direction has not been subsequently developed by other film theorists. Kaja Silverman, *The Subject of Semiotics* (New York: Oxford University, 1983), 24. I would modify this explanation, first, by stating that while film theory has generally followed the minor tradition through the figures Barthes, Derrida, and Benveniste, the visual arts have typically payed more attention to Peirce's model and its iconic dimension. Edwin Panofsky originated the term "iconology" to refer to the conceptual content of works of art. Henri Focillon proposes "iconography" as a more thoroughly formal and stylistic method, in Focillon, *La Vie des Formes* (1934), reprinted as *The Life of Forms*, trans. Charles Hogan and George Kubler (New York: Zone Books, 1948), 23. This privilege is perhaps logical given art history's conventional preoccupation with iconology and iconography—terms that predated Peirce. Although there appear to be few purely Peircian media semioticians, important aspects of his model *have* been assimilated into the French model. Clearly, Peirce remains a potentially useful basis for the semiotic analysis of media. Some writers like Ellen Seiter have made suggestive applications of Peirce's iconographic breakdown in the analysis of television. Ellen Seiter, "Semiotics and Television," *Channels of Discourse: Television and Con-*

temporary Criticism (Chapel Hill: University of North Carolina, 1987), 19–24. Unfortunately Seiter also invokes Ellis's and Altman's glance theory—an analytical move that deemphasizes image in favor of the audio and verbal texts, in favor of linguistic signs that television constantly uses to anchor the image. Seiter, *Channels of Discourse, 25–27.*

It is an over-simplification to say that film-semiotics has been wholly Saussurian-based. However, many of the manifestations in the field, through the decades of the 1970s and 1980s, are traceable back to Saussure primarily through the influential and seminal work of Christian Metz. Christian Metz, *Film Language* (New York: Oxford University, 1968). One valuable outcome of Metz's work are the self-criticisms that result from his exhaustive interrogation of central precepts. In addition to the major concepts that resulted from the language analogy in *Film Language*, Metz recognized the problems inherent in a simplistic film-language analogy—problems spelled out in well-known essays "The Cinema: Language or Language System." His schematization of the "Grand Syntagmatique," in *Film Language*, 145–146, resulted from the apparent fittingness of linguistic analysis in film theory and proved very influential in subsequent analyses. Yes, film like language could be segmented into parts. These segmented parts could then be arranged to produce meanings (a second requirement of language). But no, the rules for this arrangement, the syntax and grammar of film, was not apparently stateable in a precise way for all cases. Metz, 31–91. The cinema, in short, is a "loose system." Seiter describes television as even looser, since it is comprised of "weaker codes." Seiter, 29. Television and film appear meaningful only when physically seen and produced, not when explained or interpreted according to a preexisting syntax.

35. Metz, *The Imaginary Signifier*, 3–16.

36. Early critics of Metz's film as language position, both cineastes and aestheticians, may have been threatened by the somewhat arrogant scientific claim and style of the newly announced field as Turim suggests. Maureen Turim, "A Project of Deblocage," *Semiotica*, 50, nos. 1–2, (1984), 181–190. Later critics, like Dudley Andrew, would attack semiology's formalism, impracticality, and its demphasis on the role of interpretation in the production of meaning. Dudley Andrew, *Concepts in Film Theory* (New York: Oxford University, 1984), 57–74. These developments included the infusion of new theoretical master codes as part of a synthesis within the tradition of French semiology. Ideological explications were imported to semiology from Althusser, and psychoanalytic treatments from Lacan.

37. Roland Barthes, *Mythologies* (New York: Hill and Wang, 1957), and "Introduction to the Structural Analysis of Narrative," *Image-Music-Text* (New York: Hill and Wang, 1977), 79–124. Emile Benveniste gives the narratological equation a political and social dynamic by describing the goals that speakers inherently have in influencing hearing subjects in *Problems in General Linguistics* (Coral Gables: University of Miami, 1971), 209. In short, the privileged linguistic-based semiotic equation had been opened up to include important extra-textual determinations, specifically, psychological and social factors. But analyses of these factors were still typically based to some degree on the model of language. That is, rather than articulating the ways that film was like language, semiology now shifted its concerns to the ways that *ideology is constructed through transactions of language*, or the way that history is effaced by language, or the way that subjectivity is created by textual discourse. In each case, semiology was still language-based, but language was now seen as psychic, sexual, and political.

The language-based semiological model was opened up to include other factors, the cultural and ideological. However, this shift still did not enable film semiology to deal with the pervasive aspects of nonverbal communication in film and television. A different kind of theorization is needed. Invoking the names and insights of Freud or Lacan and their views on language does not solve semiology's linguistic limitations either. Psychoanalysis opened up semiology to the human subject, but it was to a clearly conjectural and speculative human subject. Saussure's linguistic model, although quickly assimilated and radically developed in the last two and a half decades, has continued to lead film and television theory into a prisonhouse of language. Confined by its own linguistic presuppositions, it is doubtful whether subsequent conceptual machinations, poststructuralist or otherwise, will allow critical theory with semiological origins to account convincingly for nonverbal semiosis.

38. The leading English-language organ of the French semiotic tradition in film was

Screen. Published in England, *Screen* was influential during the 1970s and 1980s in synthesizing and disseminating language-based semiology, complete with its Lacanian and Althusserian manifestations.

39. Heath, *Questions of Cinema*, 204.

40. This metaphor Heath derives from the Russian formalists V. M. Volosinov and L. S. Vygotsky among others. It is curious, however, that when faced with the inadequacy of both image and written language as a model for film, Heath shifts to speech as a driving metaphor. It is interesting, because in an evolutionary and biological sense, speech is a particularized manifestation of a previously existing languagelike system. That is, speech depends upon the existence of an abstractive capability, one that can image and model alternative worlds outside of the signifying subject. As the semioticians cited below describe, speech, like language, follows the advent of human semiosis and modeling in both the evolutionary and developmental schema.

41. A later example of the persistence of, and allegience to, the cinematic-language model is found in a suggestive article by Daniel Milo, "The Culinary Character of Cinematic Language," *Semiotica*, 58, nos. 1–2 (1986), 83–99. In a stimulating account of how cinematic language has a culinary character (how film language is like cooking), Milo's stated objective is to keep alive the study of cinematic language by making it more rational and useful. Food, perhaps the most visceral form of human consumption (certainly when compared to media), is reduced to rationality and language. Milo's analysis articulates the two sides of the linguistic faculty operative in film, the "productive" aspect, and the "perceptive" aspect. He concludes that the foodlike morphemes (the words) of cinema-language are "reality, photographed, and tape recorded" (94). It is ironic that, twenty years after Metz introduced linguistic based semiology, a follower of that tradition concludes that "reality-units" comprise the raw material of film-language—not the frame, not the shot, not the sequence. It is as if Bazin, an old-theory patriarch discarded by the semiological movement, has been reincarnated in semiology's second generation. Even apart from this theoretical inversion, however, the question remains, if cinema is like food, why persist in calling it a language? Food, like the filmic and televisual image, is polysemous, multisensory, and to a great extent nonverbal. Like food, the image is consumed without naming. Language, it seems, is the burden that even recent semiologists carry into analysis.

42. A view back on the early critics of Metz might elicit the response "of course, film-and-language is a mute issue." Bill Nichols, for instance, challenged the privileged model of language in analyses in "Style, Grammar, and the Movies," in *Movies and Methods*, Bill Nichols, ed. (Berkeley: University of California, 1976), 607–628. The legacies of Saussure and Metz, however, continue in figures like Heath and to a lesser extent in Foucault despite historical precedents critical of the master-model of language.

43. A distinction has been proposed by Thomas Sebeok, "Ecumenicalism in Semiotics," in *A Perfusion of Signs*, Thomas Sebeok, ed. (Bloomington: Indiana University Press, 1977), 182, and further developed by John Deely, "Pars Pro Toto," in *Frontiers in Semiotics*, John Deely, ed. (Bloomington: Indiana University Press, 1986), viii–xvii , that describes word-based ("glottocentric") semiology of Saussure as the "minor tradition." Sebeok and Deely contrast this relatively recent semiological discourse (typically embraced in literary and aesthetic circles) to what they describe as the "major tradition" of semiotics (which includes many other disciplines including science and philosophy). The latter tradition, primarily French in origin, dominated literary and aesthetic studies for several decades. Film study, from which many contemporary critical theorists of television come, has traditionally privileged French theory and still frequently betrays an allegience to the idea that semiosis is a language-based operation.

44. The major tradition of semi-*otics*, as opposed to semi-*ology*, claims as its founder John Locke, "Division of the Sciences," *An Essay Concerning Human Understanding*, reprinted in Deely, *Frontiers in Semiotics*, 3–4, and the philosopher C. S. Peirce as its most extensive and comprehensive theorist. See "Logic as Semiotic: The Theory of Signs," *The Collected Papers of C. S. Peirce* (Cambridge: Harvard University Press, 1931), in Deely, 4–23. In establishing the historical precedents for this tradition of semiotics, the writers trace the origins of the "science of signs" back to medieval scholastics like John Poinsot, *Tractatus De Signis*, trans. and ed. John by Deely (Berkeley: University of California, 1985), and to

Greek philosophy and medicine (including Alcmaeon, Hippocrates, and Galen). For a historical summary of this and subsequent developments see T. A. Sebeok, "Semiotics and Its Cogeners," in Deely, *Frontiers in Semiotics,* 255–263. Seen in this light, semiotics has been applied to signs in a wide range of institutions and traditions: from legal proceedings, to genetics, to the discourses of economics.

45. This major tradition of semiotics, centered around the theories of Locke, Peirce, and Morris, has helped to stimulate an integrated science of communication. It is a framework that includes not just word-centered, man-made semiotic systems, but also animal, endo-semiotic, and genetic-semiotic systems. "Zoosemiotic" systems are described in T. A. Sebeok, "The Notion of Zoosemiotics," and "Talking With Animals: Zoosemiotcs Explained," in Deely, *Frontiers in Semiotics,* 74–75, 76–82. Plant semiosis is described in M. Krampen, "Phytosemiotics," *Semiotica,* 36, 3-4, (1981), 187–209. Following Sebeok, Deely, and others, then, all living forms, plants and animals, selectively perceive their environments and work to process signs. The survival of animal and plant forms, in fact, depends upon successful decoding and interpretation of signs.

46. Ferdinand de Saussure, *Course in General Linguistics* (New York: McGraw-Hill, 1966), 16.

47. Arthur Asa Berger, *Media Analysis Techniques* (Beverly Hills/London: Sage Publications, 1982), 14, 17. Obviously this privileged linguistic model for semiosis, derived from Saussure's minor tradition, is of dubious value to plant and animal systems. Such systems are comprised of subjects that can neither speak nor write, yet are semiotic nevertheless. Since humans are by nature *also* biological—an animal species—it follows that linguistic semiotics can only *incompletely* account for humankind's broader biological sign systems. Research indicates that all living systems are involved in patterns of nonverbal communication and in the perpetual interchange of messages. If one views human semiosis within its broader biological context—as a perceptual, neurological, and genetic process—then semiotics, both human and cultural, would more accurately be seen as a modeling of the outside world.

48. Thomas A. Sebeok, "On the Phylogenesis of Communication, Language, and Speech," *Recherche Semiotiques/Semiotic Inquiry* 5, 4 (1985). The work of Gregory Bateson for example, counters positions that place an evolutionary (and thereby hierarchical) overvaluation on language. This Bateson and Sebeok do by demonstrating that nonverbal communication did not atrophy, was not displaced by evolution, but has instead flourished alongside language in humans. Gregory Bateson, "Redundancy in Coding," in *Animal Communication,* T. A. Sebeok, ed. (Bloomington: Indiana University, 1968), 614–626. Biologist François Jacob, following Von Uexkull, demonstrated how all living organisms model their universe and selectively perceive their environment as a biological necessity. François Jacob, *The Possible and the Actual* (Seattle: University of Washington, 1982).

49. Evolutionary evidence seems to indicate that because of the rapid increase of brain size long before the acquisition of speech in Homo sapiens, a modeling capacity existed in its predecessor Homo habilis. Stephen Gould and Elisabeth Vurba argue from this that speech and linguistic communication were "exapted" (not adapted) from a previously existing languagelike modeling capacity. Stephen J. Gould and Elisabeth S. Vurba, "Exaption—A Missing Term in the Science of Form," *Paleobiology* 8, 1 (Winter 1982), 4–15.

50. It is necessary, then, to shift from these perspectives to the question of film and television semiotics per se. Given the semiotic critique of semiology, what then are the possibilities for a viable nonverbal based semiotics of television?

51. The important work of Michel Foucault shows convincingly that even sexuality and pleasure can be understood as social constructions and political forms of oppression. Michel Foucault, *The History of Sexuality,* trans. Robert Hurley (New York: Pantheon, 1978). Following Foucault, excessive visuality can be considered in the same manner that one thinks about sexuality: as a social practice. Both the image and sexuality clearly involve physical pleasure. Foucault states: "The central issue . . . is not to determine whether one says yes or no to sex, whether one formulates prohibitions or permissions. . . . *What is at issue,* briefly, *is* the over-all 'discursive fact,' *the way in which sex is 'put into discourse'* " (italics mine; 11). The discourse of sex, rather than the practice of sex, is pivotal for Foucault in attacking the "regime of power-knowledge-pleasure" that dominates culture. In this light,

poststructuralist theory assumes its radical stance as a analytical-verbal practice—as a discursive interrogation—that deconstructs the falsely perceived immanence, pleasure, and universality of sex. Regardless of its emancipatory aims, though, this *discursification of pleasure* is locked into a world of semantic radicality. Poststructuralism has, in effect, replaced experience with the conspiracies of discourse. The phenomenal world is a deception.

52. John Fiske, *Television Culture* (New York: Methuen, 1987).

53. See especially the work of Paul Ekman and his colleagues. "Universals and Cultural Differences in Facial Expressions of Emotion," in Nebraska Symposium on Emotion 1971, James K. Cole, ed. (Lincoln: University of Nebraska Press, 1972), 207–283, is but one of many pieces of research that describe these conditions.

54. It has become commonplace, in addition, to assume that language is the prerequisite for all things human. Humanists elevate the species by describing its unique signifying abilites (the species can rationalize, express, create, etc.). Critical theorists frequently do a similar thing, but for entirely different reasons. Discourses and social codes dominate the reality and functions of human subjects. These views do not, however, alter the fact that humans also enact an extensive array of nonverbal communicative strategies. These simultaneous nonverbal signifying channels do not result because humans are distinct from animal species, but precisely because they continually utilize and engage their abilities as biological beings. The neural and biological capacities of the human subject give them a rich and complex array of nonverbal semiotic behaviors. Although constantly intertwined with language, these behaviors are persistently present and unavoidable—in the semiosis of television and in the televisual image.

55. Consider the fact that both sex and imagery have biological components and functions, in addition to encoded meanings. Just as with sex, consuming images can be pleasurable without first naming those images. Reconsidering sexualities and visual pleasures as political discourses is a useful and important strategy. But extrapolating, on the basis of this kind of discursivity, to sweeping definitions about the human ability to experience is shortsighted.

56. There has been, after all, much that is poststructural in the pages of this study. Media semiology has found itself in an awkward position. For heuristic efficacy, it has had to modify its commitment to the cinema-as-language model of semiosis. The linguistic paradigm continues to evolve. Some writers shifted to cinema as (inner) speech, others to cinema as food-reality units. *Screen* championed cinema language as ideology. Metz and others shifted to cinema as dreamwork. The poststructuralist tradition continues to allow an aversion to the image by viewing cultural operations, experience, and media imagery as discourse. Each manifestation still assumes that language is a master code, a structuring principle, or a determining foundation for speech, ideology, dreams. As long as each model *overdetermines language* in this way, it limits its explanatory power to only part of the cinematic and televisual phenomenon. In practice, such theories are trapped in a Saussurian-based cul-de-sac of language.

57. George Gilder, *Life After Television* (New York: Norton, 1992). Many of the ideas, for example, that issued from the stage at the information superhighway summit—a spectacle staged by the Academy of Television Arts and Sciences in Los Angeles on January 11, 1994—invoked the very terms with which Gilder had described the new technologies. *Life After Television,* and *Microcosm*, the scientific precursor upon which the latter book was based, shared an unabashed advocacy of television's nemesis—the telecomputer—which they marshal from a sociotechnological foundation hybridized from free-market economics, cybernetics, and material physics.

58. That is, against the law of macroscosm in Newtonian physics—the "many body" problem in which increased system complexity leads to exponentially growing, and finally incapacitating, inefficiencies—Gilder contrasts his law of microcosm.

59. Gilder's vision of entrepreneurship and interactivity is a fundamental form of xenophobia. His call to abandon both broadcast television and HDTV development in favor of the American-controlled telecomputer is couched as a last-ditch plea for the United States to save itself from Japanese technological and therefore economic dominance (*Life After Television,* 12–13) Our relation to the Japanese is couched in us versus them terms. Our need to attack the top-down model of inefficient American corporatism is cast in military

terms. Connectivity, a mythos based on integration, can then be used strategically as a form of proprietary exclusion and unabashed nationalism. Yes, there is still time to beat the Japanese, but only if we act and act now. We still own the patents that can strike Japan's Achilles heel and lead America back to global dominance. For all of its talk of decentralization, then, the book also romanticizes nationalism and ownership. Technologies are, clearly, proprietary.

60. Neil Postman claims that "pictures have little difficulty in overwhelming words and short-circuiting introspection." *Amusing Ourselves to Death*, 103.

61. The best example and defense of critical theory's apologetic—one that justifies its existence in opposition to American mass communications—is found in the introduction to Robert Allen, *Channels of Discourse: Reassembled*, (Chapel Hill: University of North Carolina Press, 1987). John Fiske also elaborates and justifies the new ideologically and semiotically informed field in relation to traditional methods of television study in *Reading Television* (New York: Methuen, 1980), and in *Television Culture* (New York: Routlege, 1987). While critical in some ways, Fiske acknowledges and explicates the viability of developments in content analysis through explications of the work of George Gerbner and his colleagues. Journals of mass communications, like the journal of *Broadcasting and Electronic Media,* on the other hand typically review current work in critical and cultural studies as undisciplined. When a mass-communications scholar coauthors a book with a critical theorist, however, such critics really show their colors. A review by Michael R. Real of Daniel Dayan and Elihu Katz's *Media Events* in the *Journal of Broadcasting and Electronic Media* 37, 1 (Winter 1993), 123–126, reads more like a conspiracy theory and exercise in mind reading. While praising the contributions of communications scholar Katz, the reviewer castigates those components of the book he imagines coming from French semiotician Dayan.

62. Systematic stylistic description is not just an issue of critical theory, it is a requisite, though often overlooked, issue for broadcasting science as well. It is clear that television reception, along with many other forms of interpersonal communication, involves simultaneous and sometimes contradictory engagement through scores of semiotic and perceptual channels (for example, spatial, tactile, olfactory, kinesic). Research from a physiological perspective has shown that the verbal strata may be the least important of the simultaneously used human communication channels. See the citations (especially Deely, Vurba, Bateson, and Sebeok) in the earlier discussion of nonverbal semiosis.

63. See for example John E. O'Connor's categorization of Erik Barnouw's *Image Empire* as a "history of television as an industry and an artform." *American History, American Television: Interpreting the Video Past* (New York: Unger, 1985). Few of the contributors in O'Connor's anthology utilize Barnouw's method and approach.

64. I am especially thankful to theorist Teshome Gabriel for demonstrating the extent and significance of cultural hybridity in the new multiethnic communities in Los Angeles.

65. Mary Rourke, "Fashion," *Los Angeles Times Magazine*, January 2, 1993, 25.

66. Scott Cooper is quoted in Stuart Hall, "Cultural Studies and Its Theoretical Legacies," *Cultural Studies*, Lawrence Grossberg et al., eds. (New York: Routlege, 1992), 294.

67. "O Cool, Cool World," *Spy*, January 1994, 6.

68. Michael Eisner's keynote address to promotion executives at the PROMAX-BDA Conference in Orlando, Florida, June 1993, as quoted in Mike Freeman, "Eisner on the Info Highway: Slow Down," *Broadcasting and Cable*, June 21, 1993, 26.

Bibliography

"A-53 D Digital Special Effects." Promotional brochure. Abekas, Redwood City, Calif., 1988.

"Abekas: Tools for Creative Expression." Promotional brochure. Abekas, Redwood City, Calif., 1989.

Adler, Richard. "Introduction: A Context for Criticism." In *Television as a Cultural Force,* edited by Richard Adler, 1–16. New York: Praeger, 1976.

Ailes, Roger. "The Truth About Tabloid TV." *TV Guide*, May 7, 1993.

"All Bent Out of Shape." *Post*, September 18, 1989.

Allen, Robert, editor. *Channels of Discourse*. Chapel Hill: University of North Carolina Press, 1987.

Allen, Robert and Douglas Gomery. *Film History: Theory and Practice*. New York: Alfred A. Knopf, 1985.

Allen, Stuart. "Lighting for Television: Faster Filmstocks Are Changing the Ways that Cinematographers Approach Their Work." *Film and Video* (June 1990): 47.

Altman, Rick. "Television/Sound." In *Studies in Entertainment: Critical Approaches to Mass Culture*, edited by Tania Modleski, 39–54. Bloomington: Indiana University Press, 1986.

Alvarado, Manuel and John Thompson, editors. *The Media Reader*. London: British Film Institute, 1990.

Andrew, Dudley. *Concepts in Film Theory*. New York: Oxford University Press, 1984.

———. *The Major Film Theories*. New York: Oxford University Press, 1976.

Ang, Ien. *Watching Dallas: Soap Opera and the Melodramatic Imagination*. New York: Methuen, 1985.

Antin, David. "Video: The Distinctive Features of the Medium." In *Video Art: An Anthology*, edited by Beryl Korot and Ira Schneider, 174–183. New York: Harcourt, Brace, Jovanovich, 1976.

Arbuthnot, Lucie and Gail Seneca. "Pre-Text and Text in *Gentlemen Prefer Blondes*." *Film Reader* 5 (1982): 13–23.

Armes, Roy. *On Video*. New York: Routlege, Chapman, and Hall, 1988.

Bahktin, Mikhail. *The Dialogic Imagination*. Edited by Michael Holguist. Translated by Caryl Emerson amd Michael Holquist. Austin: University of Texas, 1981.

Balfour, Ian. "The Playhouse of the Signifier." *Camera Obscura* (May 1988): 155–168.

Baltrusaitis, Jurgis. *Aberrations: An Essay on the Legend of Form*. Cambridge: MIT Press, 1989.

Barnouw, Erik. *Tube of Plenty: The Evolution of American Television*. New York: Oxford University Press, 1975, 1982.

Barthes, Roland. "Introduction to the Structural Analysis of Narrative." In *Image-Music-Text*, edited by Roland Barthes, 79–124. New York: Hill and Wang.

———. *Mythologies*. New York: Hill and Wang, 1957.

———. *Writing Degree Zero*. New York: Hill and Wang, 1968.

Bateson, Gregory. "Redundancy in Coding." In *Animal Communication*. Edited by T. A. Sebeok, 614–626. Bloomington: Indiana University Press, 1968.

Baudrillard, Jean. *The Ecstasy of Communication*. Translated by Bernard Schutze and Caroline Schutze. New York: Semiotext(e), 1987.

———. *Forget Foucault*. New York: Semiotext(e), 1987.

———. *Jean Baudrillard: Selected Writings*. Edited by Mark Poster. Stanford, Calif.: Stanford University Press, 1988.

———. *Simulations*. New York: Semiotext(e), 1983.

Baudry, Jean-Louis. "The Apparatus: Metapsychological Approaches to the Impression of Reality in the Cinema." *Camera Obscura* (Fall 1976): 104–128.

———. "Ideological Effects of the Basic Cinematographic Appartus." *Film Quarterly* 28, no. 2 (Winter 1974–75). Reprinted in *Narrative, Apparatus, Ideology: A Film Theory Reader*, edited by Robert Rosen, 289–295. New York: Columbia University Press, 1986.

Bazin, André. *What Is Cinema?* Volumes I and II. Translated by Hugh Gray. Berkeley: University of California Press, 1971.

Beal, R. R. "RCA Developments in Television." *Journal of Society of Motion Picture Engineers* (August 1937), 121–143.

Beardsley, Monroe and William K. Wimsatt. "The Intentionalist Fallacy." In *Problems in Aesthetics*, edited by Morris Weitz, 347–360. New York: Macmillan, 1987.

Belton, John and Elizabeth Weiz, editors. *Film Sound: Theory and Practice*. New York: Columbia University Press, 1985.

Benjamin, Walter. "The Work of Art in the Age of Mechanical Production." In *Illuminations*. Translated by Harry Zohn. New York: Schoken Books, 1969. Reprinted in John Hanhardt, editor, *Video Culture: A Critical Investigation*, 27–52. Rochester, N.Y.: Visual Studies Workshop, 1986.

Benveniste, Emile. *Problems in General Linguistics*. Coral Gables, Fla.: University of Miami, 1971.

Bennett, Tony, Susan Boyd-Bowman, Colin Mercer, and Janet Woollacott, editors. *Popular Television and Film: A Reader*. London: British Film Institute, 1981.

——— and Janet Woollacott, editors. *Bond and Beyond: The Political Career of a Popular Hero*. New York: Methuen, 1987.

Berger, Arthur Asa. *Media Analysis Techniques*. Beverly Hills, Calif.: Sage Publications, 1982.

Berger, John. *Ways of Seeing*. London: British Broadcasting Corp. and Penguin Books, 1972.

Blair, Iain. "*Needful Things*: Producer Jack Cummins, Director Fraser Heston and Cinematographer Tony Westman Bring Their Production of the Stephen King Novel to the Pacific Northwest," *Film and Video*, April 1993.

Blonsky, Marshall, editor. *On Signs*. Baltimore, Md.: Johns Hopkins University Press, 1985.

Blumenthal, Sydney. "The Syndicated Presidency," *The New Yorker* (April 5, 1993), 42–43.

Boddy, William. *Fifties Television: The Industry and Its Critics*, Urbana: University of Illinois Press, 1990.

Bogle, Donald. *Toms, Coons, Mulattoes, Mammies, and Bucks: An Interpretive History of Blacks in American Film*. New York: Viking, 1973.

Boorstin, Daniel. *The Image: A Guide to Pseudo-Events in America*. New York: Atheneum, 1962.

Bordwell, David. *Making Meaning: Inference and Rhetoric in the Interpretation of Cinema*. Cambridge: Harvard University Press, 1989.

———. *Narration in the Fiction Film*. Madison: University of Wisconsin Press, 1985.

———, Janet Staiger, and Kristin Thompson. *The Classical Hollywood Style: Film Style and the Mode of Production to 1960*. New York: Columbia University Press, 1985.

Boss Film Studios. Print advertisement. *Millimeter* (October 1989): 127.

Bourdieu, Pierre. *Distinction: A Social Critique of the Judgement of Taste*. Cambridge: Harvard University Press, 1984.

Boyle, Dierdre. "Guerrilla Television." In *Transmissions*, edited by Peter D'Agostino, 203–214. New York: Tanam Press, 1985.

Brecht, Berthold. *Brecht on Theater*. Translated by John Willet. New York: Hill and Wang, 1964.

Brown, Lester L. *Television: The Business Behind the Box*. New York: Harcourt, Brace, Jovanovich, 1971.

Brown, Rich. "Chancellor: 'Too Much Vox Populi," *Broadcasting*, February 1, 1993, 8.

Browne, Nick, editor. *American Television: New Directions in History and Theory*. New York: Harwood Academic, 1994.

———, editor. *Cahiers du Cinéma: 1969–1972, The Politics of Representation*. Cambridge: Harvard University Press, 1990.

———. "The Political Economy of the Television (Super) Text." *Quarterly Review of Film Studies* 9, no. 3 (Summer 1984): 174–182.

———. "Race: The Political Unconscious of American Film." *East-West Film Journal* 6, no. 1 (January 1992): 5–16.

Bruce, Bryan. "Pee-Wee Herman: The Homosexual Subtext." *Cine Action* 9 (Summer 1987), 3–6.

Bunish, Christina. "The Search for Realism: Directors David Steinberg, Ron Dexter, and Bob Eggers Face the Challenges of Capturing Reality." *Film and Video* (September 1990): 68.

Burgoyne, Robert. "The Cinematic Narrator." *Journal of Film and Video* 42, no. 1 (Spring 1990): 3–16.

Burns, Bill, Timothy Carlson, and Jerry Lazar, "Failed Shows, Few Replacements Fueled Shakeup at CBS," *TV Guide*, December 9, 1989, 51.

Caldwell, John. "Non-verbal Semiotics in Film and Video." Paper for the International Summer Institute of Semiotic and Structural Studies, Evanston, Illinois, 1986.

———. "Performing Style: Industrial Strength Semiotics." In *American Television: New Directions in History and Theory*. Edited by Nick Browne. New York: Harwood Academic, 1994.

———. "Public-Private Interventions: Technological Competence, Liveness and Participation TV." Paper presented at the Society for Cinema Studies Conference, New Orleans, February 1993.

———. "Salvador/ Noriega: Convulsions of Topicality." *Jump-Cut* 37 (1992): 15–29.

———. "Television Negotiates the Gaze: The Televisual Documentary." Paper presented at the Ohio University Film Conference, November 1990.

Camera Obscura 17, special issue entitled "Male Trouble" (May 1988).

Capra, Frank. *The Name Above the Title: An Autobiography*. New York: MacMillan Company, 1971.

Cawelti, John. *Adventure, Mystery, and Romance*. Chicago: University of Chicago Press, 1976.

Carter, Bill. "Fall TV Schedules Go Back on Basics." *New York Times*, May 25, 1991, 480.

Cerone, Daniel. "Game Shows: The Price Isn't Right." *Los Angeles Times*, August 28, 1991, F1, 5–6.

Chatman, Seymour. *Story and Discourse: Narration in Fiction and Film*. Ithaca: Cornell University Press, 1978.

"Cinema/Sound." *Yale French Studies* 60 (1980)

"Close-ups: Michael Oblowitz." *Millimeter* (February 1989): 196.

Comer, Brooke Sheffield. "Music Video That Looks Like Film." *American Cinematographer* (September 1986): 95.

Comolli, Jean-Louis. "Technique and Ideology: Camera, Perspective, Depth of Field." In *Narrative, Apparatus, Ideology*. Edited by Philip Rosen. Translated by the British Film Institute, 421–443. New York: Columbia University Press, 1986.

Cook, Philip, Douglas Gomery, Lawrence Lichty, editors. *American Media: The Wilson Quarterly Review*. Washington: Wilson Center Press, 1989.

Corrigan, Timothy. *A Cinema Without Walls: Movies and Culture After Vietnam*. New Brunswick: Rutgers University Press, 1991.

Cosgrove, Kevin. "Regis' Recipe for a Healthy Life," *TV Guide*, March 6, 1993, 8–11.

Custen, George. *Bio/Pics: How Hollywood Constructed Public History*. New Brunswick: Rutgers University Press, 1992.

D'Agostino, Peter, editor. *Transmission: Theory and Practice for a New Television Aesthetics*. New York: Tanam, 1985.

"DaVinci." Print promotional brochure, 1989.

Davis, Douglas, and Allison Simmons, editors. *The New Television: Public/Private Art*. Cambridge, Mass.: MIT Press, 1977.

Davis, Helen. "Designer's Role Within the Graphic Environment." *Computer Pictures* (August–September 1989): S20.

Davis, Mike. *City of Quartz*. New York: Vintage, 1990.

Dayan, Daniel. "The Tutor Code of Classical Cinema." *Movies and Methods*, edited by Bill Nichols, 438–451. Berkeley: University of California Press, 1976.

———— and Elihu Katz. *Media Events: The Live History of Broadcasting*. Cambridge: Harvard University Press, 1992.

Debord, Guy. *Society of the Spectacle*. Detroit: Black and Red, 1970, 1977.

Deely, John. "Pars Pro Toto." In *Frontiers in Semiotics*. Edited by John Deely, viii–xvii. Bloomington: Indiana University Press, 1986.

Delta 1. Print advertisement. *Millimeter* (October 1989): 76.

Deming, Robert. "The Television Spectator-Subject." *Journal of Film and Video* 37 (Summer 1985): 48–63.

Dempsey, John. "More Mags Will Fly in the Fall: Too Much of a Good Thing?" *Variety*, April 12, 1989, reprinted in *Variety: Broadcast-Video Sourcebook I*, edited by Marilyn Matelski and David Thomas, 27. Boston: Focal Press, 1990.

DuBrow, Rick. "Emmy's 'Peak's' Season." *Los Angeles Times*, August 3, 1990, F1.

————. "No. 1 CBS Has Its Eye on Middle-Age Viewers," *Los Angeles Times*, May 21, 1993, F21.

Dyer, Gillian. *Advertising as Communication*. New York: Methuen, 1982.

"Eclipse CVE 200." Promotional brochure. DSC Corporation, Gainesville, Fla., 1989.

Eco, Umberto. *Theory of Semiotics*. Bloomington: Indiana University Press, 1976.

Eisenstein, Sergei. *Film Form: Essays in Film Theory*. Edited and translated by Jay Leyda. New York: Harcourt, Brace, 1942.

————. *Film Sense*. Edited and translated by Jay Leyda. New York: Harcourt, Brace, 1942.

Ellis, Jack. *The Documentary Idea: A Critical History of English Language Documentary Film and Video*. Englewood Cliffs, N.J.: Prentice-Hall, 1989.

Ellis, John. *Visible Fictions: Cinema, Television, Video*. London: Routlege and Kegan Paul, 1982.

"Electronic Imagery." *American Cinematographer* (June 1987): 89.

"Encore." Print promotional brochure. Encore, 1989.

Enzensberger, Hans Magnus. *The Consciousness Industry*. New York: Seabury Press, 1974.

Estrin, Eric and Michael Berlin. Public address on screenwriting. California State University, Long Beach, Calif., March 1990.

Everett, Anna. "The Emancipatory Use of Video in the Classroom," unpublished ms., UCLA, December 11, 1992.

"Every Picture Tells a Story." Promotional brochure. Quantel, 1989.

Ewen, Stuart. *All-Consuming Images: The Politics of Style in Contemporary Television*. New York: Basic Books, 1988.

Feuer, Jane. "The Concept of Live Television: Ontology as Ideology." In *Regarding Television,* edited by E. Ann Kaplan, 12–21. Los Angeles: The American Film Institute, 1983.

————. "Melodrama, Serial Form, and Television Today." *Screen* 25, no. 1 (January–February 1984). Reprinted in *The Media Reader*, edited by Manual Alvarado and John O. Thompson, 253–264. London: British Film Institute Press, 1990.

————, Paul Kerr, and Tise Vahimagi, editors. *MTM: Quality Television*. London: British Film Institute, 1984.

Field, Syd. *The Art of Screenwriting*. New York: Dell, 1979.

Fields-Meyer, Thomas, and Richard L. Meyer. "Bedrock Values." *New York Times*, March 27, 1993, A19.

Finnerman, Gerald Perry. "*Moonlighting*: Here's Looking at You Kid."*American Cinematographer* (April 1989): 70–71.

Fisher, Bob. "*Cagney and Lacey*: The New Look in L.A." *American Cinematographer* (January 1987): 88.

Fishbein, Leslie. "*Roots*: Docudrama and the Interpretation of History." In *American History, American Television*. Edited by, John E. O'Connor, 279–305. New York: Ungar, 1983.

Fiske, John. "British Cultural Studies and Television." In *Channels of Discourse*, edited by Robert Allen, 254–289. Chapel Hill: University of North Carolina Press, 1987.

———— and John Hartley. *Reading Television*. New York: Methuen, 1978.

————. *Television Culture*. New York: Methuen, 1987.

Flitterman-Lewis, Sandy. "All's Well That Doesn't End—Soap Opera and the Marriage Motif." In *Private Screenings: Television and the Female Consumer*, edited by Lynn Spigel and Denise Mann, 217–226. Minneapolis: University of Minnesota Press, 1992.

————. "Psychoanalysis, Film, and Television." In *Channels of Discourse*, edited by Robert Allen, 172–210. Chapel Hill: University of North Carolina Press, 1987.

————. "The Real Soap Operas: TV Commercials." In *Regarding Television*, edited by E. Ann Kaplan, 84–96. Los Angeles: The American Film Institute, 1983.

Focillon, Henri. *La Vie des Formes*. Paris: Presses Universitaires de France, 1934. Reprinted as *The Life of Forms*. Translated by Charles Hogan and George Kubler. New York: Zone Books, 1989.

Foisie, Geoffrey. "Fox Hounds ABC-TV." In *Broadcasting and Cable*, June 14, 1993, 65.

Foucault, Michel. *Discipline and Punish*. Translated by Alan Sheridan. New York: Random House, 1977, 1979.

————. *The History of Sexuality*. Volume 1. Translated by Robert Hurley. New York: Pantheon, 1976.

Ford, Andrea. "Suspected Gang Members Held in Slayings of 2 Men." *Los Angeles Times*, January 6, 1993, B1, B4.

Foster, Hal, editor. *The Anti-Aesthetic: Essays on Postmodern Culture*. Port Townsend, Wash.: Bay Press, 1983.

————. *Recodings: Art Spectacle and Cultural Politics*. Port Townsend, Wash.: Bay Press, 1985.

"Fourteen in the Fast Lane: Profiling the Chief Programmers for the Leading Cable Networks," *Broadcasting and Cable*, June 7, 1993, 62.

Frank, Reuven. "Television News: Chasing Scripts, Not Stories." *Broadcasting and Cable*, March 8, 1993, 12.

Freeman, Mike. "Co-op Advertising: Trick or Treat." *Broadcasting and Cable*, June 14, 1983, 56.

————. "Eisner on the Info Highway: Slow Down," *Broadcasting and Cable*, June 21, 1993, 26.

————. "Entertainment Tonight Turns 3,000." *Broadcasting and Cable*, May 8, 1993, 30.

————. "MCA Taps Big-Screen Producers for TV." *Broadcasting and Cable*, March 8, 1993, 23–24.

Freud, Sigmund. *A General Introduction to Psychoanalysis*. New York: Washington Square Press, 1962.

Friedman, James "Live Television: Ceremony, (Re)presentation, Unstructured and Unscripted Events." Paper presented at the Screen Studies Conference, Glasgow, Scotland, June 1993.

————. "The Race for the Presidency: (Re)Presentation, Participation and the Televisual Voter." Paper presented at the Society for Cinema Studies Conference, February 1993.

Friedman, Jeffrey, editor. *Milestones in Motion Picture and Television Technology: The SMPTE 75th Anniversary Collection.* White Plains, N.Y.: Society of Motion Pictures and Technical Engineers, 1991.

Gabriel, Teshome. "Towards a Critical Theory of Third World Films." In *Questions of Third Cinema*, edited by Paul Willeman, 30–51. London: British Film Institute, 1990.

Gans, Herbert. *Popular Culture and High Culture: An Analysis and Evaluation of Taste.* New York: Basic Books, 1974, 1975.

Garland, David. "East to End Our Lives: Etching the Edges of U.K. and U.S. Soap Operas." Paper presented at SCS conference in Los Angeles, May 1991.

Gerbner, George. "Persian Gulf War, The Movie." In *Triumph of the Image: The Media's War in the Persian Gulf—A Global Perspective*, edited by Hamid Mowlana, George Gerbner, and Herbert Schiller, 243–248. Boulder: Westview Press, 1992.

————, L. Gross, N. Signorelli, M. Morgan, and M. Jackson-Beeck, "The Demonstration of Power: Violence Profile No. 10." *Journal of Communication* 29, 3 (1979): 177–196.

"Getting the Word Out," *Broadcasting*, November 16, 1992, 90.

Giles, Dennis. "Television Reception." *Journal of Film and Video* 37 (Summer 1985): 12–25.

Gillette, Frank. "Masque in Real Time." In *Video Art: An Anthology*, edited by Beryl Korot and Ira Schneider, 218–219. New York: Harcourt, Brace, Jovanovich, 1976.

Gitlin, Todd. *Inside Prime Time.* New York: Pantheon, 1983.

————. "Postmodernism Defined, At Last." *Dissent* (Winter 1989). Reprinted in *Utne Reader*, 34 (July/August 1989): 52–61.

————, editor. *Watching Television.* New York: Pantheon, 1986.

Gledhill, Christine. "Genre." In *The Cinema Book,* edited by Pam Cook, 58–112. New York: Pantheon, 1985.

Goffman, Erving. *Frame Analysis: An Essay on the Organization of Experience.* Boston: Northeastern University Press, 1974, 1986.

————. *Gender Advertisements.* New York: Harper and Row, 1979.

Gorbman, Claudia. *Unheard Melodies: Narrative Film Music.* Bloomington: Indiana University Press, 1987.

Gould, Stephen J. and Elisabeth S. Vurba. "Exaption—A Missing Term in the Science of Form." *Paleobiology* 8, 1 (Winter 1982): 4–15.

Gradus, Ben. *Directing the Television Commercial.* Los Angeles: Directors Guild of America, 1981.

Greenberg, Clement. "Modernist Painting." In *Art and Literature* 4 (Spring, 1965): 193–201. Reprinted in *Modern Art and Modernism*, edited by Francis Franscina and Charles Harrison, 5–10. New York: Harper and Row, 1982.

Gross, Larry. "Out of the Mainstream: Sexual Minorities and the Mass Media." In *Remote Control: Television, Audiences and Cultural Power*, edited by Ellen Seiter, Hans Borcher, Gabrielle Kreutzner, and Eva-Maria Warth, 130–149. New York: Routlege, 1989.

Grossberg, Lawrence, Cary Nelson, and Paula Treichler, editors. *Cultural Studies.* New York: Routledge, 1992.

Guynn, William. "The Ambivalent Spectator: *Je Sais Bien, Mais Quend Meme*." Paper presented at the Ohio University Film Conference on Documentary, November 9, 1990.

Hagen, Charles. "The Power of a Video Image Depends upon Its Caption," *New York Times*, May 10, 1992, 32.

Hall, Stuart. "Cultural Studies and Its Theoretical Legacies." In *Cultural Studies*, edited by Lawrence Grossberg, Cary Nelson, and Paula Treichler, 277–285. New York: Routlege, 1992.

———. "Encoding/Decoding." In *Culture, Media, Language*, edited by S. Hall, D. Hobson, A. Lowe, and P. Willis, 128–139. London: Hutchinson, 1980.

——— and T. Jefferson, editors. *Resistance Through Rituals*. London: Hutchinson, 1978.

Hamamoto, Darrel. *Nervous Laughter: Television Situation Comedy and Liberal Democratic Ideology*. New York: Praeger, 1990.

Handy, Bruce. "A Guide to Postmodern Everything." *Spy*, April 1988. Reprinted in *Utne Reader* 34 (July/August 1989): 53–69.

Hanhardt, John, editor *Video Culture: A Critical Investigation*. Rochester: Visual Studies Workshop, 1986.

"Hard News with Soft Edges." *Post*, October 19, 1989, 44.

Harries, Dan. "Fringe Benefits/ The Subculture of Film Cult(ure)." Paper presented at the Society for Cinema Studies Conference, Los Angeles, May 26, 1991.

Heath, Stephen. *Questions of Cinema*. Bloomington: Indiana University Press, 1981.

Hebdige, Dick. *Subculture: The Meaning of Style*. New York: Methuen, 1979.

Heuring, David. "The Street: Shooting Video with an Eye to Film." *American Cinematographer* (June 1988): 73.

Hilliard, Robert L. *Writing for Television and Radio*. Belmont, Calif.: Wadsworth, 1981.

Hoberman, J. "Sex, Drugs, and Dreadlocks." *Premiere*, August 1993, 49–51.

Holsinger, Ralph L. *Media Law*. New York: Random House, 1987.

Houston, Beverle. "Viewing Television: The Metapsychology of Endless Consumption." *Quarterly Review of Film Studies* 9, no. 3 (Summer 1984): 183–195.

Hovland, Carl, Arthur A. Lumsdaine, and Fred D. Sheffield. *Experiments in Mass Communication*. Princeton: Princeton University Press, 1949.

Hunt, Albert. *The Language of Television: Uses and Abuses*. London: Eyre Methuen, 1981.

Huyssen, Andreas. "Mass Culture as Woman: Modernism's Other." In *Studies in Entertainment: Critical Approaches to Mass Culture*, edited by Tania Modleski, 188–208. Bloomington: Indiana University Press, 1986.

"If You're Into Page Turns, Don't Turn This Page." Print advertisement. *Video Systems*, September 1987.

Intintoli, Michael. "The Study of Televisual Production: A Research Report." *Visual Anthropology* 1, no. 2 (1988): 192–193.

Jackson, Eugene. "Monday Memo." *Broadcasting and Cable*, June 14, 1993, 98.

Jacob, François. *The Possible and the Actual*. Seattle: University of Washington Press, 1982.

Fredric Jameson. *The Political Unconscious: Narrative as a Socially Symbolic Act*. Ithaca: Cornell University Press, 1981.

———. "Postmodernism and Consumer Society." In *The Anti-Aesthetic: Essays in Postmodern Culture*, edited by Hal Foster, 111–125. Port Townsend, Wash.: Bay Press, 1983.

———. *Signatures of the Visible*. London: Routlege, 1990.

Jarvik, Laurence and Nancy Strickland. "Cinema Very TV." *California* (July 1989): 198–200.

Jenkins, C. F. "Radio Photographs, Radio Movies, and Radio Vision." *Journal of Society of Motion Picture Engineers* (May 1923), 78–88.

Jenkins, Henry III. *Textual Poaching: Television Fans and Participatory Culture*. New York: Routlege, 1992.

Jessell, Harry A. "New Wave Newscasts Anchor WSVN Makeover: Ex-Affiliate Finds a New Niche." *Broadcasting*, October 12, 1992, 24.

Joyrich, Lynne. "Critical and Textual Hypermasculinity." In *Logics of Television: Essays in Cultural Criticism*, edited by Patricia Mellencamp, 156–172. Bloomington: Indiana

University Press, 1990.

Kaminsky, Stuart. *American Television Genres*. Chicago: Nelson-Hall Publishers, 1985.

Kaplan, E. Ann, editor. *Regarding Television: Critical Approaches, An Anthology*. Frederick, Md.: University Publications of America/American Film Institute, 1983.

————. *Rocking Around the Clock: Music Television, Postmodernism, and Consumer Culture*. New York: Methuen, 1987.

Kay, Jeff. "Sex, Mud, and Rock and Roll." *Los Angeles Times*, November 9, 1989, F1, F2.

Katz, Elihu, Tamar Liebes, and Lili Berko. "On Commuting Between Television Fiction and Real Life." *Quarterly Review of Film and Video* 14, nos. 1–2 (1992): 157–178.

Kirschenbaum, Jill. "Dynamic Duo: Director-Editor Teams Look for Acceptance." *Millimeter* (October 1989): 147.

Kellner, Douglas. "Television, Mythology, Ritual." *Praxis: Art and Ideology*, Part 2, 6 (1982): 133–156.

Kernan, Lisa. "Consuming Production: Reflexivity Remystified in the Television 'Making-Of' Documentary." Paper presented at the Society for Cinema Studies Conference, University of Pittsburgh, May 2, 1992.

Kennedy, Lisa. "Risque Business: Fox Stoops to Conquer." *Village Voice* April 10, 1990, 37.

Kim, L. S. "The Erasure of Difference and the Denial of Ethnicity: The Ethnic Domestic in Television." Paper presented at the Screen Theory Conference, Glascow, Scotland, June 1993.

"King: Video Blurs the Line in Beating Trial, Experts Say." *Los Angeles Times*, February 14, 1993, A34.

Kleid, Beth. "Escape Routes: NBC Lathers up Prime Time with Two Soapy Mini-Series." *Television Times*, August 22, 1993, 15.

Kleinhans, Chuck. "Working-Class Film Heroes: Junior Johnson, Evel Knievel and the Film Audience." In *Jump-cut: Hollywood, Politics and Counter-Cinema*, edited by Peter Steven, 64–82. New York: Praeger, 1985.

————, John Hess, and Julia Lesage. "After Cosby/ After the L.A. Rebellion: The Politics of Transnational Culture in the Post Cold War Era." *Jump Cut* 37 (1992): 2–4.

Knobloch, Susan. "Narrating the Staged: Time and Televisual Rhetoric in a Corporate Audio-Biography of Motown Records." Unpublished paper, UCLA. March 24, 1993.

Krampen, M. "Phytosemiotics." *Semiotica* 36, nos. 3–4 (1981): 187–209.

Krauss, Rosalind. "Video: The Aesthetics of Narcissism." In *New Artists Video*, edited by Gregory Battock, 43–64. New York: Dutton, 1978.

Kubey, Robert, and Mihalyi Csikszentmihalyi. *Television and the Quality of Life: How Viewing Shapes Everyday Experience*. Hillsdale, N.J.: Lawrence Erlbaum Associates, 1990.

Kuhn, Thomas. *The Structure of Scientific Revolutions*. Chicago: University of Chicago Press, 1962.

Kuney, Jack. *Television Directors on Directing*. New York: Praeger Publishers, 1990.

Larner, Stevan. "Beauty and the Beast: God Bless the Child," *American Cinematographer* (April 1989), 71.

Lasch, Christopher. *The Culture of Narcissism: American Life in an Age of Diminishing Expectations*. New York: W.W. Norton, 1978.

Lasswell, H. D. *Propaganda Techniques in the World War*. New York: Alfred A. Knopf, 1927.

Lazarsfeld, Paul, Bernard Berelson, and Hazel Gaudet. *The People's Choice*. New York: Columbia University Press, 1948.

————, and R. K. Merton. "Mass Communication, Popular Taste, and Organized Social Action." In *Communication of Ideas*, edited by L. Bryson, 95–118. New York: Harper and Row, 1948.

Lesage, Julia. "Artful Racism, Artful Rape: Griffith's Broken Blossoms." *Jump Cut* 26 (1981). Reprinted in *Jump Cut: Hollywood, Politics,and Counter-Cinema*, edited by Peter

Steven, 247–268. New York: Praeger, 1985.

———. "The Human Subject—You, He, or Me? (Or The Case of the Missing Penis." *Jump Cut* 4 (November–December 1974). Reprinted in *Screen* 16, no. 2 (Summer 1975): 73–82.

Lauretis, Teresa de. *Alice Doesn't: Feminism, Semiotics, Cinema*. Bloomington: Indiana University Press, 1984.

Lévi-Strauss, Claude. "The Structural Study of Myth." *Structural Anthropology*. Translated by Claire Johnson and Brooke Grundfest Schoepf. New York: Basic Books, 1963.

Lewis, Lisa, editor. *Adoring Audience: Fan Culture and Popular Media*. New York: Routlege, 1992.

Lewis, Mildred. "Paradigms of Race: Televisuality and Otherness, An Examination of *Roc*." Unpublished ms., UCLA, March 1993.

Lichty, Lawrence. "Watergate, the Evening News, and the 1972 Election." In *American History/American Television*, edited by O'Connor, 232–255. New York: Ungar, 1983.

Lippman, John. "All Webs Spinning Record Coin." *Variety*, November 8, 1989, 47.

———. "Networks Push for Cheaper Shows." *Los Angeles Times,* February 19, 1991, D1, 10.

———. "Networks Tap into Television Production." *Los Angeles Times*, June 4, 1993, D1, D5.

———. "Tangled TV Webs Begin to Unravel." *Los Angeles Times*, August 8, 1991, F17.

Lipsitz, George. *Time Passages: Collective Memory and American Popular Culture*. Minneapolis: University of Minnesota Press, 1990.

Littlejohn, David. "Thoughts on Television Criticism." In *Television as a Cultural Force,* edited by Richard Adler, 147–174. New York: Praeger, 1976.

Locke, John. "Division of the Sciences." In *Frontiers in Semiotics*, edited by John Deely, 3–4. Bloomington: University of Indiana Press, 1986.

Lyotard, François. *The Postmodern Condition: A Report on Knowledge*. Translated by G. Bennington and B. Massumi. Minneapolis: University of Minnesota Press, 1984.

MacCabe, Colin. *High Theory/Low Culture: Analysing Popular Television and Film*. New York: St. Martin's Press, 1986.

MacDonald, J. Fred. *Blacks and White on TV: African Americans in Television Since 1948*. Chicago: Nelson Hall, 1992.

———. *One Nation Under Television*. New York: Pantheon, 1990.

MacNeill, Alex. *Total Television: A Comprehensive Guide to Programming, 1948 to Present*. New York: Penguin, 1991.

Mander, Jerry. *Four Arguments for the Elimination of Television*. New York: Morrow Quill, 1978.

Mann, Denise. "The Spectacularization of Everyday Life: Recycling Hollywood Stars and Fans in Early Television." In *Private Screenings: Television and the Female Consumer*, edited by Lynn Spigel and Denise Mann, 41–70. Minneapolis: University of Minnesota Press, 1992.

Marc, David. *Demographic Vistas: Television in American Culture*. Philadelphia: University of Pennsylvania, 1984.

Maslow, A. H. *Motivation and Personality*. New York: Harper and Row, 1954.

Matelski, Marilyn, and David Thomas. *Variety: Broadcast Video Sourcebook I*. Boston: Focal Press, 1990.

Marin, Rick. "Reading Wild Palms." *TV Guide*, May 15, 1993, 16.

Matthias, Harry, and Richard Patterson. *Electronic Cinematography: Achieving Photographic Control Over the Video Image*. Belmont, Calif.: Wadsworth, 1985.

Mattleart, Michele. *Women, Media, Crisis: Femininity and Disorder*. London: Comedia Publishing, 1986.

McGorry, Ken. "Animate Objects." *Post* (September 1989): 118–120.

McClellan, Steve. "Grabbing the Grazers in a Crowded Field." *Broadcasting*, February 1, 1993, 14.

McLuhan, Marshall. *Understanding the Media: The Extensions of Man*. New York: McGraw-Hill, 1964.

―――, and Quentin Fiore. *The Medium Is the Massage: An Inventory of Effects*. New York: Bantam Books, 1967.

Mellencamp, Patricia. "TV Time and Catastrophe: Or Beyond the Pleasure Principle of Television." In *Logic of Television: Essays in Cultural Criticism*, edited by Patricia Mellencamp, 240–266. Bloomington: Indiana University Press, 1990.

―――. "Prologue." In *Logic of Television: Essays in Cultural Criticism*, edited by Patricia Mellencamp, 1–13. Bloomington: Indiana University Press, 1990.

Metz, Christian. *Film Language: A Semiotic of the Cinema*. Translated by Michael Taylor. New York: Oxford University Press, 1974.

―――. *The Imaginary Signifier*. Bloomington: Indiana University Press, 1976.

Milne, Tom, editor. *Godard on Godard*. New York: DeCapo, 1972.

Milo, Daniel. "The Culinary Character of Cinematic Language." *Semiotica* 58, 1–2 (1986): 83–99.

Minh-ha, Trinh T. "(Un)Naming Cultures." *Discourse: Journal for Theoretical Studies in Media and Culture* 11, no. 2 (Summer 1989): 5–18.

―――. *Woman, Native, Other*. Bloomington: Indiana University Press, 1989.

Mitchell, Kim. "Fortunately, They're Not in the Demographic." *Multichannel News*, March 8, 1993, 6.

Modleski, Tania. *Feminisim Without Women: Culture and Criticism in a "Postfeminist" Age*. New York: Routlege, Chapman, and Hall, 1991.

―――. *Loving With a Vengeance: Mass-Produced Fantasies for Women*. New York: Metheun, 1982.

―――, editor. *Studies in Entertainment: Critical Apporaches to Mass Culture*. Bloomington: Indiana University Press, 1986.

Molino, Jean. Introduction. In Henri Focillon, *The Life of Forms in Art*, 9–30. New York: Zone Books, 1989.

Moore, Frazier. "From Schools to Truck Stops, 'Place-Based' Media Flourish." *TV Times*, March 1993, 7.

Morley, David. *The "Nationwide" Audience*. London: The British Film Institute, 1980.

Morrison, Patt. "Image Isn't Everything." *Los Angeles Times Magazine*, February 21, 1993, 14.

Morse, Margaret. "The Ontology of Everyday Distraction." In *Logics of Television: Essays in Cultural Criticism*, edited by Patricia Mellencamp, 193–221. Bloomington: Indiana University Press, 1990.

―――. "Sport on Television: Replay and Display." In *Regarding Television: Critical Approaches, An Anthology*, edited by E. Ann Kaplan, 44–66. Frederick, Md.: University Publications of America/ American Film Institute

―――. "The Television News Personality and Credibility." In *Studies in Entertainment: Critical Approaches to Mass Culture*, edited by Tania Modleski, 55–79. Bloomington: Indiana University Press, 1986.

Moshavi, Sharon D. "Niche Cable Networks Attract Advertisers of Same Genre." *Broadcasting and Cable*, March 8, 1993, 47.

Mowlana, Hamid, George Gerbner, and Herbert Schiller, editors., *Triumph of the Image: The Media's War in the Persian Gulf—A Global Perspective*. Boulder: Westview Press, 1992.

Mulvey, Laura. "Visual Pleasure in the Narrative Cinema." *Screen* 16, no. 3 (Autumn 1975). Reprinted in *Narrative, Apparatus, Ideology*, edited by Phillip Rosen, 198–209. New York: Columbia University Press, 1986.

Murphy, Mary. "Tsk, Tsk, Tori," *TV Guide*, May 8, 1993, 19.

Naficy, Hamid. "Mediawork's Representation of the Other: The Case of Iran." In *Questions of Third Cinema*, edited by Paul Willemen, 227–239. London: British Film Institute, 1990.

Newcomb, Horace, and Robert S. Alley. *The Producer's Medium*. New York: Oxford University Press, 1983.

———, editor. *Television Criticism: The Critical View*. 4th ed. New York: Oxford University Press, 1987.

———. *Television: The Most Popular Art*. New York: Anchor Books, 1974.

Newton, Jim. "U.S. Loses Bid to Question Koon on Manuscript." *Los Angeles Times*, March 26, 1993, A1, A18.

Nichols, Aiden. "The Vindication of the Icons." In *The Art of the God Incarnate*. New York: Paulist Press, 1980.

Nichols, Bill. *Ideology and the Image, Social Representation in Cinema and Other Arts*. Bloomington: Indiana University Press, 1981.

———. "Style, Grammar, and the Movies." In *Movies and Methods*, edited by Bill Nichols, 607–628. Berkeley: University of California, 1976.

"Now It Costs Less to Get More in 3D Video Effects: Including Page Turns for Under $20,000." Advertisement by Microtime. *Videography* 14, no. 10 (October 1989), 6.

Nowell-Smith, Geoffrey. "Minnelli and Melodrama." *Screen* 18, no. 2 (Summer 1977): 112–118.

"O Cool, Cool World," *Spy*, January 1994, 6.

Ober, Josephine. "Cover Story: Team Work on *Covington Cross*." In *Camera*, edited by Eastman Kodak. Spring 1992–1993, 3–4.

O'Conner, John E., editor *American History, American Television: Interpreting the Video Past*. New York: Ungar, 1985.

O'Neill, Edward R. "The Seen of the Crime: Policing the Domestic Subject." Paper presented at the Console-ing Passions Conference, USC, April 4, 1993.

Owen, B., and S. Wildman. *Video Economics*. Cambridge, Mass.: Harvard University Press, 1992.

"Paintbox in the DPC." Print promotional brochure. *Quantel*, Fall 1989.

"Pastiche." Print promotional brochure. Electronic Graphics Corp., 1989.

Peary, Gerald, and Roger Shatzkin, editors. *The Modern American Novel and the Movies*. New York: Ungar, 1978.

Penley, Constance. "The Cabinet of Dr. Pee-Wee: Consumerism and Sexual Terror." *Camera Obscura*, May 17, 1988. Reprinted in Constance Penley, *The Future of Illusion: Film, Feminism, and Psycholanalysis*. Minneapolis: University of Minnesota Press, 1989.

Peirce, Charles Sanders. *Collected Papers*. Vol. 2. Cambridge: Harvard University Press, 1931–1958.

———. "Logic as Semiotic: The Theory of Signs." In *Frontiers in Semiotics*, edited by John Deely, 4–23. Bloomington: Indiana University Press, 1986.

Pike, Kenneth. *Language in Relation to a Unified Theory of the Structure of Human Behavior*. Glendale, Calif.: Summer Institute of Linguistics, 1954.

Poinsot, John. *Tractatus De Signis*. Translated and edited by John Deely. Berkeley: University of California Press, 1985.

Postman, Neil. *Amusing Ourselves to Death: Public Discourse in the Age of Show Business*. New York: Penguin, 1985.

"Postproduction switcher, 8200 C." Print promotional brochure. Crosspoint Latch Corporation, Union, N.J., 1989.

Price, Jonathan. *Video Visions: A Medium Discovers Itself.* New York: The New American Library, 1972.

"Rad TV." *Village Voice,* April 10, 1990, 32–42.

"Real Time Color Corrector." Print promotional brochure. Colorgraphics Systems Corp., 1989.

Renov, Michael. Opening Address. Paper presented at Ohio University Film Conference on Documentary, Athens, Ohio, November 1990.

Review of "Dead End." *American Cinematogapher* (January 1987): 94.

Rochlin, Margy. "The Unblinking Eye: David Lynch and Mark Frost Chronicle America." *International Documentary* (Fall 1990): 11.

Rodman, Howard, N. "Small Screen Shooters: Four Distinguished Cinematographers Discuss the Craft of Shooting Film for Episodic Television." *Millimeter,* (August 1988): 143.

Rosen, Phillip, editor. *Narrative Apparatus, Ideology.* New York: Columbia University Press, 1986.

Rosenthal, Alan, editor. *New Challenges for Documentary.* Berkeley: University of California Press, 1988.

Sahagun, Louis, and Michael Kennedy. "FBI Puts Blame on Koresh for Cultists' Deaths." *Los Angeles Times,* April 21, 1993 A1, A13.

Saussure, Ferdinand de. *Course in General Linguistics.* New York: McGraw Hill, 1966.

S.C. "*Twin Peaks* Caps ABC Season." *Broadcasting,* April 23, 1990, 35.

Schafer, Richard. "Choice of Transfers: Film to Tape." *American Cinematographer* (September 1986): 97–99.

Schatz, Thomas. *Hollywood Genres: Formulas, Filmmaking, and the Studio System.* New York: Random House, 1981.

Scholes, Robert. "Narration and Narrativity in Film." In *Film Theory and Criticism,* edited by Gerald Mast and Marshall Cohen, 390–404. New York: Oxford University Press, 1985.

Schwoch, James. *The American Radio Industry and Its Latin American Activities, 1900–1939.* Urbana: University of Illinois Press, 1990.

———, Mimi White, and Susan Reilly. *Media Knowledge: Readings in Popular Culture, Pedagogy, and Critical Citizenship.* Albany: State University of New York Press, 1990.

"Screening the War: Filmmakers and Critics on the Images that Made History," special issue of *International Documentary* (Spring 1991): 20–25.

Scully, Sean. "For Some, Interactive Future is Now," *Broadcasting and Cable,* June 14, 1993, 77–78.

———. "Low-Cost High Tech Is Graphic Equalizer." *Broadcasting and Cable,* June 14, 1993, 58.

Sebeok, Thomas "The Notion of Zoosemiotics." In *Frontiers in Semiotics,* edited by John Deely, 74–82. Bloomington: Indiana University Press, 1986.

———. "On the Phylogenesis of Communication, Language, and Speech." *Recherche Semiotiques/ Semiotic Inquiry* 5, no. 4 (1985): 361–367.

———. *A Perfusion of Signs.* Bloomington: Indiana University Press, 1977.

———. "Semiotics and Its Cogeners." *Frontiers in Semiotics,* edited by John Deely, 255–263. Bloomington: Indiana University Press, 1986.

Seiter, Ellen. "Semiotics and Television." In *Channels of Discourse,* edited by Robert Allen, 19–24. Chapel Hill: University of North Carolina, 1987.

———, H. Borcher, G. Kreutzner, and E. Warth, editors. *Remote Control: Television, Audiences and Cultural Power.* New York: Routlege, 1989.

Shaefer, Dennis, and Larry Salvato. *Masters of Light*. Berkeley: University of California Press, 1986.

Siebert, Fred S., Theodore Peterson, and Wilbur Schramm. *Four Theories of the Press*. Urbana: University of Illinois Press, 1963.

Shiver, Jube, Jr. "U.S. Jobs and Better TVs?: HDTV Universal Standards—Political and Economic Pressures May Color Decision." *Los Angeles Times*, May 8, 1993, D1–D2.

Silverman, Kaja. *The Accoustic Mirror*. Bloomington: Indiana University Press, 1988.

———. *The Subject of Semiotics*. New York: Oxford University Press, 1983.

"Six Levels of Video in a Single Pass." Print advertisement. *Video Systems*, Crosspoint Latch Corporation, October 1989, 136.

Smith, Beretta. "Arming Youth: Video for the Revolution." Unpublished ms., UCLA, December 1, 1992.

Solomon, Charles. "'Rangers,' 'Bonkers!' Not Kiddie Fare." *Los Angeles Times*, September 6, 1993, F11.

Spigel, Lynn. *Make Room for Television*, Chicago: University of Chicago Press, 1992.

———, and Denise Mann, editors. *Private Screenings: Television and the Female Consumer*. Minneapolis: University of Minnesota Press, 1992

Squiers, Carol, Ken Burns, Murray Fromson, Fadwa El Guindi, Ralph Arlyck, Ivan Zatz-Diaz, Barbara Trent and Joanne Dororshow. "Screening the War." *International Documentary* (Spring 1991): 20–25.

Stalter, Katharine. "Developing Quality Television: Executives Work with Producers and the Networks to Improve Programming." *Film and Video* (January 1993): 45–48.

———. "Working in the New Post Environment." *Film and Video* (April 1993): 100.

Stam, Robert. *Subversive Pleasures: Bakhtin, Cultural Criticism, and Film*. Baltimore: Johns Hopkins University Press, 1989.

Stemple, Guido H. "Content Analysis." In *Research Methods in Mass Communications*, edited by Guido Stempel III and Bruce H. Westley, 119–131. Englewood Cliffs, N. J.: Prentice-Hall, 1981.

Stewart, Susan. *On Longing: Narratives of the Miniature, the Gigantic, the Souvenir, the Collection*. Baltimore: Johns Hopkins University Press, 1984.

Stockler, Bruce. "Seducing Reality: Documentaries Mix Truth and Fashion." *Millimeter* (May 1988): 48.

Struzzi, Diane. "Scandals, Rumors, and Rock 'N' Roll: On the Edge with Margeotes, Fertitta, and Weiss." *Millimeter* (October 1989): 126.

Taylor, Ella. *Primetime Families*. Berkeley: University of California Press, 1989.

Tichenor, Phillip. "The Logic of Social and Behavioral Science." In *Research Methods in Mass Communications*, edited by Guido Stempel III and Bruce H. Westley, 10–28. Englewood Cliffs, N.J.: Prentice-Hall, 1981.

Thomas, Laurie and Barry Litman. "Fox Broadcasting Company: Why Now?: An Economic Study of the Rise of the Fourth Broadcast Network." *Journal of Broadcasting and Electronic Media* 35, no. 2 (Spring 1991): 139–157.

Thompson, Kristin. *Ivan the Terrible: A Neoformalist Analysis*. Princeton: Princeton University Press, 1981. Chapter 9 reprinted in *Narrative Apparatus, Ideology*, edited by Phillip Rosen, 130–142. New York: Columbia University Press, 1986.

"Thomson Digital Image/TDI print ad. *Video Systems*, June 1993, 70.

Timberg, Bernard and David Barker, editors. "The Evolution of Encoding Research." *Journal of Film and Video* 41, no. 2 (Summer 1989): 3–14.

Todorov, Tzvetan. *Mikhail Bahktin and the Dialogical Principle*. Translated by Wlad Godzich. Minneapolis: University of Minnesota Press, 1984.

Tragardh, Lars, and Paul Meyers. "Graphics: Softward Update/ An Industry in Flux." *Video Systems* (December 1987): 30.

Turim, Maureen. "A Project of Deblocage." *Semiotica* 50, 1-2 (1984): 181–190.

Venturi, Robert, Denise Scott Brown, and Steven Izenour. *Learning from Las Vegas*. Cambridge: MIT Press, 1972.

Vianello, Robert. "The Power Politics of 'Live' Television." *Journal of Film and Video* 37 (Summer 1985): 26–40.

Viera, David. *Lighting for Film and Electronic Cinematography*. Belmont, Calif.: Wadsworth, 1993.

"Viet Vet Kovic May Challenge Dornan." *Los Angeles Times,* January 5, 1990, A3.

Wagner, John. "Absolute TV." *Quarterly Review of Film and Video* 14, 1–2, (1992): 93–102.

Walkerdine, Valerie. "Video Replay: Family, Films, and Fantasy." In *Formations of Fantasy*, edited by Victor Burgin, James Donald, and Cora Kaplan. London: Methuen, 1986. Reprinted in *The Media Reader,* edited by Manuel Alvarado and John Thompson, 339–358. London: British Film Institute, 1990.

Ward, Larry. *Electronic Moviemaking*. Belmont, Calif.: Wadsworth, 1990.

Wells, Jeffrey. "Is It the Reel Thing: Big Name Directors Try to Bring Film Magic to Coke Ads," *Los Angeles Times*, February 17, 1993, F1, F6.

White, Mimi. "Ideological Analysis and Television." In *Channels of Discourse: Reassembled*, edited by Robert Allen, 161–202. Chapel Hill: University of North Carolina Press, 1987, 1992.

———. *Tele-Advising: Therapeutic Discourse in American Television*. Chapel Hill: University of North Carolina Press, 1992.

———. "Television—A Narrative, a History." *Cultural Studies* (1990): 282–300.

———. "Women, Memory, and Serial Melodrama: Rewriting/Reviewing History." Paper presented at Console-ing Passions Conference, USC, April 2–4, 1993.

———. "Women with a Past: Television's Recent Historical Fictions." Unpublished ms. Northwestern University, April 1993.

Whitney, Dwight. "Look Who's In the 'In' Crowd." *TV Guide*, September 21, 1968, 20–23.

Williams, Raymond. "Interview with Raymond Williams." In *Studies in Entertainment: Critical Approaches to Mass Culture*, edited by Tania Modleski, 3–17. Bloomington: Indiana University Press, 1986.

———. *Raymond Williams on Television: Selected Writings*. New York: Routlege, 1989.

———. *The Sociology of Culture*. New York: Schocken, 1981.

———. *Television, Technology, and Cultural Form*. New York: Schocken, 1975.

Williams, Scott. "Doing a Little Fall Arithmetic," *TV Times*, July 11, 1993, 7.

Williamson, Judith. *Decoding Advertisements: Ideology and Meaning in Advertising*. London: Marion Boyars, 1978.

Wilson, Anton. *Anton Wilson's Cinema Workshop*. Los Angeles: American Society of Cinematographers, 1983.

Winston, Brian. "Great Artist of Fly on the Wall: The Griersonian Accommodation and Its Destruction." In *Visual Explorations of the World: Selected Papers from the International Conference on Visual Communications*, edited by Jay Ruby and Martin Taureg, 190–204. Aachen, GDR: Edition Herodot im Rader Verlag, 1987.

———. *Misunderstanding Media*. London: Routledge and Kegan Paul, 1986.

———. "Television at a Glance." *Quarterly Review of Film Studies* 9, no. 3 (Summer 1984): 256–261.

Wollen, Peter. *Raiding the Icebox: Reflections on Twentieth-Century Culture*. Bloomington: Indiana University Press, 1993.

———. *Readings and Writings: Semiotic Counter-Strategies*. London: Verso, 1982.

———. *Signs and Meanings in Cinema*. Bloomington: Indiana University Press, 1969.

Wood, Peter H. "Television as Dream." In *Television a Cultural Force*, edited by Richard Adler, 17–35. New York: Praeger, 1976.

WRAL-TV5, Raleigh, North Carolina, Advertisement. "Situations Wanted Technical." *Broad-*

casting, November 30, 1992, 57.

"The Writers of *thirtysomething.*" *thirtysomething stories*. New York: Pocket Books, 1992.

Wurtzel, Alan. *Television Production*. New York: McGraw-Hill, 1979.

Yost, Scott D. "Viewpoint: Triumphs of the Tube." *Emmy* (August 1993): 19.

Zettl, Herbert. *Sight, Sound, Motion: Applied Media Aesthetics*. Belmont, Calif.: Wadsworth, 1973.

Zoglin, Richard. "The Big Boy's Blues," *Time*, October 17, 1988, 59.

Index

About the Author

John Thornton Caldwell is Chair of the Radio/TV/Film Department at California State University, Long Beach. His work has been published in *Cinema Journal, Jump-Cut,* and *American Television*, broadcast on WGBH, WNED, WTTW, and SBS-Australia, and screened in festivals in Berlin, Paris, New York, Los Angeles, and Amsterdam. He is the recipient of awards in film/video from the Regional Fellowships, state arts councils, and The National Endowment for the Arts.